THE GENE ILLUSION

Genetic research in psychiatry and psychology under the microscope

THE GENE ILLUSION

Genetic research in psychiatry and psychology under the microscope

Jay Joseph

PCCS BOOKS
Ross-on-Wye

First published in 2003

PCCS BOOKS
Llangarron
Ross-on-Wye
Herefordshire
HR9 6PT
UK
Tel +44 (0)1989 77 07 07
enquiries@pccsbks.globalnet.co.uk
www.pccs-books.co.uk

The Gene Illusion
**Genetic research in psychiatry and psychology under the
microscope**

British Library Cataloguing in Publication Data.
A catalogue record for this book is available from the British Library.

ISBN 1 898059 47 0

Cover design by Denis Postle.
Printed by Bookcraft, Midsomer Norton, UK.

CONTENTS

ACKNOWLEDGEMENTS

Professor Steve Baldwin of the University of Teesside, my friend and colleague, was tragically killed in the March, 2001 train crash in Selby, England. Steve was always available to provide helpful advice and support. I miss him dearly, and I dedicate this book to his memory. I would like to thank the many people who helped make this book possible. In doing so, I want to make clear that the views expressed in this book are entirely my own, and do not necessarily reflect the opinions of those who helped me along the way. In addition, I take full responsibility for any errors found in the text. The origins of this book lie in my 1998 doctoral dissertation analyzing, or some would say 'deconstructing,' schizophrenia twin and adoption studies. Several papers based on this research have been published over the past few years. I thank Raymond C. Russ, Editor of the *Journal of Mind and Behavior*, and David Cohen, Editor of *Ethical Human Sciences and Services*. They believed in my work and were instrumental in helping me get started in the publishing arena. Several people read one or more chapters and provided important feedback. I am thinking in particular of Ty Colbert, Leon J. Kamin, Jonathan Leo, and Ken Richardson. I also thank Ty Colbert, David Jacobs, Jonathan Leo, Robert W. Morgan, Loren Mosher, Benno Müller-Hill, Sarah Norgate, Alvin Pam, Stanton Peele, Ken Richardson, and Al Siebert for providing feedback along the way. I also thank my family for the support they provided. I thank Leon Kamin and Pekka Tienari for allowing me to publish their personal communication, and Mark Largent for allowing me to quote from his unpublished manuscript. Daniel Ramseyer, Annette Schoch, Ingrid Stapf, and especially Esther Rosen helped me greatly with German translations. I am also grateful to the University of California library system and the people who work hard to keep its treasures easily accessible for research. Finally, I want to acknowledge Pete Sanders and Maggie Taylor-Sanders of PCCS Books for publishing this book, and for their willingness to let me sneak in some late changes, additions and updates, as well as the Editors of the PCCS Critical Psychology Division, Craig Newnes and Guy Holmes.

For Adam

CHAPTER ONE

Introduction

In 1925 psychiatrist Abraham Myerson, who was writing at the height of the eugenics movement's influence the belief in the overriding importance of genes, observed, 'We often hear of hereditary talents, hereditary vices, and hereditary virtues, but whoever will critically examine the evidence will find that we have no proof of their existence.'[1] The evidence of Myerson's era consisted of family pedigrees, preconceived notions, and prejudice. Today it consists mainly of family studies, adoption studies, studies of twins reared together, studies of twins reared apart, and molecular genetic research. In spite of the widespread view that the results of these investigations converge on the importance of genetics, there are serious problems associated with these methods. And as we shall see in Chapter 10, molecular geneticists have failed to find postulated genes for the major psychiatric disorders. Thus, Myerson's 1925 observation is more relevant to today's evidence than is commonly believed.

In 1996 twin researcher Irving Gottesman wrote, 'no educated person . . . can be oblivious to the fact we are in the midst of a genetic revolution.' He added that the younger generation of genetic researchers could hardly imagine 'the uphill battle that had been fought for the past 45 years.'[2] With what weapons was this 'battle' fought? Gottesman identified the 'old-fashioned strategies' of family, twin, and adoption studies. And he is correct that these methods helped pave the way for the ascendancy of the genetic position as articulated by the fields of behavior genetics and psychiatric genetics.

Newspapers often report claims that a specific gene has been linked to a psychiatric disorder, trait, or behavior, although most of these claims are subsequently found to have no merit. Several books have been published popularizing behavior genetic research, which include William Wright's *Born That Way*, Lawrence Wright's *Twins*, Hamer and Copeland's *Living With Our Genes*, and Judith Harris's *The Nurture Assumption*. Most have focused on the results of twin and adoption studies which, it is claimed, call forth a radical re-evaluation of the 'nature-nurture' debate on the side of nature.

This book provides an alternative view of this body of literature. Far from establishing the importance of genes, we will see that family, twin, and adoption studies are plagued by researcher bias, unsound methodology, and a reliance on unsupported theoretical assumptions. It examines research methods, theories, and specific studies in the only way they should be examined — at their roots. Special attention is devoted to twin research because of its central position in the case for genetics.

Outline of the chapters

Each chapter of this book can be read independently, and any particular chapter (other than Chapters 3 and 4) could be skipped without missing the basic argument. Chapters 7 and 8 consist mainly of in-depth analyses of published studies and may prove difficult reading for some people. They could be skimmed by readers who are already in general agreement with the conclusions I draw at the ends of these chapters.

Chapter 2 looks into the history of twin research, for a simple reason: the full story of this history has never been told. Unfortunately, twin researchers cannot be relied upon to provide details on the more unsavory aspects of this history. In fact, the origin of twin research as a tool of eugenics, 'racial hygiene,' and Nazism is rarely mentioned by twin researchers. Beginning with Galton, I discuss the various ways that twins have been used for research purposes, as well as some of the methodological problems as discussed by critics. We will see that interest in twin research waned in the late 1940s and 1950s, but began a revival in the 1960s that has continued up to the present day.

Chapter 3 is devoted to an analysis of the theoretical underpinnings of the most commonly used method in twin research, the 'classical twin method.' The twin method compares reared-together identical and reared-together same-sex fraternal twins, and its proponents claim that the greater correlation or concordance of identical twins points to the operation of genetic factors on the trait in question. However, this finding depends on the validity of a critical theoretical assumption which holds that identical and fraternal twins experience equal environments. Interestingly, most people — including many leading twin researchers — understand that identical twins' physical and social environments are much more similar than fraternals. Nonetheless, twin researchers continue to uphold the validity of the twin method on the basis of a new definition of the equal environment assumption. I shall, however, demonstrate that this new definition is as untenable as its predecessor. Thus, contrary to the views expressed in countless textbooks and popular reviews, identical-fraternal comparisons cannot be relied upon to tell us anything about genetic influences on psychological trait differences.

Chapter 4 turns to studies of twins whom the investigators have called 'reared-apart.' Although these types of investigations might appear to eliminate the environmental factors confounding the twin method, we will see that they all suffer from important methodological problems, not the least of which is the fact that few pairs were actually reared apart without knowledge of and contact with each other. I focus on the well-known Minnesota studies because they have been instrumental in strengthening the genetic position. I also look at cases of individual pairs who have been reported by journalists. Although their stories are interesting, there are many reasons why they don't tell us anything important about genetics.

In Chapter 5, I explain the reasons why heritability figures are discussed only briefly in this book. The heritability concept was designed for use in agriculture to predict the results of a program of controlled breeding. Unfortunately, it has been falsely promoted as a 'nature-nurture ratio' for the relative influences of genes and environment on particular traits.[3] As I argue in this chapter, there is little reason to believe that heritability estimates can serve this function.

Chapter 6 begins a two-part series looking into the evidence supporting the widely held view that schizophrenia is a genetic disorder. The genetic basis of schizophrenia is nowadays so widely accepted that the question is almost never debated. Amazingly, however, the evidence supporting this position is extremely weak. The main evidence, we will discover, is derived from biased, methodologically unsound, and environmentally confounded research. In this chapter, I point to trends and data from the schizophrenia twin studies to further illustrate the problems with the twin method.

The schizophrenia adoption studies of the 1960s and 1970s are largely responsible for closing the 'genetics of schizophrenia' debate. Chapter 7 consists of an in-depth critical analysis of these studies, which were performed in the United States, Denmark, and Finland. We will discover that the schizophrenia adoption studies were plagued by methodological problems and bias. Most importantly, all schizophrenia adoption studies were performed in regions having eugenic sterilization laws in the era when adoptees were placed. This makes it unlikely that 'hereditarily tainted' children were placed into the same types of homes as other adoptees. Therefore, the basic assumption of adoption studies — that experimental and control adoptees had the same chance of being placed into available adoptive homes — was violated. On the basis of the serious problems with twin and adoption studies, I call for reopening the debate on the role of genetics in schizophrenia and other psychiatric disorders.

Chapter 8 examines twin and adoption studies of criminal and antisocial behavior as an example of how genetic research has been carried out in the area of human behavioral differences. Genetic theories of criminality have been regaining the foothold they had before they were discredited by their association with Nazism and German 'criminal biology.' In keeping with one of the main themes of this book, I demonstrate that the reported higher identical twin concordance for criminality can be plausibly explained on environmental grounds. Like schizophrenia, adoption studies have been put forward as important evidence in support of the role of genetics. Also like schizophrenia, there are several invalidating features of these studies, which I document.

In Chapter 9, I examine the argument that intelligence (as allegedly measured by standardized IQ tests) has an important genetic component. Studies cited in support of this position are looked at only briefly, and emphasis is placed on the intelligence tests themselves. Specifically, we will see that assumptions about the genetic inferiority of lower classes and subordinated races are *built into* IQ tests. Thus, it is astonishing that anyone who knows how these tests are constructed could argue that the lower IQ scores of blacks versus whites, or working class versus upper class, are the result of genetic differences.

Chapter 10 examines the failure of molecular geneticists to find genes for psychiatric disorders and psychological trait variation. The belief that these genes exist derives from the results of family, twin, and adoption studies. As I argue throughout this book, these studies do not show what they are believed to show, and this is being borne out in the failure to find genes which in most cases are unlikely to exist. Unfortunately, the general public is sometimes led to believe that genes have been found. The sensational accounts of the original 'findings' are followed by brief reports of the inevitable failures to replicate. Behavior genetics and psychiatric genetics have arrived at a blind alley as they attempt to show us what they believed all along actually exist. Unlike family, twin, and adoption studies, in this case unsound methodology and glossed-over implausible assumptions cannot produce the desired results of finding actual genes.

In Chapter 11, I summarize and integrate the main arguments made in the preceding chapters, and propose that the entire body of evidence produced by behavior genetics and psychiatric geneticists be re-evaluated. I note further that biased and unsound research is not limited to these fields, but is a widespread problem. Finally, I provide an optimistic view of human progress; a view which stands in stark contrast to the bleak forecasts of most human genetic investigators.

Family studies

Although twin and adoption studies are a major focus of this book, family studies are sometimes cited in support of the genetic position, and therefore deserve mention. The family (or consanguinity) method of study constitutes a systematic attempt to determine whether a condition clusters in families, thereby laying the basis for the possibility of finding a genetic component with other methods. Family studies locate persons affected with a particular trait or condition and attempt to determine whether their biological relatives are similarly affected more often than members of the general population or a control group. If a condition is found to cluster or 'run' in families, it is said to be familial. Note that 'familial' is not the same as 'genetic.' Unfortunately, many people view these terms as being synonymous, when in fact they are not. As most genetic researchers now acknowledge, the aggregation of a condition in families is consistent with genetic or environmental causation. Psychiatric geneticists Steven Faraone and Ming Tsuang, for example, noted that a family study can provide only 'the initial hint that a disorder might have a genetic component,' because 'disorders can "run in families" for nongenetic reasons such as shared environmental adversity, viral transmission, and social learning.' They concluded that 'Although family studies are indispensable for establishing the familiality of disorders, they cannot, by themselves, establish what type of transmission.'[4] However, this has not always been the prevailing view.

The first schizophrenia family study was published in Germany during World War I by Ernst Rüdin, and the most influential study of this type was published by Franz J. Kallmann in 1938. (Kallmann, as we will see, also studied twins.) Most of the early family studies were authored by strong proponents of the genetic position, and most did not perform blind diagnoses. Kallmann believed that the familiality of schizophrenia *proved* that the condition was genetic in origin: 'The principal aim of our investigations was to offer conclusive proof of the inheritance of schizophrenia and to help, in this way, to establish a dependable basis for the clinical and eugenic activities of psychiatry.'[5] Psychologist Alvin Pam noted Kallmann's faulty logic and commented further on the family study method,

> The most serious breach in inductive logic committed by Kallmann was his use of kinship concordance rates to determine genetic transmission of psychopathology. We have already noted that no family inheritance study can control for environment in human research; such data, therefore, are nowhere near 'suggestive' — they are at best inconclusive and at worst misleading . . . This inferential limitation holds with respect to any consanguinity finding, even if the design and technique employed in the investigation were scientifically impeccable.[6]

Today, most behavior genetic and psychiatric genetic researchers acknowledge that family studies cannot establish the existence of genetic factors, and have cited twin and adoption studies as the primary evidence in favor of the genetic basis of schizophrenia and other conditions.

In spite of the formal pronouncement that family studies cannot be used as evidence of genetic transmission, some researchers attempt to use the results from these studies in support of genetic conclusions. For example, in Russell Barkley's authoritative 1998 handbook for the diagnosis and treatment of attention-deficit hyperactivity disorder (ADHD), he wrote, 'Family aggregation studies find that ADHD clusters among biological relatives of children or adults with the disorder, strongly implying a hereditary basis to this condition.'[7] And schizophrenia researchers Frangos and colleagues concluded their 1985 family study by writing, 'We consider the results to provide evidence that even narrowly defined schizophrenia . . . has a genetic component . . .'[8] Whereas

Faraone and Tsuang correctly viewed the results from family studies as providing only an 'initial hint' that genetic factors might be operating, investigators such as Barkley and Frangos et al. believed that these results provide evidence in favor of genetics.

Even before the first family studies were performed, patterns found in the family pedigrees of affected individuals were seen as definitive proof of genetic influences. The case of pellagra provides us with an example. Pellagra is a disease that ravaged poor people in the southern part of the United States during the first half of the twentieth century. Before then, it had been known in southern Europe for almost 200 years. The often fatal disease, still found among the world's poor, is characterized by severe skin rash, gastrointestinal problems, and mental disturbance. Between 1730 and 1930 pellagra claimed over half a million lives, including tens of thousands of poor blacks and whites in the southern United States. On the basis of the pioneering work of Joseph Goldberger and others in the early part of the twentieth century, it was firmly established that pellagra is caused by a dietary deficiency linked to malnutrition among the southern poor. In other words, pellagra was (and is) a disease of hunger and poverty. But it has not always been viewed this way.

The pioneers of the American eugenics movement believed that pellagra had a hereditary basis. According to a 1916 article by Charles Davenport, 'Pellagra is not an inheritable disease in the sense in which brown eye color is inheritable. The course of the disease does depend, however, on certain constitutional, inheritable traits of the affected individual.'[9] Davenport based his conclusions on the results of his family pedigrees, then known as 'eugenical family studies.'[10] Yet even before the advent of eugenical family studies, Davenport believed that 'many physical, mental and moral traits have been proved to have an hereditary basis, and it seems probable that in practically all there is an hereditary factor of more or less importance.' He saw eugenical family studies as 'afford[ing] the means of studying this [proven] hereditary factor.'[11]

In support of his position, Davenport produced 15 pages of pedigrees of families affected by pellagra. These diagrams showed that pellagra clustered in families at a rate far greater than would be expected in the general population. Although Davenport mistakenly believed that the condition was communicable, his fundamental position was that 'the constitution of the organism must be held to be the principal cause of the diversity which persons show in their reaction to the same disease-inciting factors. This constitution of the organism is a racial, that is, hereditary factor.'[12] In Davenport's paper one searches in vain for any mention of the fact that most pellagrins lived in dire poverty. Conspicuously absent from his study are words such as 'poor,' 'poverty,' 'hunger,' or 'malnutrition,' in spite of the fact that several researchers had pointed to nutritional factors in pellagra and that there were no reported cases of health care providers, working in close contact with pellagrins, who had developed the disease. Above all, pellagra was known to be correlated with poverty and the consumption of corn.

That Davenport and others were blinded by hereditarian views and failed to pay serious attention to the possible environmental confounds in pellagra family pedigrees is hardly an original observation; indeed, it is a classic example of the potential fallacy of reaching conclusions about genetic factors when studying people who share a common environment as well as common genes.[13] This is true for virtually all psychiatric conditions, psychological traits, behaviors, and medical conditions. Thus, it is now generally agreed that the results from family studies are explainable on either genetic or environmental grounds.

Terms used in the book

I would like to define some terms that will be used throughout the book. I use **human genetics** as a shorthand for 'behavior genetics and psychiatric genetics,' although several aspects of human genetic research do not fit into these categories. I frequently discuss the underlying **assumptions** of genetic research. In this sense, an assumption is something that is taken to be true. The conclusions of behavior genetic research are based on many questionable assumptions, some of which the researchers identify and discuss, and many others that are not identified. **Concordance rates** are used in psychiatric twin studies. Twins are said to be concordant for a condition when both members of the pair are affected, and *discordant* when only one is affected. Unless otherwise noted, all discussions of twins in this book refer to pairs reared together in the same home. **Correlation coefficients** quantify the degree of relationship or association between two variables. A positive correlation is expressed as a coefficient ranging from 0.0 to 1.0. A 'relationship' means that two occurrences vary together, not necessarily that they are similar. In addition, correlation says nothing about the cause of the association. A classic example is that ice-cream sales increases are often correlated with an increase in the crime rate, which does not mean that eating ice cream causes people to commit criminal acts. The correlation is spurious because warm weather contributes to the causes of both higher ice-cream sales and a higher crime rate. Although the position that correlation does not imply cause is a basic statistical principle, it is frequently misunderstood or ignored by people attempting to use correlational data to imply cause and effect. When I say that a study is **methodologically** unsound, I mean that it violates a set or system of methods and principles used in psychological or psychiatric research.

The word **environmental** refers to all non-genetic factors influencing a trait or contributing to the causes of a condition. Examples of environmental factors include family or social milieu, viruses, prenatal complications, air and water pollution, and socioeconomic status. When discussing environmental and genetic influences, it should be emphasized that they frequently interact. For example, human height is influenced by heredity, but the social environments of adults in western societies who are three feet tall and those who are six feet tall will be vastly different.

Another word I employ frequently is **confound**. A basic study tests the effect of a factor or factors (the independent variable) on another factor (the dependent variable). The investigators attempt to eliminate (control for) factors other than the independent variable which could influence (confound) the results. For example, a serology laboratory takes precautions to keep foreign substances out of its samples, since contamination could lead to false results. In human genetic research, conclusions favoring genetic factors are dependent upon being able to control for environmental confounds. Human genetic investigators usually state that the results of family studies are confounded by environmental factors, but that twin and adoption studies are not.

In discussions of psychiatric adoption studies, the terms **index group**, **experimental group**, and **control group** are used frequently. I use 'index group' and 'experimental group' interchangeably in reference to subjects identified because they are, or because they are related to, people diagnosed with the condition under study. Control groups consist of people matched with members of the index (experimental) group, but who are not diagnosed with the condition under study. The original investigators' conclusions are drawn from comparisons between the index group and the control group. The **genetic theory of schizophrenia** refers to the view that although environmental factors might be important, genetic factors are equally if not more important.

It is also important to distinguish between genetic influences on a **trait** versus genetic influences on **trait differences**. A trait is a distinguishing characteristic or quality of a person. Clearly, there are many inherited physical and mental traits that distinguish humans from other species. Behavior genetic research takes this basic fact as a starting point and seeks to establish that genetic factors contribute greatly to differences among people for particular human traits such as personality and intelligence. As behavior geneticist Robert Plomin has written, 'It is critically important to recognize that behavioral genetic research is limited to the investigation of the genetic and environmental origins of individual differences within a species, not species-typical development.'[14] Still, the critical differentiation between genetic influences on behavior versus genetic influences on behavioral *differences* is often conflated. In a 1994 article in *Science*, Plomin added to this problem by writing, 'Just 15 years ago, the idea of genetic influence on complex human behavior was anathema to many behavior scientists.'[15] While hardly anyone in 1979 doubted that genes influence human behavior, many disputed the contention that genes were involved in human behavioral *differences*.

I also discuss various aspects of the way that the **eugenics** movement impacted genetic research. This movement was founded by Francis Galton in the nineteenth century. Eugenics refers to the belief that human beings can be improved by means of selective breeding. 'Positive eugenics' refers to measures used to increase the reproduction of those seen as genetically 'fit,' whereas 'negative eugenics' refers to attempts to curb or eliminate the reproduction of the 'unfit.' Examples of negative eugenic measures include marriage restrictions, laws against miscegenation (racially mixed marriages), sterilization, and in the case of the Nazi Germany, genocide. The eugenics movement was very active and influential in the United States in the first four decades of the twentieth century, until the crimes of the Nazis in the name of eugenics and 'racial hygiene' demonstrated what the logic of eugenics could ultimately lead to. The mainstays of eugenic research were family and twin studies. Although we have seen that family studies are no longer seen as conclusive evidence for genetic factors, twin studies remain 'the workhorse of behavioral genetics.'[16] The use of twins in an attempt to prove that heredity is important has a long and yet largely unknown history. In the next chapter, I attempt to give the reader a better sense of this history.

Notes

[1] Myerson, 1925, p. 23.
[2] Gottesman, 1996, p. 52.
[3] Hirsch, 1997.
[4] Faraone & Tsuang, 1995, pp. 88–9.
[5] Kallmann, 1938a, p. xiv.
[6] Pam, 1995, p. 19.
[7] Barkley, 1998, p. 36.
[8] Frangos et al., 1985, p. 385.
[9] Davenport, 1916, p. 13.
[10] Davenport & Laughlin, 1915.
[11] Ibid., p. 3.
[12] Davenport, 1916, p. 4. Davenport mistakenly believed that blacks were constitutionally less susceptible to pellagra than whites.
[13] See Chase, 1975, 1980.
[14] Plomin, 1996, p. 30.
[15] Plomin, 1994, p. 187.
[16] Plomin & DeFries, 1998, p. 64.

CHAPTER TWO

Twin Research: Misunderstanding twins from Galton to the 21st century

Why shall we only deprive these persons, of no use to society or even for themselves, the ability of reproduction? Is it not even kinder to take their lives? This kind of dubious reasoning will be the outcome of the methods proposed today.

Swedish member of Parliament Carl Lindhagen speaking in opposition to a proposed eugenic sterilization law in 1922.[1]

That man should be one of the few species that produce an abundant supply of identical twins is so fortunate that it might be regarded as a direct intervention of Providence!

Eugenicist J. A. Fraser-Roberts in 1934.[2]

The history of twin research is usually told by twin researchers. These typically brief accounts are written from the point of view that twin studies are a valuable tool for assessing the influence of genetic factors on particular traits, and tend to portray the history as a linear path from necessarily crude methods and uncertain biology to the scientifically precise twin studies of the current period. Because one of the central themes of this book is the problem of drawing genetic inferences from the results of twin research, I will focus more closely on the theoretical and methodological shortcomings of this research. This chapter is a relatively brief overview of a large body of work, and I will cover only what I view as the most important points.

The story cannot be told without an integration of the social and political views and motivations of twin researchers, as well as the social and political environments in which their research was carried out. As we will see, most pioneers of twin research believed strongly in the importance of heredity, which often led them to advocate the use of selective breeding programs for humans. To leave this out is to fail to understand the history of twin research. It would be like writing a history of the U.S. Civil War without mentioning slavery.

Twin research from 1870 to 1924

Galton

The history of twin research usually starts with Galton, which is where I also shall begin. Francis Galton (1822–1911), born to a well-off Birmingham, England family, was a pioneer of the fields of statistics and psychology, and is widely recognized as the father of twin research. Galton was also the founder of the eugenics movement and coined the word 'eugenics,' meaning 'well-born.' In his 1869 book *Hereditary Genius* he wrote that 'it would be quite practical to produce a highly-gifted race of men by judicious marriages during several consecutive generations.'[3] As for the others, Galton wrote in 1883 that 'There exists a sentiment, for the most part quite unreasonable, against the gradual extinction of an inferior race.'[4]

In *Hereditary Genius* Galton demonstrated that British judges, commanders, scientists, and the members of other highly regarded professions tended to produce highly achieving offspring. Although Galton attributed the success of fathers and sons to heredity, the resemblance also can be explained by the vastly superior socioeconomic status of these fathers. Galton's awareness of this objection led him to seek other methods to demonstrate the importance of heredity. In an article published in 1875 (and reproduced in a revised form in his 1883 book, *Inquiries into Human Faculty*), Galton discussed the need for

> some new method by which it would be possible to weigh in just scales the respective effects of nature and nurture, and to ascertain their several shares in framing the disposition and intellectual ability of men. The life history of twins supplies what I wanted.[5]

Galton sent 600 inquiries to persons he knew to be twins or to their relatives, but received only 159 responses.[6] From these responses he obtained 35 cases of twins 'of close similarity' and 20 cases where twins displayed 'sharply contrasted characteristics, both of body and of mind.' Contrary to popular belief, Galton did not compare identical and fraternal twins for the purpose of determining the existence of hereditary influences, and thus he cannot be credited with inventing what we now know as the twin method.[7] Galton did, however, believe that there were two types of twins. In his view, one type 'is derived from a separate ovum, while the other is due to the development of two germinal spots in the same ovum.'[8]

Galton viewed his 35 closely similar pairs as being alike because of their shared heredity, and noted that he had not heard of a case where 'growing dissimilarity' was the result of 'firm freewill of one or both of the twins.' He argued that physical illness or accidents were the only circumstances, apart from heredity, that would have a 'marked effect on the character of adults.'[9] Of his 20 dissimilar pairs, Galton found that none was reported to have become more similar due to having received identical nurture. He concluded, 'There is no escape from the conclusion that nature prevails enormously over nurture when the differences of nurture do not exceed what is commonly to be found among persons of the same rank of society and in the same country.'[10]

Clearly, Galton came to important conclusions on the basis of what we today would consider sketchy, responder-biased, and second-hand reports, and like many subsequent twin researchers he studied twins for the purpose of proving what he already believed to be true. Galton saw the histories of his similar twins as evidence of the importance of their common heredity, or as he put it, 'the clocks of their two lives move regularly at the same rate, governed by their internal mechanism.'[11] Galton based his conclusion on reports in which twins were said to have become sick and recovered at the same time. He viewed cases describing twins who simultaneously became ill from contagious diseases as evidence of an 'intimate . . . constitutional resemblance,'[12] when today we could say that close physical proximity was the likely cause. In another case, both members of a pair of female twins developed 'the defect of not being able to come downstairs quickly' at the age of 20, which Galton also attributed to heredity.[13]

Turning to behavioral and psychological traits, Galton's explanations become even more implausible. He described a twin pair who 'make the same remarks on the same occasion' and 'begin singing the same song at the same moment,' as if they were genetically programmed to do so independently of one another.[14] Galton's belief in genetic predestination is illustrated by his interpretation of the history of a pair of French twins.[15] François and Martin were 50-year-old twin railroad contractors living two leagues apart from each other in small French towns. On January 15 someone robbed the box where the brothers had deposited their savings. On the night of January 23 both twins, though living separately, were said to have been awakend at 3 a.m. by the same dream, at which point they shouted, 'I have caught the thief! I have caught the thief! They are doing mischief to my brother!' Martin attacked his grandchild and would have strangled him had he not been stopped. He went outside and attempted to drown himself in the River Steir, but was prevented from doing this by his son. Martin was then was picked up by gendarmes and was taken to an asylum, where he died three hours later. According to the account, François had calmed down by the morning of January 24 and attempted to solve the robbery. Then,

> By a strange chance, he crossed his brother's path at the moment when the latter was struggling with the gendarmes; then he himself became maddened, giving way to extravagant gestures and using incoherent language (similar to that of his brother). He then asked to be bled, which was done, and afterwards, declaring himself to be better, went out on the pretext of executing some commission, but really to drown himself in the River Steir, which he actually did, at the very same spot where Martin had attempted to do the same thing a few hours earlier.[16]

Galton apparently viewed these twins to have been constitutionally programmed at birth to attempt to drown themselves in the same river on the same day. According to the story, Martin had regained his composure until he came across his twin brother who was struggling with the gendarmes. It was only then that he began to make 'extravagant gestures' and use 'incoherent language . . . similar to that of his brother.' Clearly, there was a strong bond between these twins which likely manifested itself in similar behavior, and explains François's reaction to what was happening to his twin brother. This story describes the mutual interaction and association of twins, which Galton completely misunderstood as evidence that the genetically set 'clocks of their lives' moved at the same rate.

In studying twins, Galton saw the opportunity to strengthen his case for a hereditary behavioral blueprint stamped on people from conception. He interpreted his case histories as showing that twins who closely resembled each other in childhood were 'keeping time like two watches, hardly to be thrown out of accord except by some physical jar.'[17] Although Galton was aware of the possibility that common upbringing could be a factor, he was blind to the influences of identity confusion and emotional closeness. Galton saw the twins' histories as evidence that heredity has mapped out a behavioral life plan for each person in the same way that it might influence, for example, baldness, menstruation, or menopause. He was arguing that psychological traits are under as much hereditary control as physical traits, failing to recognize that twins' behavioral similarity could be the result of their similar environment and the emotional bond of the pairs. In another example of Galton's failure to understand the behavior he was reporting, he wrote that 'no less than nine anecdotes have reached me of a twin seeing his or her reflection in a looking-glass, and addressing it in the belief it was the other twin in person.'[18] Galton did not understand that these stories could describe the identity confusion of twins who — because of similar appearance, mutual interaction, and societal expectations that they see themselves as two halves of one whole — found it difficult to become individuals separate from each other. 'Identity formation,' wrote Ricardo Ainslie in his 1985 book on twinship, 'is often considered the cornerstone of any discussion of the psychology of twinship. The idea that twins encounter difficulties in the process of identity formation is as pervasive in scientific writings on twinship as it is in popular culture.'[19]

From a historical perspective, the most important aspect of Galton's twin study was his failure to understand the causes of the observed behavioral similarities of twins. This error, as we will see, was repeated by subsequent twin researchers similarly blinded by a belief in the importance of eugenics and genetics.

After 1875 Galton did not perform twin research and he turned his attention to other questions. It would be another 30 years before the publication of the next twin study, this time by an American investigator. Nevertheless, Galton is rightly considered the father of eugenics and twin research — and these 'fatherhoods' are by no means independent of each other. Interestingly, Galton was one of the first to mention reports of adoptees in the context of genetic research. Using stories supplied to him, in 1865 Galton attempted to show that 'a wild, untamable restlessness is innate with savages.'[20] His evidence?

> I have collected numerous instances where children of a low race have been separated at an early age from their parents, and reared as part of a settler's family, quite apart from their own people. Yet, after years of civilized ways, in some fit of passion, or under some craving, like that of a bird about to emigrate, they have abandoned their home, flung away their dress, and sought their countryman in the bush, among whom they have subsequently been found living in contented barbarism, without a vestige of their gentle nature.[21]

Thorndike

The first twin study to use psychological tests was published in 1905 by the noted American psychologist E. L. Thorndike, who sought to assess the relative influences of nature and nurture on mental abilities. Thorndike selected 50 pairs of same-sex twins from the New York City school population. He was interested in using twins in order to test whether training and common environment had an important role in mental development. If this were the case, he reasoned, then (1) twins should grow more alike as they get older, and 13 to 14-year-old twins should be much more alike than 9 to 10-year-old twins; (2) the similarity of ordinary siblings not more than 4–5 years apart should resemble the similarity of twin pairs; and (3) if training influences mental traits, twins should be more similar for abilities that require training as opposed to those less subject to training. Thorndike believed that his data did not support the effects of training, and concluded that he had found 'conclusive evidence' that 'mental likenesses . . . are due, to at least nine-tenths of their amount, to original nurture.'[22]

Thorndike did not compare correlations between identical and fraternal twins. He believed that 'The evidence . . . gives no reason for acceptance of the hypothesis of two distinct types of twins.'[23] He viewed the evidence as suggesting that all twins were the result of two separate ova. Thorndike's data revealed a unimodal distribution of the twins' test scores, but he expected that if there were two types of twins, there would be a bimodal distribution. Although by this time many scientists had come to accept the theory of two types of twins, the authority of Thorndike (and later Fisher[24]) delayed the widespread acceptance of this idea until the mid-1920s.

In reading through Thorndike's monograph, one is struck by the certainty of his belief in the importance of genetics on the basis of what we would today view as weak evidence. Like Galton, Thorndike used twins as a backdrop for expressing his pre-existing belief in the importance of heredity:

> The facts are easily, simply and completely explained by one simple

hypothesis: namely, that the natures of the germ cells — the conditions of conception — cause whatever similarities and differences exist in the original natures of men, that these conditions influence body and mind equally, and that in life the differences in modification of body and mind produced by such differences as obtain between the environments of present-day New York City public school children are slight.[25]

Also like Galton, who acknowledged that some would doubt his conclusion 'that nurture should go for so little,'[26] Thorndike discussed his and others' 'repugnance to assigning so little efficacy to environmental forces as the facts of this study seem to demand.'[27]

Following Thorndike's study, almost two decades would pass before twins would again be used in psychological research. In the 1920s, a convergence of events led to a rapid growth in twin research.

Origins of the twin method
It became firmly established in the mid-1920s that there are two types of twins: identical (also known as monozygotic, MZ, one-egg, duplicate, or uniovular) and fraternal (also known as dizygotic, DZ, two-egg, or binovular) twins, although several investigators had been certain of this years earlier.[28] In 1924 Hermann W. Siemens of Germany became the first to propose that the similarities of reared-together identical and reared-together fraternal twins be compared as a way of determining the role of genetic factors for diseases or traits.

In his book *Die Zwillingspathologie* (Twin Pathology), Siemens presented the results of a study using this new method, which came to be known as the 'classical twin method,' or 'twin method.'[29] He invented the twin method in the context of the study of racial biology (rassenbiologie), family biology, and twin biology. According to Siemens the main problem with racial pathology research (rassenpathologische Forschung) was that its findings were not widely accepted due to the existence of plausible alternative explanations. The new method of comparing the similarity of identical and fraternal twins offered, according to Siemens, a more definitive research method in support of racial hygienic ideas.[30] Siemens, a member of the famous German industrialist family, was a leading figure in the German racial hygiene movement in the early to middle parts of the twentieth century, and was later a supporter of the Nazi regime (see below).

Three contemporary behavior geneticists have challenged the claim that Siemens deserves full credit as the discoverer of the twin method. In 1990 Rende, Plomin, and Vandenberg[31] argued that American psychologist Curtis Merriman should share this distinction on the basis of his 1924 twin study published in *Psychological Monographs*.[32] However, the authors acknowledged that, unlike Siemens, Merriman 'did not indicate that the comparison between identical and fraternal twin resemblance could be used to assess hereditary influences on a trait.'[33]

Merriman believed that identical twins should show a greater resemblance than fraternals, but his observation was made in the context of demonstrating the existence of two types of twins. As Rende et al. acknowledged, Siemens is usually recognized in Europe as the discoverer of the twin method. But this view was also held by early American twin researchers. For example, in 1934 twin researcher Paul Wilson named Siemens, Dahlberg, and Holzinger as the developers of the twin method,[34] and in 1937 Newman, Freeman, and Holzinger wrote, 'Siemens in 1924 published a book entitled *Zwillingspathologie*, in which he proposed a new method of studying pathology and human heredity in general.'[35] It should be added that the Italian twin historian Luigi Gedda named Siemens as deserving 'credit for opening a new approach to twin studies.'[36] I am not aware of any twin researcher naming Merriman as a discoverer of the twin method prior to Rende and colleagues' 1990 article. Having raised Merriman to the status of a co-founder of the twin method 65 years after the fact and contrary to the position of all other twin researchers, in 1997 behavior geneticists Plomin, DeFries, McClearn, and Rutter named Merriman as the *sole* discoverer of the method: 'The first real twin study in which identical and fraternal twins were compared in order to estimate genetic influence was conducted in 1924 (Merriman, 1924).'[37] Interestingly Plomin, as a co-author of the 1990 Rende and colleagues article, had written that Merriman *did not* study twins for this purpose (see above).

Siemens: the unknown pioneer

Hermann Werner Siemens (1891–1969) holds the distinction of being one of the most unknown and unheralded inventors of a widely used research technique. Contemporary twin researchers rarely mention him, and his 103-page founding document of the twin method, *Die Zwillingspathologie*, has not been translated into English. One explanation for twin researchers' reluctance to talk about Siemens is the nature of his views on racial hygiene and German National Socialism.

The German Society for Racial Hygiene (Gesellschaft für Rassenhygiene), the world's first eugenics organization,[38] was founded in 1905 by Alfred Ploetz, Anastasius Nordenholz, and Ernst Rüdin. Its goals were to promote the reproduction of the genetically 'fit,' and to prevent the reproduction of those viewed as unfit. The racial hygienic and eugenics movements elsewhere shared similar goals and ideas, but many in the German movement preferred the name 'racial hygiene' because they wanted to stress the racial nature of their movement.[39] The German racial hygiene movement grew slowly through the years of World War I and emerged in the mid-to-late 1920s as an influential force in German intellectual circles. In 1916 Siemens published the first edition of his book *Grundzüge der Vererbungslehre, Rassenhygiene und Bevölkerung-spolitik* (Foundations of Genetics, Racial Hygiene, and Population Policy). This was followed by at least 12 more editions spanning more than 35 years.

In 1924 an English edition of Siemens's book was published under the title *Race Hygiene and Heredity*.[40] In this book, Siemens argued strongly for the importance of genetic influences and called for the implementation of racial hygienic measures. He warned that European civilization was in danger because 'fertility among the inferior is greater than that among those of more average capacity,'[41] and claimed that Rome and Greece had fallen because 'care was not taken to maintain a sufficient fertility among the "capable," who were the real girders of civilization.'[42] Siemens believed that the state must take action in order to stop the reproduction of the 'unfit':

> It cannot be denied that the slightly feeble-minded, those who lack self-control and those with feeble will, which make up the largest contingent of confirmed criminals, tramps and prostitutes, represent a certain danger from the standpoint of race hygiene, for through their lack of inhibition they tend to multiply rapidly. The sterilization of such pathological persons at their own request should, therefore, as soon as possible, become a matter of legal regulation . . . To render pathological natures permanently harmless and to prevent the reproduction of other miserable creatures should consciously become the goal of our courts.[43]

It is unclear whether Siemens's desire for the courts to act was a call for the compulsory sterilization of 'pathological persons.' This was published in the atmosphere of Weimar Germany but, as we will see, the atmosphere changed radically after the National Socialist seizure of power in 1933. Siemens invented the twin method in order to create more credible evidence in favor of the importance of heredity and to help eugenicists and German racial hygienists identify the carriers of genetic defects, so that these 'miserable creatures' would be convinced not to reproduce, or would be prevented from reproducing.[44]

After the Nazis took power in 1933, most German racial hygienists called for compulsory eugenic sterilization. For Siemens, Hitler's regime transformed his racial hygienic ideas from a 'utopian dream' into state policy. Siemens's forward to the 1937 (eighth) edition of his book on racial hygiene contains this passage:

> Since the National Socialist seizure of power the political goals that we, the racial-hygienists, are in favor of, have now become a part — and not the least important one — of the German government program. 'Racial hygiene as a utopian dream' became 'Racial hygiene as political program' . . . Our future will be governed by racial hygiene — or it will not exist at all.[45]

Later in the eighth edition, Siemens discussed how Galton's ideas were being put into practice in Germany:

> Galton already saw the possibility of integrating racial-hygienic ideals — just like a new religion — into the national conscious. The national

[völkische] state, however, is now called on to be really serious about it. According to its *Führer,* it is the obligation of the national state 'to declare children as a people's most precious commodity' so that 'it will one day be considered reprehensible to withhold healthy children from the nation.' [46]

Here, Siemens praised the 'Führer' while dropping the mask of the apolitical scientist. He spoke of his field by its proper designation: 'Racial hygiene as political program.'

Siemens lived in the Netherlands during the Nazi period, having been named the chair of dermatology at the University of Leiden in 1929,[47] so the views expressed in his writings cannot be attributed to the pressures of living under the regime. Although postwar editions of his book on racial hygiene and population policy removed statements in support of Hitler's policies, the 1952 edition of Siemens's book 'continued to advocate the forcible sterilization of inferior stocks.'[48] In the 1930s and early 1940s Siemens was a co-editor of Rüdin's *Archiv für Rassen- und Gesellschaftsbiologie* (Archive for Racial and Social Biology). The *Archiv* began publication in 1904 as the journal of the Society for Racial Hygiene. By the mid-1930s, it had become an official organ of the Nazis' Reich Committee for Public Health.[49]

Developments in twin research, 1924–1945

Early criticism of the twin method

The twin method rests on three basic theoretical assumptions which have changed little since the mid-1920s: (1) there are two types of twins, identical and fraternal, and researchers can reliably distinguish between them; (2) the results from identical-fraternal comparisons can be generalized to the single-born population; and (3) the environments of identical and fraternal twins are roughly equal.[50] The last assumption is known today as the 'equal environment assumption' and will be examined in detail in Chapter 3. The questionable validity of this assumption has been the most common criticism of the twin method, and it is therefore useful to look at the writings of some early critics as a prelude to a more in-depth discussion.

Early critics of the twin method pointed out that identical twins experience a more similar environment than fraternals, which might cause identicals to be more similar. Because this objection often was raised by people who believed strongly in the importance of heredity, it was sometimes argued that identical twins created more similar environments for themselves on the basis of their more similar inherited personalities and temperament. For example, in 1930 American eugenicist S. J. Holmes wrote that 'the environmentalist may contend that ordinary, or fraternal, twins by virtue of their differences soon get into different relations with their environment and come to experience very unlike reactions from their associates.'[51] In the same year, British medical statistician Percy

Stocks argued that the equal environment assumption 'may involve a fallacy' because

> many dizygotic twins are very different in general body build, healthiness, tastes and temperament so that they naturally tend to subject themselves, or be subjected, to differences in nurture to a greater degree than monozygotic twins who have usually the same needs, tastes and inclinations and are rarely seen apart during childhood.[52]

In 1934 Wilson conducted his own twin study and concluded that 'The greater dissimilarities in environment of the fraternal pairs may be attributed in part to the influence of their different hereditary makeup.'[53] Wilson therefore believed that the assumption of equal environments was 'unjustified.'[54] In the same year, Helmut von Bracken published the results of a study which found that identical twins experience a higher degree of attachment than fraternals. He attributed this to identical twins' greater genetic similarity, which reduces 'slight misunderstandings which are an important cause of conflicts between brothers and sisters . . .'[55] Von Bracken also noted, 'where identicals live together, there is a characteristic tendency towards uniformity which must inevitably exert a powerful causal influence upon the properties of such twins.'[56] Although von Bracken was the first person to study differing levels of attachment and association among identical and fraternal twins, in his 1934 article he did not discuss the implications his findings might have for the twin method. In 1941 American psychologist R. S. Woodworth performed a critical review of the literature and concluded that twins reared in the same home 'may have very different effective environments, for the reason, at bottom, that they differ genetically.'[57] According to Woodworth, 'we cannot derive much information from a comparison of the results from the two classes of twins.'[58]

In a 1940 summary of his ten years of twin research, University of California researcher Harold Carter wrote that 'the assumption that the nurture influences are approximately equal for fraternal and identical twins . . . seems untenable to anyone who has had much contact with twins in their own social environment.'[59] He further observed that

> Identical twins obviously like each other better; they obviously have the same friends more often; they obviously spend more time together; and they are obviously treated by their friends, parents, teachers, and acquaintances as if they were more alike than fraternal twins are.[60]

We have seen that most of the early critics believed that the twin method was flawed because the greater genetic dissimilarity of fraternal twins led them to create more dissimilar environments than identical twins. Interestingly, in Chapter 3 we will discover that contemporary twin researchers *uphold* the validity of the twin method on the basis of the very same argument.

Developments in North America

Apart from Germany, the United States was the leader in twin research in the 1920s and 1930s.[61] I have already mentioned Merriman's 1924 study. In 1925 Lauterbach published his twin study of intelligence.[62] Like Merriman, he did not make direct identical-fraternal correlational comparisons, but instead gathered information in order to answer other questions about the features and abilities of twins. One of his findings was that, in accord with Merriman, twins were not intellectually handicapped when compared with non-twins. This was seen as an important question, as it related to whether IQ twin studies are generalizable to the single-born population.

Two additional early IQ twin studies were published by the Canadian Wingfield in 1928 and the American Hirsch in 1930.[63] Both attempted to make identical-fraternal comparisons, although critics have noted that these studies suffered from inadequate methods of separating identical and fraternal twins.[64] Wingfield believed, as he wrote in *Eugenics Review*, that 'intelligence is an inherited trait.'[65] Hirsch claimed that his data showed that 'heredity was about five times as significant as environment in determining IQ differences between twins . . .'[66] and concluded, 'We could not better end our essay than by quoting from Professor Conklin,' who wrote that civilization was threatened by 'hereditarily primitive men,' and that 'either the responsibilities of life must be reduced and the march of civilization stayed, or a better race of men, with greater hereditary abilities, must be bred.'[67] Four years earlier, Hirsch had written a lengthy monograph in which he contrasted the abilities of the 'Nordic,' 'Alpine,' and 'Mediterranean' 'races' of Europe, while advocating eugenic measures. According to Hirsch, 'It would really be worse than miscegenation for millions of our citizens of many of our Natio-Racial groups to breed freely among themselves or to blend with others of our Natio-Racial groups. Both of these are contingencies that will *naturally* occur unless intelligence and foresight intervene.'[68]

In the 1930s Carter published twin studies of personality, vocational interests, and emotional traits,[69] and Rosanoff and colleagues studied schizophrenia, criminality, and delinquency.[70] Aaron Rosanoff was an early American psychiatric twin researcher and eugenicist who served on the committee on insanity of the American Breeders Association.[71] In the 1938 edition of his *Manual of Psychiatry and Mental Hygiene*, Rosanoff, although recognizing that social and economic factors influence criminal behavior, wrote,

> In all probability, even under ideal social and economic conditions, there will still be many cases of antisocial personalities in which rehabilitation or a tolerable social adjustment will prove unattainable. For such cases, *permanent segregation* and, possibly, *sterilization* will have to be arranged, for the double purpose of protection of society and eugenic effect.[72]

Rosanoff wrote that 'for mental disorders which are determined solely by hereditary factors the only prevention is through eugenics,'[73] although 'the problem is more complicated in connection with mental disorders which are of complex and variable etiology.'[74] Given that Rosanoff was a leading voice in psychiatry for eugenics and sterilization in the 1920s and 1930s (particularly in California), it is intriguing that Edward Shorter, in his 1997 *A History of Psychiatry*, cited Rosanoff as an example of a benevolent early psychiatric geneticist. According to Shorter, 'Rosanoff was not a marginal right-wing figure, as some historians of psychiatry have tried to characterize these early psychiatric geneticists.'[75]

For most eugenicists, a demonstration that conditions or traits run in families was sufficient evidence to implicate heredity.[76] As twin researcher and eugenicist Horatio H. Newman wrote in his 1932 book, *Evolution, Genetics, and Eugenics*, the three main 'methods of research in eugenics' were 'The pedigree method,' 'The statistical method,' and 'The twin method.'[77] Today, twin studies are the most frequently cited evidence in support of genetic influences because the results of family studies are *now* recognized as explainable on environmental as well as on genetic grounds.

Twins reared apart

Although there is controversy about who invented the twin method, according to contemporary twin researchers Bouchard and Pedersen there is no controversy about who invented the twins reared-apart method, because 'apparently, no one knows.'[78] Galton had received a report on a reared-apart identical pair, but did not mention this in his published writings.[79] The mid-1920s witnessed the first published reports of identical twins reared apart. The first reported pair, 'B' and 'J,' were separated at two weeks of age and were brought up in different U.S. states. These twins were first reported in 1922 by Paul Popenoe, who was also a prominent champion of compulsory eugenic sterilization in the United States and was a co-author of *Applied Eugenics*. This was followed by an investigation by the subsequent Nobel laureate geneticist and prominent eugenicist, H. J. Muller.[80] B and J have thus gone into history as the 'Popenoe-Muller pair.' Muller found that the twins scored similarly on IQ tests, but on other psychological tests their scores were no more similar than randomly selected individuals. Other pairs were subsequently identified, and in 1937 the first account of a study of several reared-apart twins was published by Newman, Freeman, and Holzinger (see Chapter 4). Various single-case studies of reared-apart pairs concordant or discordant for psychiatric disorders have been published over the years. Reports of schizophrenia diagnoses among these pairs are discussed in Chapter 6.

The Co-twin Control Method

Another method of studying twins is known as the 'co-twin control method,' which was developed in the late 1920s by Gesell and

Thompson.[81] This method performs an environmental manipulation on one member of an identical twin pair in order to test the effect of the manipulation. For example, it might be possible to test the efficacy of a particular teaching method by applying it to one twin and not the other. Because genetic factors are controlled for on the basis of the genes shared by identical twins, it is reasoned that test scores would reveal whether the method improved learning. The co-twin control method differs from other types of twin research because it is utilized to investigate environmental influences, whereas most other types of twin studies are used to detect genetic influences. Still, a finding that environmental influences do not lead to large behavioral differences can be used to support the genetic position, and Gesell, like other pioneer twin researchers, was influenced by eugenic ideas.[81] The co-twin control method is not used widely today for several reasons, which include ethical problems and its history of abuse (see below).

Twin research in Germany

Germany was the world center of twin research in the late 1920s and 1930s. It is commonly known that biology and genetics played a major role in the 'Third Reich' and helped provide a 'scientific' justification for the murder of mental patients and, ultimately, the Holocaust. What is not well known is that the intellectual seeds of the misuse of genetics were sown long before Hitler was named Chancellor in January, 1933.

One of the early investigators interested in twin research was Wilhelm Weinberg (1862–1937), who also served as chairman of the Society for Racial Hygiene's Stuttgart branch.[82] Weinberg was a popularizer of the 'differential method,' which uses a simple calculation to estimate the number of identical pairs that should be found in a given population of twins. He also created the first 'morbidity tables' used in medical statistics, as well as an age-correction method subsequently used by several twin researchers. In a 1938 obituary written by F. J. Kallmann, it was said that Weinberg's most well-known publications 'deal with the frequency of taint in the families of tuberculosis and schizophrenia patients, the relationship between tuberculosis infection and heredity, the racial importance of accurate fertility and mortality statistics, and the consideration of selective factors and age distribution in biological studies.'[83] According to Stern, Weinberg worked alone without colleagues or students, although other investigators (such as Rüdin) sought his advice from time to time. 'Only when he was old,' wrote Stern, 'did a new generation again begin to explore the areas in which he had achieved much, a nearly lone worker.'[84]

In 1928 Hans Luxenburger published the first schizophrenia twin study. Another study performed in Germany was published by Johannes Lange in 1929, which appeared in English the following year under the title *Crime and Destiny* (the German title of Lange's book, *Verbrechen als Schicksal*, translates as 'Crime *as* Destiny'). Lange studied 13 identical and 17 fraternal pairs and found a much higher identical twin concordance rate for criminality. He concluded that genetic factors 'play a predominant part'[85] in criminality, and that 'We must try to make it impossible for human beings with positive criminal tendencies to be born.'[86]

1933 was a pivotal year in world history and also began the golden age of twin research in Germany. Hitler's regime gave a prominent place to biology and genetics. Indeed, Rudolf Hess called the National Socialist movement 'applied racial science.'[87] It was therefore natural that Germany would take the lead in racial hygienic measures, and in July, 1933, the regime passed the 'Law for the Prevention of Genetically Diseased Offspring.' This law (which was not officially abolished until 1988[88]) established genetic courts having the power to involuntarily sterilize people labeled feebleminded, schizophrenic, manic-depressive, epileptic, genetically blind of deaf, or alcoholic. The law was justified on the basis of genetic studies, which included twin research. In an official commentary on the new sterilization law, Gütt, Rüdin, and Ruttke wrote that natural laws applicable to plants and animals also applied to humans, and that this view was 'fully and completely . . . confirmed during the last three decades both through family research and through the study of bastards and twins.'[89] The commentary included family pedigree charts and illustrations showing how the sterilization procedure is surgically performed.[90]

The focus of Nazi-era genetic studies included criminality, epilepsy, manic depression, tuberculosis, mental retardation, and abnormal personality. These studies were performed in order to help justify the regime's racial hygienic measures. The authors of two criminal twin studies published in 1936, Heinrich Kranz and Friedrich Stumpfl, both called for the compulsory sterilization of criminals. According to Kranz,

> One could ascertain so far on the basis of twin concordance rates that have been found that the imbecile criminals are undesirable in terms of racial hygiene [rassenhygienisch unerwünscht]; furthermore, some types of criminal psychopaths are borderline psychotics and severe alcoholics. All of these are already being recorded to a large extent through the sterilization law.[91]

Kranz wrote that the genetic impairment of sex offenders 'can hardly be questioned,' while adding that 'the castration law is simultaneously fulfilling the racial hygienic task (rassenhygienische Aufgabe).'[92] Stumpfl went on to work in the Innsbruck *Intitut für Erb- und Rassenbiologie* (Institute for Genetics and Racial Biology) in Nazi occupied Austria from 1939–1945.[93]

Twins were regarded as a precious resource in Hitler's Germany. As the prominent racial hygienist Eugen Fischer observed, twin research was the single most important tool in the racial hygiene field.[94] In 1936, summer camps for 'Aryan' twins were established in Germany for the purpose of scientific study. As described in 1939 by Germany's leading twin researcher, Otmar Freiherr von Verschuer, 'the latest development of the methodology of twin research' are 'the so-called "twins camps"':

Twin children of Berlin and other towns who were in need of a holiday were brought together in two camps, for a 6 weeks' summer holiday. Several doctors, psychologists, and trained nurses lived with the children during this time, kept them under constant observation, and took notes. Every experience of the twins from early morning until bedtime, the daily rhythm of life, the succession of moods, the reaction to difficulties, to successful and unsuccessful experiences and to conflicts were exactly noted. This kind of observation was supplemented by experimental psychological tests introduced unobtrusively into the daily life. These investigations have led to essentially new conclusions about the hereditary nature and development of normal psychological tendencies, about which we cannot give a detailed account here.[95]

According to the Reich Research Council records, twin camp research was funded until November, 1944.[96]

A visitor to Germany in the Winter of 1939–1940 wrote of twin studies which 'oblige some ten thousand pairs of twins, as well as triplets and even quadruplets, to report to a scientific institute at regular intervals for all kinds of recordings and tests.'[97] He described the setting where twins were tested:

I was particularly interested in their laboratory for the study of the inheritance of behavior and mental capacities. For this purpose the twins were placed in two identical rooms, separated by a narrow corridor for the observer, who had a free view over both rooms through big windows in the walls . . . If the observer wants a permanent record of the behavior of the twins, he can operate two concealed movie cameras with the lenses nicely camouflaged in the wall paper.[98]

Rüdin

The psychiatric genetics field was born in Germany in the early twentieth century. The founder of the movement was the Swiss-German psychiatrist Ernst Rüdin (1874–1952), who (as we have seen) was also one of the founding members of the German Society for Racial Hygiene. The primary goal of psychiatric genetics was to establish the genetic basis of psychiatric conditions for the purpose of promoting racial hygienic measures for those diagnosed with these conditions. German psychiatric genetics played a central role in training people to conduct twin studies, and researchers from other European countries came to Munich to study under Rüdin and his associates at the Genealogical-Demographic Department of the Kaiser-Wilhelm Institute of Psychiatry. Among these students were several people who went on to have long and influential careers in psychiatry, such as Franz Kallmann (Germany and the United States), Eliot Slater (UK), Eric Strömgren (Denmark), and Erik Essen-Möller (Sweden).

Today, one can still find positive references to the 'pioneering work' of Rüdin's 'Munich School.'[99] Some contemporary psychiatric geneticists claim that the findings of Rüdin and his associates were merely *used* by the Nazis to justify murder.[100] This position overlooks the role of German scientists as collaborators with the Nazi regime and as people who provided the intellectual justification for many of its practices.[101] To understand this more clearly, we must look at the views and deeds of Rüdin himself.

Rüdin was the founder of psychiatric genetics and has been referred to as one of its 'great pioneers.'[102] We have seen that Rüdin was a popularizer of the Nazi eugenic compulsory sterilization law, under which 300,000–400,000 people were forcibly sterilized.[103] In 1934 Rüdin (like Siemens) recognized that 'only through the political work of Adolf Hitler, and only through him has our more than thirty-year-old dream become reality: to be able to put race hygiene into action.'[104] In 1935 Rüdin proposed that the sterilization law be extended to include 'valueless individuals . . . all who were socially inferior psychopaths on account of moral confusion or severe ethical defects.'[105] Rüdin thereby remained true to the views he had put forward at the beginning of his career in 1903, when he wrote that in order to obtain 'biologically fit members of the race,' it would be necessary to promote 'maximum propagation of those who are healthy, robust and . . . ethically superior,' while preventing 'the weak, ill, unfit, and morally reprehensible from reproduction by artificial selection . . . by instruction and by private and government force.'[106] Thus, before the passage of compulsory eugenic sterilization laws anywhere in the world, the 29-year-old Rüdin was calling for the use of 'private and government force' to prevent the reproduction of the 'unfit.'

According to his biographer, Rüdin 'played a major role in the propagation of racial hygiene doctrines in the "Third Reich."'[107] Hitler awarded Rüdin the Goethe medal for art and science in 1939, which was accompanied by a telegram from Nazi Interior Minister Wilhelm Frick, which read, 'To the indefatigable champion of racial hygiene and meritorious pioneer of the racial-hygienic measures of the Third Reich I send . . . my heartiest congratulations.'[108] In the same year Rüdin, who by now had joined the Nazi party,[109] gave a talk which ended with a long quotation from the 'Führer, spoken at the 1937 Party Day in Nuremberg.' Rüdin ended his speech by calling on psychiatrists and neurologists to 'not only hear [Hitler's] serious words, but to take them to heart and act accordingly!'[110] Rüdin's university lecture notes help illuminate his racial ideas. He instructed his students that 'Nordics [are] tall, slim, narrow, and long-headed. Blue-eyed, creative, brave, heroic spirit, self-restrained, cautious, self- respecting, [and] honest.' Rüdin's notes for 'Jews' read, 'racial mixture, predominantly Near Eastern, oriental and negroid. Characterized by persuasiveness and importunacy [*sic*]; early maturity, quickness of understanding but not creative; *thiebhaft* [*sic*] (tendency to follow baser instincts with emphasis on sexual crimes); dishonest; tendency to agitate (Karl Marx, Kurt Eisner).'[111]

In the late 1930s the regime took 'racial hygiene' and eugenic sterilization to its logical conclusion and instituted an extermination campaign against mental patients and 'defectives.' This Nazi extermination campaign was euphemistically called 'euthanasia' (the program was code-named 'T4'), and resulted in the murder of approximately 70,000 people.[112] It is known that Rüdin discussed 'euthanasia' as a type of 'therapeutic reform.'[113] A memorandum on euthanasia co-authored by Rüdin and others suggests that Rüdin supported the murder of mental patients:

> Even the euthanasia measures will meet with general understanding and approval, as it becomes established and more generally known that, in each and every case of mental disease, all possible measures were taken either to cure the patients or to improve their state sufficiently to enable them to return to work which is economically worthwhile, either in their original professions or in some other occupation.[114]

In 1942 Rüdin stressed 'the value of eliminating young children of clearly inferior quality,'[115] and in the same year he published an article celebrating the tenth anniversary of National Socialist rule, while the Holocaust was already underway:

> The results of our science had earlier attracted much attention (both support and opposition) in national and international circles. Nevertheless, it will always remain the undying, historic achievement of Adolf Hitler and his followers that they dared to take the first trail-blazing and decisive steps toward such brilliant race-hygienic achievement in and for the German people. In so doing, they went beyond the boundaries of purely scientific knowledge. He and his followers were concerned with putting into practice the theories and advances of Nordic race-conceptions . . . the fight against parasitic alien races such as the Jews and Gypsies . . . and preventing the breeding of those with hereditary diseases and those of inferior stock.[116]

Rüdin spoke favorably about Nazi racial laws which led to a 'progressive reduction of Jewish influence and especially the hindrance of further intrusions of Jewish blood into the German gene pool.'[117] Rüdin saw World War II as being the result of 'the Jewish-plutocratic and Bolshevik directed forces,' and regretted that the racial hygienic agenda had been slowed by the war.[118] This setback, he wrote, 'will only spur us on to multiply our racial hygienic efforts after the victory.'[119] Readers may wonder why someone of Rüdin's stature would put these views into publication, but the answer is clear: at the time these lines were written, Rüdin, like most Germans, expected Germany to win the war.

Nonetheless, Rüdin's contemporary supporters continue to claim that he opposed the Nazi T4 euthanasia program and that he helped Jews and was an opponent of anti-Semitism. Gottesman and Bertelsen claimed that Rüdin 'was not directly involved' in the euthanasia program, and that when he found out about it 'in 1940, he declared it to be murder.'[120] Gottesman and Bertelsen also claimed that Rüdin, 'rationalizing after the war,' stated that his reason for joining the Nazi party in 1937 was the necessity 'to preserve his life's work, his research institute, and its funding, and to protect his staff, including Jews, for as long as possible.'[121] Many Nazi criminals rationalized their actions after the war. In common language it is called trying to save your skin. Former Munich student Erik Strömgren wrote about a 1936 funeral for a Jewish researcher where, braving bad weather, 'everybody' at the institute attended because they 'wanted to pay him the last tribute.' The implication was that Rüdin and his colleagues were not anti-Semitic.[122]

Luxenburger

Rüdin's history is discussed here as an important example; there were many other figures in German psychiatry who held similar views and performed similar functions. Some were prominent twin researchers. Rüdin's associate Hans Luxenburger (1894–1976) published the first schizophrenia twin study in 1928 and wrote extensively on twin research and psychiatric genetics throughout the 1930s, yet the unsavory aspects of his background are left unmentioned by contemporary twin researchers.

For Luxenburger, the 'core significance of human twin and multiple research' was to provide a scientific basis for the 'unity of racial hygienic and medical thinking.'[123] Luxenburger supported and helped implement the 1933 eugenic sterilization law, although he viewed the sterilization of identified 'unfit' people as an inadequate measure to achieve racial hygienic goals, *before* Hitler's rise to power. At the twenty-fifth annual conference of Bavarian psychiatrists held in 1931, Luxenburger discussed calls for sterilization.

> Firstly, occasional mistakes are unavoidable, but they can be tolerated in view of the grave crisis in which our race currently finds itself. It is clear that we cannot fully prevent the transmission of recessive hereditary properties through sterilization; but they will certainly be considerably contained, and it is impossible to see why one should sit back and do nothing only because a radical eradication of degenerate hereditary properties is still impossible today.[124]

Apparently even in 1931, compulsory sterilization was for Luxenburger only a temporary method to 'contain' 'degenerate hereditary properties' until a future time when they could be 'radically eradicated.'

During this period Luxenburger published articles expressing his outrage over the lack of legal support to eugenic practices. In 1931 Luxenburger wrote about the 'misdeed' of allowing 'the most dangerous people . . . the improved or cured mentally ill' from helping to create the next generation. 'We are forced,' he lamented, 'to watch without the ability to act.'[125] In another article he expressed misgivings about therapeutic interventions with psychiatrically diagnosed people because 'if the germ cells are left untouched,' the 'hereditarily ill' recovered patients 'become a serious danger for the race' because they will pass along their predispositions to their offspring.[126] Luxenburger warned that 'If psychiatric practice succeeds in widely preventing the manifestation of a mental illness or in the reduction of the time spent in an asylum by the mentally ill, this would undoubtedly lead to a disaster for the race.'[127] He concluded by stating that eugenics will support therapeutic interventions only on the assurance that, in addition to the health of the individual, the 'health of the race' is also considered.[128]

Luxenburger's hard-line racial hygienic views are seen clearly in a 1931 article in *Eugenik*. There, he demanded the sterilization of 'every securely diagnosed schizophrenic' before discharge from an institution, if the person was of an age where they could father a child or become pregnant. Additionally, he demanded that people diagnosed with schizophrenia who do not necessarily need to be institutionalized should be institutionalized if they do not submit to sterilization.[129] Turning his attention to other diagnoses, Luxenburger wrote that every person who (after being tested) is judged to be hereditarily feebleminded 'must be sterilized.'[130] Not only the mentally retarded but also their entire families were candidates for racial hygienic measures. According to Luxenburger, 'The biological, social, and cultural standard among the families of the hereditarily feebleminded (erblich Schwachsinniger) is in most cases so low that the complete elimination (völlige Ausschaltung) of these families from the population would be a clear gain.'[131] He wrote that every criminal 'whose criminality is mostly hereditary' has to be 'stopped from procreation,' eventually through force, even though the 'social causes of criminality have to be studied and completely eliminated.'[132] Regarding 'sexual perversion,' Luxenburger warned that 'attempts to heal the sexually perverse, especially exhibitionists or homosexuals, through marriage or normal sexual activity without prior sterilization are not only ineffective, they are a virtual crime against the race.'[133]

Luxenburger worried that the procreation of the 'unfit' would lead to a situation where 'the fit will be overtaken and annihilated' by a 'huge army of dull inferiors.' It is 'the responsibility of eugenics,' he wrote, 'to make sure that this horrible vision will never become a reality in our culture.'[134] The situation was so grave, Luxenburger warned, that urgent measures were required:

> The preservation of the fit by means of elimination of the unfit [Ausmerze des Untüchtigen] — the highest goal of eugenics — will never be fully achieved. The eugenicist is in the position of a surgeon who, with the intention of getting rid of a cancerous growth, will prefer to cut off some of the healthy substance in order to be sure to get rid of the disease.[135]

In a chilling forecast of events to come, in the early 1930s Luxenburger wrote that doctors would soon become 'the executors of the eugenic will of the nation.'[136] Although Luxenburger saw the 'elimination of the unfit' as the 'highest goal of eugenics,' he stopped short of calling for the 'destruction of lives unworthy of living,' proposed by Hoche and Binding in 1920. In 1931 Luxenburger wrote that the discussion of 'social euthanasia' (killing the 'unfit') 'is unworthy of a civilized people.'[137]

In a discussion of how the 1933 sterilization law should be carried out, Luxenburger argued that medical and racial hygienic interventions should be thought of as operating on a continuum.[138] For Luxenburger, 'the healthy identical co-twin of a schizophrenic is genetically a schizophrenic,' and 'We should demand an extension of the definition of sickness according to the Law for the Prevention of Genetically Diseased Offspring [the 1933 sterilization law] to apply to both identical twins, including the twin who is not ill.'[139] Thus, from Luxenburger's racial hygienic perspective, there was no difference between a person diagnosed with schizophrenia and his or her 'well' identical co-twin. Luxenburger listed four indications that an individual should be the subject of racial hygienic measures: (1) the person manifests a condition deemed hereditary; (2) the person is the identical twin sibling of an affected person; (3) for recessive conditions, the person has two affected parents; or (4) the person manifests latent symptoms of the (supposedly) hereditary condition. Luxenburger was particularly concerned that these 'latent' individuals would escape sterilization because they were not clinically identified.[140]

Luxenburger's position that the identical twin of a person diagnosed with schizophrenia is also 'schizophrenic' demonstrated his view of individuals more as genotypes than as people. That other biological relatives were not similarly targeted (as we will see Kallmann attempted to do) was due only to Luxenburger's inability to definitively identify them as carriers. In 1940 Luxenburger wrote that a close relationship between race and hereditary illness was 'likely,'[141] and that 'hebephrenia,' depression, and hysteria 'are rather widespread among the Jewish people living in Europe.'[142] Thus after Kristallnacht and the passage of the Nuremberg Laws, and on the eve of the Holocaust, Luxenburger found it necessary to remind his readers that Europe's Jews were the disproportionate carriers of 'hereditary illness.'

Luxenburger published several articles in the mid-1930s in von Verschuer's influential journal *Der Erbarzt* (The Genetic Doctor). This journal, which first appeared in 1934, was dedicated to the promotion of racial hygiene, the sterilization law, and the maintenance of the German 'Volk's' racial purity.[143] In a 1939 edition of the Nazi-controlled *Archiv für Rassen- und Gesellschaftsbiologie*, Luxenburger reviewed a book on racial hygiene by the Romanian eugenist G. Banu. There, he took Banu to task for not going far enough in pursuit of the racial hygienic agenda.[144] In the late 1930s Luxenburger ran afoul of the regime because he publicly criticized Julius Streicher's theory that the children of an 'Aryan' woman would not be 'Aryan' if she had previously had intercourse with a 'non-Aryan.' Still, Luxenburger was one of the Nazi era's leading psychiatric genetic and racial hygienic ideologues, and played a role in creating the climate that culminated in mass murder.

Schulz

Bruno Schulz (not to be confused with Bruno K. Schultz) was another important associate of Rüdin. Of the leading Munich psychiatric geneticists, Schulz is most often portrayed as a opponent of the National Socialist regime. It also is said that he, and not Rüdin, was the person most responsible for training students in psychiatric genetic methods. In 1991 Gottesman wrote that Schulz was a 'star member of Rüdin's Munich school.'[145] According to British twin researcher Eliot Slater, a student of Schulz's in 1934, 'Schulz's political integrity was the equal of his scientific honesty. He could not stand the Nazi ideology and never compromised with it.'[146] (Slater did admit that Klaus Conrad, another Munich twin researcher, was 'an enthusiastic Nazi.'[147]) However, it was not possible to be a public figure (and publish) under the Nazi totalitarian regime without 'compromising' with it. Of course, it would have taken tremendous courage to speak out under the conditions of Hitler's Germany, and one cannot be condemned for not doing so. Still, Slater's account describes the impossible scenario of Schulz living under a fascist regime, as an opponent, and not compromising with it. In 1997 American psychiatric geneticist and twin researcher Kenneth Kendler portrayed Schulz in a positive light, writing that 'Schulz never joined the Nazi party to the detriment of his academic career.'[148] According to Strömgren, Schulz 'was deeply disgusted with the Nazi regime.'[149]

In fact, Schulz aided the regime and supported its early racial hygienic measures. In 1934 Schulz published an article entitled 'Racial Hygienic Marriage Counseling' in *Volk und Rasse* ('People and Race'; the edition was a *Festschrift* to Ernst Rüdin on the occasion of his sixtieth birthday) in which he supported racial hygienic measures and the sterilization law.[150] *Volk und Rasse* had been taken over by Hitler's SS (Schutzstaffel) in 1933 and was used to promote racial theories and Nordic superiority.[151] The editorial board of the 1934 edition of *Volk und Rasse* containing Schulz's article (Volume 9, Number 5) featured Nazi 'Reichsminister' Walter Darré, 'Reichsführer SS' Heinrich Himmler, and other members of the SS. According to a historian, *Volk und Rasse* 'became more and more a mouthpiece for Himmler's Nordic programme.'[152] Although according to Slater's account, Schulz was an uncompromising opponent of the political regime he 'detested,'[153] Schulz 'compromised' enough to write an article on racial hygiene and sterilization which earned the warm approval of 'detested' Nazis Heinrich Himmler, Walter Darré, and associates.

Schulz wrote in this 1934 article that the law was insufficient to eliminate all 'hereditary illness,' writing, 'It is without question that the reproduction of another large number of persons who do not fall under the law are undesirable in terms of racial hygiene.' In this group he included 'diabetics . . . hereditary neurotics, certain types of psychopaths, etc.' For those whom Schulz was willing to grant a reprieve from the sterilizer's knife, he recommended 'racial hygienic marriage counseling.'[154] According to Schulz, the 'Children of someone who later becomes schizophrenic have to be considered qualitatively equal to children of obvious schizophrenics' because the 'illness' serves merely to identify the nature of the genes.[155] In 1939 Schulz published an article in von Verschuer's *Der Erbarzt* in which he wrote of empirical genetic prognostic research in the context of support for the 1933 sterilization law and other German racial hygienic statutes.[156]

Rüdin's biographer, Matthias Weber, claimed that in 1936 Schulz 'questioned Nazi sterilization laws under the protection of his mathematical apparatus, which only specialists could understand.'[157] Although Weber performed a public service by exposing some of Rüdin's criminal activities, in this case he failed to mention Schulz's public support of the sterilization law in his 1934 *Volk und Rasse* article. After the war, Schulz was a founder of the Max Planck Institute in Munich, which was the successor to Rüdin's institute.

Psychiatric geneticists Rüdin, Luxenburger, Schulz, Conrad and others studied twins and popularized twin research in order to provide what they viewed as additional evidence that people with 'undesirable' traits were 'hereditarily tainted,' and deserved to fall under the sterilizer's knife or worse. This was the main reason that they studied twins, and this is how Rüdin's Munich school should be remembered.

But the full history of the Munich school continues to be denied. According to Kendler, Luxenburger was 'an active Catholic,' who 'deeply disliked the Nazis.'[158] It is true that Luxenburger was an active Catholic, although this did not prevent him from also being an active eugenicist and sterilization proponent. The claim that Luxenburger 'deeply disliked' the Nazis, while containing an element of truth, is highly misleading. From the standpoint of educated elitists such as Luxenburger, the Nazi leaders and their plebeian followers were often looked down upon as 'uneducated rabble.' As a historian correctly noted, 'Many geneticists initially had a disdainful attitude toward nazism, as being a vulgar rabble-rousing movement.'[159] Strömgren wrote that Luxenburger 'did not hesitate to make jokes about the government and the Nazis.'[160] In the same spirit, an aristocratic Wehrmacht Field Marshal may have looked down upon (and made jokes about) 'Corporal Hitler' and his modest origins and educational background. To cite this as evidence that the same general was an opponent of Nazism, however, would be a gross distortion. Undoubtedly, many racial hygienists who recognized Hitler as enabling their 'utopian dream' to become a reality were at the same time repulsed by the plebeian origins and methods of German National Socialism. But to paint them as anti-Nazi is very far from the truth. In *A History of Psychiatry*, Shorter acknowledged that Luxenburger 'served as an apologist for racist genetics for a few years under the Third Reich,' but then mistakenly claimed that Luxenburger became a critic of the regime in subsequent years.[161] Getting into trouble for having publicly criticized the ravings of one particularly outrageous Nazi figure is not the same thing as being a critic of the regime in general.

In a 1971 account by psychologist David Rosenthal, who apart from Kallmann was the most important popularizer of psychiatric genetics in the United States, we read,

> Psychiatrists founded a research institute in Munich, Germany, in the first decade of the century, and the field of psychiatric genetics was born there, fathered by Ernst Rüdin, a man little known in our own country. From this institute emerged all of the pioneering psychiatric geneticists, some of whom are still alive and active. They concerned themselves with many psychiatric disorders, especially the new diagnostic entities, dementia praecox (schizophrenia) and manic depressive psychosis. They devised systematic research strategies based on sound principles of human genetics, and generated a large body of data, never doubting that they were dealing with genetic diseases. Unfortunately, concomitant with the development of psychiatric and other aspects of human genetics, there grew in Germany an ugly, racist view of man, in which the Aryans were deemed to be the master race. Many German scientists voiced their approval of these ideas, offering pseudoscientific statistics and generalizations in support of them, while the Nazi *Herrnvolk* [master race] went about the inhuman business of committing genocide.[162]

Thus, according to Rosenthal, there was a group of dedicated psychiatric researchers, headed by Rüdin in Munich, who performed pioneering genetic studies 'based on sound principles.' At the same time presumably *other* German scientists supported the master race idea and helped set the stage for genocide. Rosenthal's account is even more inexcusable when we realize that he was very well acquainted with the Nazi-era German psychiatric genetic literature.[163]

Auschwitz

The 'twin research' at Auschwitz conducted by Josef Mengele in the midst of the Holocaust has been ably described in Robert J. Lifton's 1986 book, *The Nazi Doctors*. Mengele was a committed member of the SS and had studied under von Verschuer in Frankfurt. Mengele's teacher was perhaps the most famous twin researcher in the world during the 1930s. According to a 1937 account by American twin researchers Newman, Freeman and Holzinger,

> The most active and influential of modern students of human genetics using this twin method is O. von Verschuer, who heads a large group of students of twinning at the Kaiser-Wilhelm Institut für Anthropologie... This group of investigators has used the twin method

for the study of anthropological, clinical, serological, and physiological problems and has contributed extensively to our knowledge of these subjects and, incidentally, to our general knowledge of twins.[164]

It is likely that Mengele was one of these students.

Mengele arrived at Auschwitz in the Spring of 1943 for the purpose of conducting studies on twins. His task was to identify twin inmates from those at the camp who had not yet been put to death, as well as to select twins from among the newly transported. Mengele frequently performed this task personally by wading into the stream of new arrivals and shouting 'Zwillinge heraus!' (twins, get out!).

Mengele performed gruesome and murderous 'experiments' on twins, at times killing them personally and performing post-mortem examinations. His technique of choice was the co-twin control method. Sometimes he would send the body parts of his victims to von Verschuer in Berlin, with whom he was in close contact.[165] Thus, *the world's 'most active and influential' twin researcher was an accomplice to the murder of twins in the name of science.* In 1942 von Verschuer wrote that 'The policies of National Socialism have finally put an end to the last threat to the race — that which comes from the Jews,'[166] and in 1944 he proclaimed that the alleged dangers posed by Jews and Gypsies had been 'eliminated through the radical-political measures of recent years.'[167]

For Mengele, von Verschuer and others, Auschwitz presented an unprecedented opportunity to conduct twin research. 'It would be a sin, a crime,' Mengele is reported to have said, 'not to realize the possibilities that Auschwitz had for twin research. There would never be another chance like it.'[168] It is estimated that 3,000 twins were 'examined' by Mengele, but only about 200 were left alive by the end of the war.

Twin research and human genetic investigation were dealt a severe blow by the crimes committed by the Nazis and their willing scientific accomplices. Still, some contemporary psychiatric geneticists persist in denying these crimes. In response to my quite modest statement in 2000 that 'history has shown that the results of genetic studies have often been used to stigmatize individuals and groups,'[169] American psychiatric geneticists Stephen Faraone and Joseph Biederman responded by criticizing me for presenting 'no data' supporting this claim, while admitting only that 'psychiatric genetic data had been used by the Nazis to justify eugenic sterilization and murder.'[170] In a similar fashion, psychiatric geneticists Zerbin-Rüdin and Kendler wrote in 1996 about the Nazis' 'possible [sic] misuse and abuse of findings from the field of psychiatric genetics.'[171] Statements such as these ignore the *collusion* of German psychiatric genetics and Hitler's regime. In their 1999 psychiatric genetics textbook, Faraone, Tsuang, and Tsuang wrote that the 'findings from psychiatric genetics were abused by scientists and politicians to kill rather than to cure.'[172] But these scientists had names, and among them was the 'great pioneer' of psychiatric genetics, Dr. Ernst Rüdin.

Twin research from 1945–1960

In the years after World War II twin research was performed by a small group of people, many of whom were motivated by eugenics, who believed that their research continued to have an important role to play in assessing the relative weight of nature and nurture. This was a difficult time for twin research, as Germany was seen as an example of what eugenics and racial hygiene could ultimately lead to. In addition, behaviorist and Freudian explanations for psychiatric disorders were widely accepted, although few psychiatrists believed that conditions such as schizophrenia had no genetic component at all. For those who believed that psychiatric disorders were strongly influenced by heredity, the flame was kept burning by researchers who had studied at Rüdin's institute in Munich.

One of these researchers was Munich-trained Eliot Slater, who published the results of his British schizophrenia twin study in 1953. Along with his colleague James Shields, Slater played an important role in promoting psychiatric genetic ideas in the English-speaking world. Another Munich graduate, Erik Essen-Möller, performed twin research in Sweden and had published his own schizophrenia twin study in 1941.

Kallmann
In the United States, the message of psychiatric genetics was brought from Rüdin and Munich in the person of Franz J. Kallmann (1897–1965). In 1946 Kallmann published his schizophrenia twin study, in which he claimed to have investigated and diagnosed 691 pairs of American twins. He reported age-corrected pairwise concordance rates of 85.8% identical, and 14.7% fraternal. Kallmann concluded that schizophrenia required a genetic predisposition 'which is probably recessive and autosomal.'[173] He had published his famous schizophrenia family study eight years earlier. This investigation was based on the families of German patients diagnosed with schizophrenia (see Chapter 6 for a critique of the schizophrenia twin studies).

Kallmann was well known in the United States and the UK because of his large twin samples, and the high concordance rates he reported. He was also controversial, in part because of the impact that his strong eugenic views may have had on the way he diagnosed twins. In most cases, Kallmann was aware of the co-twin's diagnosis and whether a twin was identical or fraternal. Lewontin, Rose, and Kamin noted the irony of the fact that, while Kallmann's data is less accepted as time goes on, a belief in the genetic basis of schizophrenia, which these data helped establish, becomes more accepted.[174]

Like his former Munich colleagues, Kallmann's primary interest was to produce evidence in support of policies that would eliminate the genes he believed caused psychiatric disorders. Kallmann's background and activities in Germany and the U.S. in the 1930s are seldom discussed by his contemporary supporters. When they are, he is typically portrayed as a 'Jewish refugee' who 'fled Hitler's Germany.'[175] These portrayals grossly misrepresent Kallmann's history.

In 1935, while still active in Germany, Kallmann called for the compulsory sterilization of people diagnosed with schizophrenia and their 'heterozygous' relatives as well. According to Kallmann,

> It is desirable to extend prevention of reproduction to relatives of schizophrenics who stand out because of minor anomalies, and, above all, to define each of them as being undesirable from the eugenic point of view at the beginning of their reproductive years.[176]

Nevertheless, Kallmann was forced to leave Germany in 1936 because he had a Jewish father. It is therefore ironic that Kallmann, ever in search of 'hereditary tainting' in others, would be forced to flee his homeland with little more than the clothes on his back and his family study manuscript, on the basis of his own 'taint.'

So perhaps, upon arriving in the United States, Kallmann's personal experiences would compel him towards a greater sensitivity to those with 'bad blood'? He hardly missed a beat. In a 1938 article in *Eugenical News*, Kallmann called for the use of 'negative eugenic measures' against the 'carriers of any mental disease' meeting certain criteria.[177] He continued,

> From a eugenic point of view, it is particularly disastrous that these patients not only continue to crowd mental hospitals all over the world, but also afford, to society as a whole, an unceasing source of maladjusted cranks, asocial eccentrics and the lowest types of criminal offenders. Even the faithful believer in the predominance of individual liberty will admit that mankind would be much happier without those numerous adventurers, fanatics, and pseudo-saviors of the world who are found again and again to come from the schizophrenic genotype.[178]

Kallmann presented data from his family study which, he claimed, showed that 'schizophrenia is a single-recessive trait and a specific genetic entity.'[179] In order to protect society, Kallmann advocated 'eugenic prophylactic measures' to deal with schizophrenia 'taint-carriers.'[180]

Another misleading account of Kallmann's history was published shortly after his death by his associate John Rainier. According to Rainer, Kallmann was a 'medical man who takes a humanitarian interest in the psychological and emotional problems of his patients.'[181] Regarding Kallmann's activities in the 1930s, Rainer wrote,

> By 1934, [Kallmann] was becoming increasingly anxious to leave Germany, where he found the political abuse of genetics for racial purposes, including sterilization laws for the mentally ill, not at all consistent either with the Hippocratic responsibility of the physician or with the genetic data on schizophrenia which he was beginning to collect. In 1936 Kallmann came to the United States . . . [182]

And Kallmann himself was not above describing his motivations in ways that suited the times. Let us compare two accounts. The first is from his 1938 schizophrenia family study. In the forward to the book describing this study, Kallmann wrote,

> The present study of the genetic and reproductive conditions in schizophrenia families was undertaken in 1929, in collaboration with the German Research Institute for Psychiatry in Munich (Prof. Dr. Rüdin) . . . *The principal aim of our investigation was to offer conclusive proof of the inheritance of schizophrenia* and to help, in this way, to establish a dependable basis for the clinical and eugenic activities of psychiatry [emphasis added].[183]

However, in 1953 Kallmann provided this account:

> As a young psychiatrist, I divided my time between clinical service and work in histopathology. On visiting days, the head nurse . . . often announced a group of waiting visitors with the statement: 'Doctor, the familial taint is here.' It was just because that nurse was so domineering that I *undertook my first family study to 'disprove' the inheritance of schizophrenia* [emphasis added].[184]

Kallmann's account therefore shifted 180 degrees, from someone who studied families in order to provide 'conclusive proof' of the genetic basis of schizophrenia in order to 'aid the eugenic activities of psychiatry,' to someone who wanted to 'disprove' the idea that schizophrenia was the result of a 'familial taint.' A parallel process occurred in Germany, where post-World War II editions of the works of leading German scientists were edited to remove positive references to Hitler and National Socialism. In the 1950s and 1960s, Kallmann wrote frequently about eugenics and was active in the American Eugenics Society until his death.[185]

   In spite of his history of collaboration with Rüdin and German racial hygienics, Kallmann has been called upon by twin researchers Gottesman and Bertelsen as a posthumous character witness for Rüdin.[186] After mentioning a few of Rüdin's relatively minor crimes while ignoring the most heinous, Gottesman and Bertelsen concluded, 'perhaps 50 years is not yet enough time to render a final evaluation' of Rüdin.[187] They ended with a lengthy quote from a 1947 letter written by Kallmann in an attempt to help Rüdin escape justice in the latter's post war 'denazification' trial (Rüdin received a 500 mark fine). In his letter, Kallmann claimed that Rüdin 'is no criminal, of course.' Amazingly, Gottesman and Bertelsen were 'content to let Kallmann have the last word for now'![188] Israeli psychiatric investigators Lerer and Segman, objecting to Gottesman and Bertelsen's 'whitewash,' wrote bluntly that Rüdin 'was not only a willing accomplice to the most abhorrent crimes against humanity but an enthusiastic theorist who provided the intellectual basis for many of these crimes.'[189]

In a rare discussion of the embarrassing aspects of their field's history and origins, twin researchers Torrey, Bowler, Taylor, and Gottesman wrote in 1994 about the popularity of twin research in Nazi Germany, and about Mengele's crimes at Auschwitz in collaboration with von Verschuer.[190] At the same time, nothing was said about Rüdin's criminal role, and Luxenburger was discussed in a favorable light. Finally, Kallmann was portrayed in his usual role as a German researcher forced to flee his homeland because of his Jewish ancestry, and his racial hygienic views are left unmentioned.

Selected topics in twin research
Zygosity determination
Zygosity determination refers to the necessity of accurately assessing whether a given pair is identical or fraternal. The ability to make this distinction is critical for twin research and is more difficult to accomplish than many people might realize. Zygosity determination by physical appearance alone can lead to mistakes. It was once thought that zygosity could be determined at birth on the basis of whether twins arrived in one or two chorion sacks, but it has been known for many years that about one-third of identical twins also have two chorions. Additionally, there is the problem of determining the number of chorions many years after the births took place.

One of the first ways that zygosity was determined was the 'similarity method' discussed by Siemens in 1927, which compared traits such as eye color, hair color, skin color, palm prints, ear form and body build.[191] Siemens believed that by comparing a sufficient number of traits, he · would be able to determine zygosity with a very high level of accuracy. In 1937 Newman and colleagues agreed with Siemens that the chances of making a mistake with this method were negligible. At the same time, several twin researchers were seeking better methods.

In 1941 Essen-Möller introduced the method of blood-typing, where samples of twins' blood were compared and an accurate determination of zygosity could be made. Blood-typing is more expensive than most other methods, and the ability to obtain samples is also required. Subsequent methods have combined blood-typing with physical characteristics, and errors are said to be made less than one time in a thousand.[192] Even more recently, zygosity has been determined very accurately by DNA analysis.[193]

Another technique is the 'questionnaire method.' Several researchers have developed questionnaires asking twins or their parents about the similarity of the twins' physical characteristics, whether they were mistaken for each other as children, and whether they were considered as 'two peas in a pod.' In general, the questionnaire method has the advantage of being relatively inexpensive and fairly accurate (about 95%). It is also used in cases where the investigators are unable to examine twins personally, such as studies where psychological data are obtained by mailed questionnaires and tests.

Interestingly, the differentiating criteria found in these questionnaires could be interpreted as a demonstration of the dramatically different childhood experiences of the two types of twins. For example, the zygosity determination questionnaire of Cohen and colleagues asked mothers of twins (identical, N = 94; fraternal, N = 61) the following questions (which are accompanied by the results): 'Does either mother or father ever confuse them?' (identical = 78%, fraternal = 10%); 'Are they sometimes confused by other people in the family?' (identical = 94%, fraternal = 15%); and 'Is it hard for strangers to tell them apart?' (identical = 99%, fraternal = 16%).[194] The irony of using dissimilar identical and fraternal childhood environments as part of the twin method's zygosity determination process was not lost on Cohen and colleagues, who wrote, 'the impact of such repeated confusion on individual twinships, or the effect of these differences between MZ and DZ twins is not known with certainty. However, such information must cast doubt upon the assumption of environmental equivalence.'[195] It is a telling indictment of the twin method's logic that a procedure used to distinguish identical and fraternal twins is based on a demonstration of the dissimilarity of their childhood environments, which, paradoxically, twin method proponents (at least until recently) have assumed does not exist.

There are several other problems associated with zygosity determination in twin studies. Some researchers have given vague descriptions of how they determined zygosity. For example, in Kallmann's large 1946 schizophrenia twin study it is stated only that zygosity was determined 'on the basis of personal investigation and extended observation' using the similarity method.[196] This means that Kallmann, a strong believer in the genetic basis of schizophrenia, made personal non-blinded diagnoses. It is reasonable to suspect bias because, when there was doubt, Kallmann might have seen concordant schizophrenic twins as identical, and discordant pairs as fraternal. When we consider that Kallmann also made the schizophrenia diagnoses, bias becomes a bigger problem.

In 1962 psychologist David Rosenthal discussed some of the problems with zygosity determination in the schizophrenia twin studies. He noted that in some cases the investigator had to make a determination for a twin who was not available or had died. If the investigator decided to eliminate questionable cases, he or she may have introduced a bias in the sampling procedure. Due to the problem of small sample sizes, Rosenthal described the tendency 'to retain as many pairs as possible by getting as much information as one can about all pairs of twins and making one's best judgment about zygosity and concordance for each pair.'[197] Because of the large size of Kallmann's sample, Rosenthal believed that it was difficult for Kallmann to personally investigate all of the pairs and that he sometimes had to rely on 'lay testimony,' which had been shown to be inaccurate much of the time. Rosenthal suggested that future investigations have one or two people assigned to make zygosity diagnoses, and two (presumably other) people assigned the task of making diagnoses. Although zygosity determination has been a problem in past twin studies, it is not a major problem today as long as investigators use modern techniques such as blood-typing or DNA analysis.

Sampling issues

In psychiatric twin studies there are three main ways in which twin subjects have been obtained: the resident hospital method, the consecutive admissions method, and the register method. In the resident hospital method the researcher examines the records of psychiatric hospital patients who have been diagnosed with the disorder in question. The researcher then attempts to discover if any of these patients are members of a twin pair. After a subject is found in this way, the researcher attempts to locate the co-twin of the affected person. The zygosity of the pair is determined, after which a diagnosis of the co-twin is made. Typically, the same researcher makes the diagnosis for both members of the pair. Sometimes a twin pair will be located because the researcher has been told of the existence of a diagnosed twin (or pair of twins) by a person connected with a particular hospital. The bias this creates is clear, as identicals' more similar physical appearance would make it more likely that hospital workers would know about a pair of concordant identical as opposed to fraternal twins. In addition, the co-twin of a hospitalized identical twin would be more easy to locate.

The resident hospital method was employed in several of the early schizophrenia twin studies (Kallmann's 1946 study being the most prominent). Since the 1960s there has been a consensus among twin researchers that this method creates a sample biased in favor of concordance. Rosenthal discussed some of the problems in obtaining samples for schizophrenia twin studies from resident hospital populations.[198] He observed that many people diagnosed with relatively mild cases of schizophrenia are never hospitalized. This skews the sample towards the most severely affected, who are known to be more concordant than those with milder conditions. In addition, the studied hospitalized twins were already diagnosed with the condition when they were seen by the investigator (although this problem exists in all studies sampling from hospital populations).

Another problem is that in countries and eras where a disorder is believed to have an important genetic component, it is more likely that an identical twin will be diagnosed with a psychiatric disorder if his or her co-twin had already been diagnosed with the disorder. Studies which make blind diagnoses independent of hospital diagnoses can reduce this bias. Luxenburger argued in his 1928 schizophrenia study that samples should be obtained from twins identified from lists of people consecutively admitted to a hospital for a specific period of time. This consecutive admissions method avoids the problems associated with selecting twins who are already known to be affected with the condition in question, although the problems with sampling from hospitals remain.

The third method identifies subjects from twin registers. Beginning in the 1940s, national registers were created in Denmark, Norway, and

Sweden for the purpose of scientific research. Although twin registers existed before this time, it is not clear whether they were created for the purpose of scientific research. The first attempt to create a national twin register for this purpose may have been 1939, when Hitler's Interior Minister Wilhelm Frick 'ordered the registration of all twins, triplets, or quadruplets born in the Reich, for the express purpose of research to isolate the effects of nature and nurture in the formation of the human racial constitution.'[199] A register allows researchers to sample from an entire population of twins and makes it easier to locate both members of a pair. If the investigator is able to locate and evaluate both twins independently of any other source, bias is reduced. However, most psychiatric twin studies check the names of those identified by the twin register with people whose name appeared on psychiatric records or registers. Thus, the old problems of relying on hospital records remained.

Criticism of the twin method by geneticists
In 1950 geneticist Bronson Price published a paper entitled 'Primary Biases in Twin Studies.' Price began with a provocative statement: 'In all probability the net effect of most twin studies has been [the] underestimation of the significance of heredity in the medical and behavior sciences.'[200] Contrary to most previous critics, Price argued that the twin method *underestimated* genetic contributions. The reason, according to Price, was the unique prenatal experience of twins. As examples, he listed the effects of mutual circulation and lateral inversions. Price concluded that the authors of individual case reports of identical twins tended to exaggerate the effects of the postnatal environment, and that interpretations of larger twin studies 'have been about as mistaken.'[201] He advised future twin researchers to determine, two or more months before birth, whether identical twins are monochorial or dichorial and to factor this information into their findings. Even so, Price believed that 'such a study might show that the twin method, as ordinarily applied, is too crude for the purposes of modern nature-nurture studies.'[202]

Geneticists Neel and Schull criticized the twin method in their 1954 textbook *Human Heredity*. They noted that concordant identical twins are more often reported and are therefore more likely to come to the investigator's attention than discordant pairs. They also cited Price's criticism of the twin method's 'primary biases.' Regarding postnatal influences, Neel and Schull discussed the assumption that identical and fraternal twins share similar environments, stating, 'It has been clearly shown that rarely is this the case,' because 'Identical twins are treated more similarly than are fraternal twins by both their families and their acquaintances.'[203] Neel and Schull concluded that 'twins have contributed little which may be extrapolated to other genetic situations,' and that 'In its present context, the twin method has not vindicated the time spent on the collection of such data.'[204]

Twin research from 1960 to 1980

The end of the 1950s was not a good time for twin research for several reasons. The memories of the Holocaust and collaboration of genetic researchers with the Nazi regime were still fresh. In addition, environmental theories were popular. Finally, we have seen that the methods and assumptions of the twin method were criticized by environmentalists and geneticists alike.

Don Jackson's critique of the schizophrenia twin studies

In 1960 a chapter containing a powerful critique of genetic studies of schizophrenia appeared in *The Etiology of Schizophrenia*. Its author was Don Jackson, then Director of the Mental Research Institute in Palo Alto, California. Jackson's chapter was entitled, 'A Critique of the Literature on the Genetics of Schizophrenia.'[205] Jackson was a well-known pioneer of family systems theory and had been a co-author of a widely discussed article outlining the double-bind theory of schizophrenia.[206] At the time of Jackson's article there were five published schizophrenia twin studies — Kallmann's (1946) report being the most influential. Although Jackson's chapter was about schizophrenia, it was clearly relevant to twin studies of other psychiatric conditions and behavioral traits and, to a lesser extent, IQ. Jackson made 11 major points in his critique, and although all did not originate with him, together they made a strong case that the twin method contained serious and invalidating flaws:

1. In discussing family (consanguinity) studies, it was noted that conditions can run in families for environmental reasons.
2. There were no genetic studies of schizophrenia in which diagnoses had been made blindly. The results of these studies were therefore susceptible to the researchers' bias.
3. There were other sources of bias in the diagnostic process, such as the unreliability of schizophrenia diagnoses and the finding that people had a better chance of being diagnosed with schizophrenia the longer they stayed in the hospital. An important sampling bias was introduced by the methods used to obtain twin subjects, which could lead to inflated concordance rates.
4. Contrary to genetic expectations, fraternal twins were more concordant than non-twin siblings, even though they have the same genetic relationship to each other.
5. Contrary to genetic expectations, female identical twins were more concordant than male identical pairs.
6. Contrary to genetic expectations, female fraternal pairs were more concordant than male fraternal pairs.
7. Contrary to genetic expectations, same-sex fraternal twins were more concordant than opposite-sex fraternals.
8. Individual case histories of reared-apart identical twins concordant for schizophrenia do not provide important evidence for genetic factors because they were few in number (two), and the pairs grew

up in similar environments and had an interactive relationship with each other.

9. Identical twins grow up in a more similar environment and are treated more similarly than fraternal pairs. Therefore, greater similarity or concordance of identical pairs could be explained by the more similar environment they experience.

10. The unique psychological bond or 'ego fusion' of identical twins contributes to their higher concordance rate for schizophrenia on the basis of association and identification. Furthermore, the nature of the identical twinship itself might create conditions leading to the identity problems often experienced by people diagnosed with schizophrenia.

11. There is a striking similarity between reports of folie à deux (shared psychotic disorder) and the case histories of identical twins concordant for schizophrenia.

Jackson's chapter had an important impact on the debate over the causes of schizophrenia and raised further doubt that the twin method measured much more than the common environment and 'ego fusion' of twins.[207] In the period following its publication, twin researchers carefully considered Jackson's observations and attempted to improve the methodology of their studies. But most failed to understand that the implication of Jackson's critique was not that the twin method was in need of methodological improvement, but rather that its logic was erroneous at the core level.

As we will see in Chapter 6, the most important twin researchers of the 1960s and 1970s were in agreement that environmental similarity and association were *partly* responsible for the identical-fraternal concordance rate difference. But how did they know that environmental influences were not *entirely* responsible for the difference? The answer is that they didn't – they only *hoped* that their studies had measured genetic influences. In one of their early collaborations, Gottesman and Shields were willing to acknowledge that the greater psychological identification of identical twins could affect concordance rates 'provided that the same proportion of potential schizophrenics are held back from overt illness by identifying with a normal twin as those who become ill by identifying with an abnormal one.'[208] Gottesman and Shields provided no evidence in support of their rhetorical attempt to balance the ledgers of the twin method. On what grounds, one might ask, did Gottesman and Shields insist on a one-to-one correspondence between those twins who became concordant for reasons of association, and those who stayed 'well' for the same reason? Could we not just as easily surmise that, for reasons of identification, there are *five* twins who become concordant for every *one* who remains well? The reasoning of Gottesman and Shields constituted little more than wishful thinking in the service of keeping a theory intact.

Today, Jackson's critique is a forgotten document in the sense that the twin method is more popular than ever in spite of the fact that *not one of*

Jackson's 11 major points has ever been refuted.[209] Twin researchers were unable to convincingly argue against the most important of Jackson's observations, and agreed with him on several points. Those attempting to argue against him typically focused on Jackson's observation (Point #10) that the unique psychological bond of identical twins not only explained high schizophrenia concordance rates, but was also an additional *cause* of 'madness.' On the occasions that defenders of the twin method have directly addressed Jackson's ideas, this is the charge most often made. Let us examine the citations in question. According to Jackson:

> The identity problems of the schizophrenic, most often stressed by psychodynamically oriented writers, could find no better nidus than in the intertwining of twin identities, in the ego fusion that in one sense doubles the ego (because the other is felt as part of the self) and in another sense halves it (because the self is felt as part of the other).[210]

> If the psychodynamic thesis is correct, if ego fusion in a particular family environment can be expected to lead to joint madness, then a plausible hypothesis — contrary to the genetical hypothesis — would be that, according to the degree of likeness in siblings, we will find an increased concordance for schizophrenia, without concern for genetic similarity.[211]

> Psychosis by association apparently requires the nidus of social isolation for its hatching ... These characteristics increase the separation of the twin-pair from the rest of the world and foster a joint ego fusion ... The attempt of one to be like the other is not dissimilar to the ego fusion of some twins.[212]

Here, Jackson discusses how the intimate bond and 'ego fusion' experienced by identical twins might weaken or blur the psychological boundaries between them. While the implication (in the phrase 'no better nidus') that an individual identical twin might be more susceptible to a schizophrenia diagnosis than a single-born individual is open to question, Jackson's concept of ego fusion is an excellent description of psychodynamic factors that could induce a schizophrenia-diagnosed twin's co-twin to receive the same diagnosis much more frequently than his or her non-twin sibling. As an alternative 'plausible hypothesis' to genetic interpretations of the twin method's results, Jackson theorized that 'According to the degree of likeness in siblings, we will find an increased concordance for schizophrenia, without concern for genetic similarity.' (Slightly modifying Jackson's position, we might say that according to the degree of *environmental similarity* among siblings, we would expect greater behavioral similarity, without concern for genetic relationship.) As Jackson demonstrated, and subsequent data have confirmed, this remains a plausible interpretation of the twin method's results (see Chapter 3).

During this period, Rosenthal published a series of critical papers on the twin method which can be viewed as damage control in response to Jackson's critique.[213] Rosenthal outlined several problems in the schizophrenia twin study data, many of which had been discussed by Jackson, and offered several suggestions of how twin studies could be improved. Like Jackson, Rosenthal acknowledged the higher concordance rate among female twins, and the higher rate among fraternal same-sex versus fraternal opposite-sex pairs (in Kallmann's and Slater's studies, the difference was statistically significant). Rosenthal also argued that, contrary to Jackson's hypothesis, the data showed that identical twins were no more likely to receive a diagnosis of schizophrenia than members of the single-born population. However, subsequent data have shown that the question remains open.[214]

Rosenthal concluded his series of papers by writing that 'the total weight of the evidence' from the schizophrenia twin studies 'still strongly favors [the genetic] hypothesis, despite their weaknesses.'[215] It would only be in later years that Rosenthal would conclude that both family studies and the twin method were 'confounded,' and that 'one can draw conclusions about them only at considerable risk.'[216]

Jackson's hypothesis has been cited repeatedly by twin researchers as proof that his entire argument was faulty; thus becoming a sort of 'straw man' in the case against Jackson. Kringlen's position is typical:

> The findings furthermore cast doubt on Jackson's (1960) 'confusion of ego identity theory' of schizophrenia . . . According to Jackson, the monozygotic twinship involves problems of 'isolation, association and identification.' This hypothesis stimulated Rosenthal (1960) to investigate if schizophrenia in fact was more frequent in twins . . . Rosenthal analyzed the twin investigations published by Luxenburger . . . and Essen-Möller and could not support this theory of Jackson.[217]

What Kringlen and others failed to realize was that their argument does not detract from Jackson's basic position, because his 'confusion of ego identity theory' does not require that identical twins develop schizophrenia at rates greater than the single-born population. Jackson argued that the confusion of ego identity, experienced most strongly by identical twins, could explain why the co-twin of a person diagnosed with schizophrenia would be more likely to receive this diagnosis than another family member with whom the twin does not share comparable levels of ego fusion. His theory centered on why the *second* twin receives a diagnosis of schizophrenia, not the first.

Although none of Jackson's 11 major points was convincingly disproved by twin researchers, in countless books and articles discussing (or rather, briefly dismissing) Jackson's thesis we find the claim that Jackson's argument does not hold up because it is alleged that twins are no more susceptible to schizophrenia than are members of the single-born population. However, Jackson's most telling point was that — among pairs with the same genetic relationship to each other — those pairs experiencing a more similar environment and closer emotional bond were consistently more concordant for schizophrenia. For Jackson, this suggested that the identical-fraternal concordance rate difference (already inflated by methodological bias) could be explained on environmental grounds. Using today's behavior genetic terminology, Jackson essentially argued that the theoretical basis of the twin method — the identical-fraternal 'equal environment assumption' — is false.

Nadir

At the outset of the 1960s, twin researchers were as isolated as they had ever been. Although Jackson's critique had dealt the twin method a severe blow, it was not quite at the point of being discredited as pseudoscience. Twin research was able to withstand another blow in 1963 when Tienari reported a 0% identical twin concordance in his Finnish schizophrenia study.[218] This contrasted dramatically with Kallmann's reported 86% rate in his 1946 study. On a more positive note for twin research, the reared-apart twin studies of Shields in 1962 and Juel-Nielsen in 1965 were published during this period. Both investigators concluded that they found evidence for important genetic influences on personality and IQ. In 1966, British psychologist Cyril Burt published the final paper of his subsequently discredited report of 53 pairs of reared-apart twins (see Chapter 4).

In the Fall of 1965 a meeting was held by leading twin researchers under the auspices of the World Health Organization (WHO). The researchers upheld the validity of twin research while calling attention to its methodological problems. They addressed the question of the greater environmental similarity of identical versus fraternal twins and concluded, 'most shared post-natal experiences of MZ twins are probably not qualitatively different from those shared by DZ partners or even by sibs.'[219] However, they did acknowledge the argument that higher identical twin concordance could be explained on environmental grounds, and 'As a theoretical argument, this is almost impossible to refute.'[220]

The WHO report mentioned the possible benefits of using the co-twin control method, although it acknowledged that 'one would not, of course, risk exposing randomly selected partners in a series of MZ twins to a serious disease,' as Mengele had in Auschwitz.[221] The misuse of the co-twin control method also can be found in the United States. In 1954 a pair of American investigators published an article entitled, 'Schizophrenia in Identical Twins One of Whom was Lobotomized.'[222] The authors, Weatherly and Deabler, wrote that it is often difficult to find suitable controls for psychiatric research, but that 'identical twins offer a unique opportunity for the study of human behavior and human disease.'[223] As they described it, 'The following study is one of schizophrenia in identical twins who received quite different treatment. One was lobotomized; the other was not.'[224] They closed by recommending that if 'psychosurgery is ultimately recommended' for the non-lobotomized twin, 'it is conjectured that similar incisions severing similar nerve pathways in nearly identical brains will produce the same gratifying results.'[225]

The year 1966 saw the publication of Gottesman and Shields's British schizophrenia twin study. The study was later presented in a more detailed format, which included case history information as well as diagnoses made by a panel of well-known schizophrenia experts.[226] Gottesman and Shields's stated purpose for performing this study was 'to find out what the effects would be on the results when care was taken to avoid, or to make provision for, the alleged sources of error and bias in the earlier "classical" studies conducted before 1953.'[227] Ironically, Gottesman and Shields controlled for most of these sources of error and bias except for the most important and controversial, that is, the argument that the greater resemblance of identical versus fraternal twins can be explained by the former's greater environmental similarity.

Twin research makes a comeback
The tide began to turn in the mid-1960s in the direction of a greater acceptance of the importance of genetic influences on psychiatric disorders. For the most part this was not, as some have put it, due to the work of 'ingenious' twin researchers.[228] Rather, it was the result of the widespread acceptance of the genetic conclusions coming out of schizophrenia adoption studies performed in Oregon and Denmark. In Chapter 7 I review these studies in great detail. Suffice it to say here that, in spite of their glaring weaknesses, the adoption studies had a tremendous impact. Schizophrenia (and by implication other psychiatric disorders) was increasingly seen as having an important genetic component. Given this shift in the genetic direction, it was natural that twin studies would be seen in a new light and would not have to carry the burden of supporting the genetic position alone. Schizophrenia twin studies could now be seen as having measured the important genetic component believed to have been established (albeit incorrectly, as we will see in Chapter 7) by the adoption studies. In 1972 Gottesman and Shields could defend the validity of the twin method by pointing to these adoption studies:

> The difference in identical vs. fraternal twin concordance rates is not due to aspects of the within-family environment that are more similar for MZ than DZ twins, although there are many such aspects. Studies of MZ twins reared apart as well as adoption and fostering studies show a markedly raised incidence of schizophrenia among relatives even when they were brought up in a different home by nonrelatives.[229]

And four years later they would write, 'The old objection that elements of the shared environment "caused" the higher concordance rates in MZ twins should have been put to rest with the results of the adoption studies and the few pairs of MZ twins "reared apart" that show similar overall concordance rates'.[230]

Thus, the validity of the twin method could now be defended on the basis of the results of other types of studies. Family studies could not play this role because they were widely seen as even more vulnerable to environmental confounds than the twin method. The twin method, at least when it looked at behavior and psychopathology, received a new lease on life in the late 1960s and has been on the ascendancy ever since.

Twin research in the 1970s

The 1970s were years of growth and controversy in twin research. Studying twins became a more respectable enterprise during this period, although the old objections remained in many quarters. In 1969 Arthur Jensen cited twin studies in his article on racial differences in IQ. This article was the subject of great controversy, and the results of twin research were part of this controversy. Leon Kamin's *The Science and Politics of IQ*, published in 1974, discussed the problems in IQ twin research at length. Kamin observed that identical-fraternal comparisons are based on the assumption that 'environmentally produced within-family variance is equal for the MZ and DZ samples,'[231] while noting that 'there is much evidence to demonstrate that in fact the environmental differences to which MZs and DZs are exposed is not equal.'[232] Like the schizophrenia studies, twin studies of IQ would be supplemented by new adoption studies in the 1970s and 1980s.

On the behavioral side, the rising popularity of genetic explanations gave impetus to twin research, and many of Jackson's arguments were ignored or forgotten (Jackson died prematurely in 1968). By 1972, Gottesman and Shields were prepared to relegate Jackson's critique to the status of a historical footnote:

> [In 1960] came the book edited by Jackson . . . *The Etiology of Schizophrenia,* containing a critique by him of the 'literature of the genetics of schizophrenia.' Even today, a decade later, the chapter is considered in many quarters to be the definitive rebuttal to genetic thinking and an excuse to ignore any developments generated by genetic hypotheses; the critique dealt especially harshly with Kallmann's (1938, 1946) findings and conclusions . . . Although a few of Jackson's points were well taken, most have since been shown to have no force.[233]

Ten years later, Jackson was viewed by Gottesman and Shields as little more than a 'vocal critic of genetic interpretations.'[234]

Although adoption studies helped breathe new life into twin research, there was a rivalry between schizophrenia twin and adoption researchers during this period. In their first schizophrenia adoption study publication, American psychiatric investigator Seymour Kety and associates wrote that the results of family and twin studies were 'inconclusive,' and that they 'fail to remove the influence of certain environmental factors.'[235] By the late 1970s Kety (and Rosenthal, as we have seen) definitively concluded that the twin method was unable to separate genetic and environmental factors. According to Kety,

> Twin studies are a more compelling form of genetic data [than family studies], but even twin studies depend on the assumption that the only thing that differentiates monozygotic from dizygotic twins is their genetic relatedness, and that environmental factors are somehow canceled out or randomized. But that is not the case. Monozygotic twins share much of their environment as well as their genetic endowment. They live together; they sleep together; they are dressed alike by parents; they are paraded in a double parambulator [sic] as infants; their friends cannot distinguish one from the other. In short, they develop a certain ego identification with each other that is very hard to dissociate from the purely genetic identity with which they were born.[236]

This view was in accord with Jackson's 1960 critique, which was read by Kety while in press and had a significant impact on his thinking.[237]

Prominent twin researchers had mixed feelings about the schizophrenia adoption studies. While generally supportive of them and quite willing to bask in the widespread acceptance of genetic theories that adoption studies helped bring about, many were unwilling to accept that the twin method's importance had been diminished. Norwegian twin researcher Einar Kringlen lamented in 1976 that 'One has the feeling that results from [schizophrenia] twin research have been completely ignored.'[238] In 1976 Gottesman and Shields published a much discussed critical review of the Danish-American studies in *Schizophrenia Bulletin*, and in a later paper Gottesman wrote, 'To state that adoption studies permit the disentangling of heredity from environment may be too pat and misleading.'[239] In their 1982 book on schizophrenia, Gottesman and Shields wrote, 'The strongest evidence implicating genetic factors in the etiology of schizophrenia comes from twin studies.'[240]

Still, this was a period of growth for twin research as well as the fledgling behavior genetics field, which had weathered the Jensen controversy. The first issue of the journal *Behavior Genetics* appeared in 1970. Twin research has been referred to as the 'Rosetta Stone' of this field.[241] Psychiatric genetics had also been given a tremendous boost from the schizophrenia adoption studies. The stage was now set for the further growth of twin research in the 1980s and 90s.

Twin research from 1980 to the present

Twin research and behavioral genetics carried their momentum into the 1980s. The decade saw the publication of reared-apart twin studies performed in the United States (Minnesota) and in Scandinavia. In particular, the Minnesota studies received a great deal of attention (see

Chapter 4). To the extent that the conclusions of these studies' authors were accepted, they helped strengthen behavior genetics and twin research in general. Several popular books on twins and twin research were published in the 1980s and 1990s.

Just as it had in the 1920s, the relatively conservative political climate in the United States and elsewhere provided a fertile ground for genetic theories and research. The election of Margaret Thatcher in 1979 and Ronald Reagan in 1980, and the subsequent cutbacks in social programs, were aided by the belief that societal problems were the result of biology, as opposed to unequal income distribution, racism, unemployment, etc. Genetic theories were used, as they historically have been used, to provide a 'scientific' justification for the inequalities of the capitalist economic system. Genetic and biological theories were also promoted by pharmaceutical companies, who were poised to earn enormous profits from the sale of psychotropic drugs. In addition, the memories of the Nazis' and German genetic researchers' use of genetic theories in the service of genocide were fading. American behavior geneticist and twin researcher David Lykken, for example, bemoaned the fact that the Nazi atrocities and the 'cruel experiments conducted by Dr. Mengele' at Auschwitz prevented 'an entire generation of civilized people' from looking at the alleged role of inherited factors in crime.[242]

Kenneth Kendler and his colleagues published psychiatric studies derived from the Virginia Twin Register. At the beginning of this work, Kendler published an article in defense of the twin method and of the equal environment assumption.[243] He argued that the twin method is not seriously biased by the greater environmental similarity of identical twins and that the method is a valid technique for the detection of genetic influences on psychiatric disorders. Thus, Kendler became the first twin researcher to make a serious attempt to provide theoretical justification for the equal environment assumption, although his analysis was curiously ahistorical and omitted many important facts. For example, in Kendler's 1983 and subsequent articles he virtually ignored Jackson's critique and failed to answer most of Jackson's most important points. Although Kendler is a leading schizophrenia researcher, he has failed to explain the fact that same-sex fraternal twins have more than double the concordance rate for schizophrenia than opposite-sex fraternals — a fact that cannot be explained on genetic grounds (see Chapter 6). Nevertheless, Kendler has become one of the world's most famous and prolific twin researchers, and his group's studies are published in leading medical and psychiatric journals.

The 1980s and 1990s also saw an increase in the types of behaviors studied by twin researchers. Predictably, most twin studies found significantly greater identical twin similarity, and their authors concluded that they had found genetic influences on the condition or trait under study.

The greater acceptance of the results of twin research continued into the 1990s. The beginning of the decade saw the publication of the results of the Minnesota study in *Science*. The results of twin studies were more frequently found in the pages of leading psychological and psychiatric journals in a way that seemed to endorse their scientific validity. More popularizations of behavior genetic research were published in the late 1990s. This field was rapidly becoming a household name, and magazine covers were devoted to the 'genetic revolution.' Concurrently, the Human Genome Project (HGP) began in 1990 and continued its work throughout the decade. In 2001 Kendler identified 'five areas of development for twin studies [that] are likely to be particularly fruitful in the coming years.' These include longitudinal twin studies, studies of twins and their offspring, the use of physical markers ('endophenotypes') to identify psychiatric conditions, improved statistical methods, and 'increased interactions between twin and gene-finding approaches.'[244] Behavior genetic and psychiatric investigators inferred the existence of genes from the results of their research, and expected that these genes would soon be identified in the laboratory. Many were by now personally involved in the search for genes. Twin research, initiated by Galton five generations earlier, entered the new millennium as strong and accepted as it ever had been.

Conclusions

In this chapter I have covered the most important aspects of the history of twin research, which include disturbing times and events. Journalist William Wright, who is supportive of twin research, wrote, 'It is certainly true that the history of twin research is one of the most appalling chapters in science, having been born in Galton's aristocratic notions of the natural worthiness of the English upper class, taken to its evil extreme by eugenicists, and too readily used by American scientists to rationalize racial injustice.'[245] But three generations of twin researchers have failed to document the 'appalling' history of their discipline.

The classical twin method was the creation of German racial hygienists and American eugenicists attempting to convince the socially and politically powerful of the necessity of eliminating 'hereditary tainting' from the population, by any means necessary. The researchers themselves already believed strongly in the importance of genetics and eugenics. But as Galton wrote in 1883 and Siemens noted in the first chapter of *Die Zwillingspathologie*, others remained unconvinced.

In a related development, most of the early pioneers of statistical methods were also prominent eugenicists. Leon Kamin once observed that eugenics needed statistical methods because 'it would be necessary to devise tests and measurements that could identify the genetically superior and inferior. Hence the interest of Galton and his followers in measuring physical and psychological differences between individuals and between races.'[246] While this is a historical fact, it would be wrong to discard statistical methods solely because of the views and motivations of their developers. So too with twin research. The legitimacy of drawing conclusions about genetics from twin research can be determined only by examining a study's methods and, most importantly, by determining the validity of its theoretical assumptions.

As an interesting parallel, the contemporary 'pro-life' movement frequently documents the eugenic views of Planned Parenthood founder Margaret Sanger in an effort to discredit abortion and birth control. For example, Sanger wrote an article in 1921 entitled 'The Eugenic Value of Birth Control Propaganda.' In it, she wrote that the 'eugenic and civilizational value of Birth Control is becoming apparent to the enlightened and the intelligent,' and that the 'campaign for Birth Control . . . is practically identical in ideal with the final aims of Eugenics.' She concluded that 'Birth Control is not advanced as a panacea by which past and present evils of dysgenic breeding can be magically eliminated,' and that 'drastic and Spartan methods may be forced upon society if it continues complacently to encourage the chance and chaotic breeding that has resulted from our stupidly cruel sentimentalism.'[247] Here, Sanger promoted birth control as a means of carrying out the eugenic program. Statements such as these were common during this period in American history, as well as in Germany.

Many people resent the pro-life movement's focus on Sanger's writings because they view access to birth control and abortion rights as important gains which must be defended. Neither birth control, statistical methods, nor twin research is discredited by its history or by the motivations of its pioneers. Still, if birth control has been misused it is important to know this. Similarly, whether or not standard conclusions drawn from twin research are correct, it remains important to know how twin research, like birth control, can be used for disreputable purposes.

The goal of this chapter has been to tell the history of twin research that twin researchers and their supporters have failed to tell. In the following two chapters we will look at the methodological and theoretical underpinnings of twin research. Only then will we be able to determine its usefulness and validity. In the process we will see more clearly that, beginning with Galton, five generations of twin researchers have misunderstood the causes of the behavior they were observing and reporting.

Notes

[1] Quoted in Broberg & Tydén, 1996, p. 104.
[2] Fraser-Roberts, 1934, p. 31.
[3] Galton, 1881, p. 1.
[4] Galton, 1883, p. 308.
[5] Galton, 1875, p. 391.
[6] Burbridge, 2001.
[7] See Bulmer, 1999; Burbridge, 2001; Rende et al., 1990.
[8] Galton, 1875, p. 392.
[9] Ibid., p. 402.
[10] Ibid., p. 404.
[11] Galton, 1883, p. 237.

[12] Galton, 1875, p. 397.

[13] Ibid., p. 396.

[14] Ibid., p. 400.

[15] Ibid., p. 399.

[16] Ibid., p. 399.

[17] Ibid., p. 402.

[18] Galton, 1883, p. 221.

[19] Ainslie, 1985, p. 50.

[20] Galton, 1995, p. 406.

[21] Ibid., p. 406.

[22] Thorndike, 1905, p. vii.

[23] Ibid., p. 44.

[24] Fisher, 1919.

[25] Thorndike, 1905, p. 9.

[26] Galton, 1875, p. 404.

[27] Thorndike, 1905, p. 10.

[28] For example, Newman, 1917.

[29] Siemens, 1924b, pp. 1–3.

[30] These views are discussed in Siemens, 1924b, pp. 1–2.

[31] Rende et al., 1990.

[32] Merriman, 1924.

[33] Rende et al., 1990, p. 283.

[34] Wilson, 1934.

[35] Newman et al., 1937, p. 19.

[36] Gedda, 1961, p. 24.

[37] Plomin, DeFries, et al., 1997, p. 81.

[38] Weindling, 1999.

[39] Proctor, 1988.

[40] Siemens, 1924a.

[41] Ibid., p. 110.

[42] Ibid., p. 98.

[43] Ibid., pp. 125–7.

[44] Galton also believed in sterilization. In a 1907 letter to his collaborator Karl Pearson, Galton wrote, 'except by sterilization I cannot yet see any way of checking the produce [sic] of the unfit who are allowed their liberty and are below the reach of moral control' (quoted in Kevles, 1985, p. 94).

[45] Forward to Siemens, 1937 (my translation).

[46] Ibid., p. 180 (my translation).

[47] Wiener, 1958.

[48] Proctor, 1988, p. 306.

[49] Weindling, 1989.

[50] Plomin, DeFries, et al., 1990.

[51] Holmes, 1930, p. 247.

[52] Stocks, 1930, p. 104.

[53] Wilson, 1934, p. 352.

[54] Ibid., p. 353.

[55] von Bracken, 1934, p. 305.

[56] Ibid., p. 306.

[57] Woodworth, 1941, p. 12.

[58] Ibid., p. 21.

[59] Carter, 1940, p. 247.

[60] Ibid., p. 247.

[61] A description of the methods, results, and flaws of the studies I discuss will not be made here, other than to point out that most studies found significantly higher identical vs. fraternal similarity, which was attributed to genetic influences.

[62] Lauterbach, 1925.

[63] Wingfield, 1928; Hirsch, 1930.

[64] Newman et al., 1937.

[65] Wingfield, 1930, p. 186.

[66] Hirsch, 1930, p. 148.

[67] Conklin, quoted in Hirsch, 1930, p. 150.

[68] Hirsch, 1926, pp. 399–400.

[69] Reviewed in Carter, 1940.

[70] For schizophrenia, see Rosanoff, Handy, Plesset, & Brush, 1934; for criminality and delinquency, see Rosanoff, Handy, & Rosanoff, 1934.

[71] Haller, 1963.

[72] Rosanoff, 1938, p. 642. Emphasis in original.

[73] Ibid., pp. 749–50.

[74] Ibid., p. 750.

[75] Shorter, 1997, p. 243. The point I wish to make here is not that Rosanoff was 'right-wing,' but rather that he, like most other twin researchers of his era, was motivated by eugenic beliefs.

[76] For example, see Davenport, 1911.

[77] Newman, 1932, pp. 444-445. Newman was a member of the Advisory Council of the journal *Eugenics* (for example, see *Eugenics, III,* 1930).

[78] Bouchard & Pedersen, 1999, p. 71.

[79] Burbridge, 2001.

[80] Muller, 1925.

[81] Gesell, 1942; Gesell & Thompson, 1929. For a discussion of Gesell's view on eugenics, see Marchese, 1995.

[82] Stern, 1962.

[83] Kallmann, 1938c, p. 263.

[84] Stern, 1962, p. 5.

[85] Lange, 1930, p. 46.

[86] Ibid., p. 242.

[87] Quoted in Proctor, 1988, p. 64.

[88] Weindling, 1999.

[89] Quoted in Weinreich, 1946, p. 33.

[90] Gütt et al., 1934.

[91] Kranz, 1936, pp. 250–1 (my translation).

[92] Ibid., p. 251 (my translation).

[93] Weindling, 1989, p. 516.

[94] Proctor, 1988, p. 43.

[95] von Verschuer, 1939, p. 79.

[96] Weindling, 1989, p. 562. Further information on German twin camps can be found in Ash, 1998.

[97] Ellinger, 1942, p. 141.

[98] Ibid., p. 141.

[99] For example, see Rosenthal, 1971a; Zerbin-Rüdin & Kendler, 1996; Faraone & Biederman, 2000.

[100] For example, see Faraone, Tsuang, & Tsuang, 1999; Faraone & Biederman, 2000.

[101] For documentation, see Burleigh, 1994; Lifton, 1986; Müller-Hill, 1998; Proctor, 1988.

[102] Faraone & Tsuang, 1995, p. 124.

103 Proctor, 1988, p. 108.
104 Quoted in Weingart, 1989, p. 270.
105 Quoted in Müller-Hill, 1998, p. 33.
106 Quoted in Weber, 1996, p. 325.
107 Weber, 1996, p. 328.
108 Quoted in Weinreich, 1946, pp. 32–3.
109 Proctor, 1988, p. 292.
110 Rüdin, 1939, pp. 16–17.
111 See Bernstein, 1945, p. 5. The author interviewed Rüdin after the war, writing, 'I am sure that Prof. Rudin never so much as killed a fly in his 74 years. I am also sure he is one of the most evil men in Germany' (p. 5).
112 Proctor, 1988, p. 177. More recent estimates have put the number of murdered mental patients at 300,000 (Peters, 2001, p. 297).
113 Quoted in Weber, 1996, p. 329.
114 Quoted in Müller-Hill, 1998, p. 46.
115 Quoted in Weber, 1996, p. 329. The word 'stressed' is Weber's.
116 Rüdin, 1942, p. 321. Quoted in Müller-Hill, 1998, p. 67.
117 Rüdin, 1942, p. 321 (my translation).
118 Ibid., p. 322 (my translation).
119 Ibid., p. 322 (my translation).
120 Gottesman & Bertelsen, 1996, p. 320. Another example of historical distortion is found in Gottesman and Bertelsen's description of Rüdin's role in the forcible sterilization of children who were the offspring of black African French occupation troops and native Germans (known pejoratively in Germany as the 'Rhineland bastards' ['Rheinlandbastarde']). According to Gottesman and Bertelsen,

> A committee on which Rüdin served considered the ('Rheinland-bastarde') question again in 1935 and recommended either sterilization after parental 'persuasion' and permission or, preferably, forced emigration . . . The only detailed report of the subsequent Gestapo action in 1937 is provided by Pommerin (1979), not translated into English (Gottesman & Bertelsen, 1996, p. 320).

The sources of this information were listed as Proctor (1988), and Weindling (1989). Gottesman and Bertelsen's description is indeed taken from Weindling and from page 113 of Proctor's book. Interestingly, they chose to ignore Proctor's description of Rüdin's activities in 1937 (in English, I might add), which is found on page 112:

> The 1933 Sterilization Law made no provision for sterilization on racial grounds. (Jews, for example, were never specifically targeted by the law.) In 1937, however, on secret orders from Hitler, the 500 or so *Rheinlandbastarde* . . . were sterilized in a joint Gestapo/genetic health court operation organized by Ernst Rüdin, Fritz Lenz, and Walter Gross, among others (Proctor, 1988, p. 112).

121 Gottesman & Bertelsen, 1996, p. 320.
122 Strömgren, 1994.
123 Luxenburger, 1934b, p. 315.
124 Quoted in Burleigh, 1994, p. 41.
125 Luxenburger, 1931a, p. 755 (my translation).
126 Luxenburger, 1931c, p. 53 (my translation).

[127] Ibid., p. 55 (my translation).

[128] Ibid., p. 55 (my translation).

[129] Luxenburger, 1931b, p. 119 (my translation).

[130] Ibid., p. 122 (my translation).

[131] Ibid., p. 121 (my translation).

[132] Ibid., p. 122 (my translation).

[133] Ibid., p. 122 (my translation).

[134] Ibid., p. 124 (my translation). Similar ideas were expressed in American medical circles. In a 1933 *New England Journal of Medicine* editorial, physicians were warned that the differential reproduction rate 'threatens to replace our population with a race of feeble-minded; we must study its causes and the sources from which it springs. If we wait too long, this viper that we have nourished may prove our undoing' (quoted in Sofair & Kaldjian, 2000, p. 316).

[135] Ibid., p. 123 (my translation).

[136] Quoted in Burleigh, 1994, p. 41.

[137] Luxenburger, 1931a, p. 755 (my translation).

[138] Luxenburger, 1934b.

[139] Ibid., p. 306.

[140] See Luxenburger, 1934b.

[141] Luxenburger, 1940a, p. 256 (my translation).

[142] Ibid., p. 257.

[143] Proctor, 1988. Luxenburger published several articles in *Der Erbarzt* (for example, see Luxenburger 1934a, 1936a, 1936b).

[144] Luxenburger, 1939. In 1937–1938, Banu held a position in the far-right anti-semitic Goga-Cuza Romanian regime (Bucur, 2002, p. 43).

[145] Gottesman, 1991, p. 4.

[146] Slater, 1971, p. 18.

[147] Ibid., p. 19.

[148] Kendler, 1997, p. 461.

[149] Strömgren, 1994, p. 406.

[150] Schulz, 1934.

[151] Weindling, 1989 p. 500.

[152] Ibid., p. 500.

[153] Slater, 1971, p. 18.

[154] Schulz, 1934, pp. 38–9 (my translation).

[155] Ibid., p. 142 (my translation).

[156] Schulz, 1939.

[157] Weber, 1996, p. 328.

[158] Kendler also wrote that 'In 1935,' Luxenburger 'was banned from future public lectures during the Nazi regime' for his remarks against Streicher (Kendler, 1997, p. 461). According to the historian Paul Weindling, however, 'Rüdin, Luxenburger and Lenz continued to hold large numbers of lectures until 1937' (Weindling, 1989, p. 502). In addition, we have already seen that Luxenburger was published in itler's Gerany as late as 1940 (as another example, see Luxemburger 1940b).

[159] Weindlig, 1993, p. 643.

[160] Strömgren, 1994, p. 406.

[161] Shorter, 1997, p. 242.

[162] Rosenthal, 1971a, p. 7.

[163] For example, see the bibliography in Rosenthal, 1970b.

[164] Newman et l., 1937, p. 19.

[165] See Lifton, 1986, for an account of von Verschuer's role as accomplice to Mengele's crimes at Auschwitz.

[166] Quoted in Müller-Hill, 1998, p. 52.

[167] Quoted in Proctor, 1988, p. 211.

[168] Quoted in Lifton, 1986, p. 358.

[169] Joseph, 2000c, p. 562.

[170] Faraone & Biederman, 2000, p. 575.

[171] Zerbin-Rüdi & Kendler, 1996, p. 336.

[172] Faraone Tsuang, & Tsuang, 1999, p. 222. The authors also wrote (p. 223) that 'Adolph Hitler and his Nazi regime began a systematic program first to sterilize and then to kill "genetically defective" people.' The role of Rüdin and other psychiatric geneticists in popularizing the sterilization law is not mentioned, in spite of Faraone and colleagues' citation of two articles (Gottesman & Bertelsen, 1996; Weber, 1996) which documented Rüdin's role. Also on page 223 they wrote, 'Contemporary researchers in psychiatric genetics are especially disturbed to learn that the Nazis used this research to justify their eugenics policies regarding the mentally ill . . .' Thus, Faraone et al. portrayed eugenic sterilization as something advocated by the Nazis, but made no mention of the role of psychiatric geneticists. The false implication is that eugenic sterilization was 'their' (the Nazi's) program, but not 'ours' (the psychiatric geneticists). Also overlooked is the fact that several psychiatric geneticists were themselves members of the Nazi party, such as Rüdin himself who joined in 1937.

[173] Kallmann, 1946, p. 321.

[174] Lewontin et al., 1984.

[175] Gottesman & Shields, 1982, p. 42. This theme is consistent in Gottesman's writings on Kallmann's background.

[176] Quoted in Müller-Hill, 1998, p. 11.

[177] Kallmann, 1938b, p. 105.

[178] Ibid., p. 105.

[179] Ibid., p. 112. For those interested in Kallmann's family study, see Kallmann, 1938a.

[180] Ibid., p. 109.

[181] Rainer, 1966, p. 414.

[182] Ibid., p. 415.

[183] Kallmann, 1938a, p. xiv.

[184] Kallmann, 1953, p. 18. At the end of this quotation, Kallmann cited his 'first family study' as 'Kallmann, 1938.'

[185] Ludmerer, 1972.

[186] Gottesman & Bertelsen, 1996.

[187] Ibid., p. 320.

[188] Ibid., p. 321. Psychiatric geneticist Erik Strömgren (1994, p. 406) denied Rüdin's criminal role by writing that Rüdin was released after the war 'following very conscientious investigations performed by the Americans.'

[189] Lerer & Segman, 1997.

[190] Torrey et al., 1994.

[191] Siemens, 1927.

[192] Lykken, 1978.

[193] See, for example, Thompson et al., 2001.

[194] Cohen et al., 1973, p. 467.

[195] Dibble et al., 1978, p. 247–8.

[196] Kallmann, 1946, p. 312.

[197] Rosenthal, 1962b, p. 121.

[198] Ibid.

[199] Proctor, 1988, p. 43. I am unaware of any twin researcher mentioning the Nazi's creation of a national twin register.

[200] Price, 1950, p. 293.

[201] Ibid., p. 343.

[202] Ibid., p. 343.

[203] Neel & Schull, 1954, p. 280.

[204] Ibid., pp. 280-281.

[205] Jackson, 1960.

[206] Bateson et al., 1956.

[207] Jackson did, however, believe that 'it seems likely that hereditary factors do play a part in at least some of the schizophrenias . . .' (1960, p. 80).

[208] Gottesman & Shields, 1966a, p. 55.

[209] For a detailed discussion of this point, see Joseph, 2001b. In this article I listed 12 points. I have dropped one here because it is redundant.

[210] Jackson, 1960, p. 66.

[211] Ibid., p. 67.

[212] Ibid., p. 69.

[213] See Rosenthal, 1960, 1961, 1962a, 1962b.

[214] Joseph, 2001b.

[215] Rosenthal, 1962b, p. 132.

[216] Rosenthal, 1979, p. 25.

[217] Kringlen, 1967a, p. 61.

[218] Tienari, 1963.

[219] World Health Organization, 1966, p. 115.

[220] Ibid., p. 115.

[221] Ibid., p. 116.

[222] Weatherly & Deabler, 1954. Although the authors did not state that the lobotomy was performed in the context of a co-twin control study, they discussed the story in a way that would encourage this type of research.

[223] Ibid., p. 265.

[224] Ibid., p. 262.

[225] Ibid., p. 266.

[226] Gottesman & Shields, 1972.

[227] Ibid., p. xvi.

[228] Neale & Oltmanns, 1980, p. 174.

[229] Gottesman & Shields, 1972, p. 318.

[230] Gottesman & Shields, 1976a, p. 371.

[231] Kamin, 1974, p. 96.

[232] Ibid., p. 97.

[233] Gottesman & Shields, 1972, p. 4.

[234] Gottesman & Shields, 1982, p. 73.

[235] Kety et al., 1968, p. 345.

[236] Kety, 1978, p. 48.

[237] See Kety, 1959.

[238] Kringlen, 1976, p. 429.

[239] Gottesman, 1978, p. 63.

[240] Gottesman & Shields, 1982, p. 71.

[241] Bouchard, 1999, p. ix.

[242] Lykken, 1995, p. 71.

[243] Kendler, 1983.

[244] Kendler, 2001, p. 1012.

[245] L. Wright, 1997, p. 33.

[246] Kamin, in Eysenck vs. Kamin, 1981, p. 91.

[247] Sanger, 1921, p. 5.

CHAPTER THREE

The Twin Method: An environmentally confounded research method

> If identical twins have the same trait people are inclined to think that this proves that the trait is hereditary. The fallacy is most easily seen if we take an imaginary example. Suppose that a student of human heredity should hail from another planet and that he should be required to use the twin method to find out whether or not people's clothes were a direct consequence of heredity. He would find that identical twins were very often dressed alike, down to the small details, and that this was uncommon with fraternal twins. He would confidently conclude that the choice of clothes was almost an exclusively hereditary trait and might even suppose them to be part of the natural skin of the human animal by the exercise of superficial reasoning.
>
> *Geneticist Lionel Penrose in 1973.*[1]

The twin method's comparison of reared-together identical versus reared-together same-sex fraternal twins has been the main way that twins have been used in genetic research. In the previous chapter we looked at the critics' objections that the twin method is biased because identical twins share more similar environments than fraternals — implying that some or all of the greater observed similarity of identical twins is explained by their greater environmental similarity. This question is central to the validity of the twin method.

Critics also have pointed to several other methodological problems with studies using the twin method, such as the lack of an adequate and consistent definition of the trait under study, the use of non-blinded diagnoses, diagnoses that were made on the basis of sketchy information, inadequate or biased methods of zygosity determination, the unnecessary use of age-correction formulas, the use of non-representative sample populations, and the lack of adequate descriptions of the methods used in some of the studies. Although these problems are important, I will focus on the main theoretical assumption of the twin method for the following reason: if all methodological problems associated with the twin method were resolved, identical twins would still be more concordant or correlate higher than same-sex fraternal twins for most traits (although the differences would not be as great as those reported in methodologically flawed studies). Twin researchers, in my opinion, are therefore correct that identical-fraternal differences are real. We will see, however, that there is little reason to accept that identical-fraternal comparisons measures much more – or anything more – than the greater environmental similarity experienced by identical versus fraternal twins.[2]

The Equal Environment Assumption (EEA)

Two definitions
The validity of drawing genetic inferences from the twin method is dependent on the 'equal environment assumption' (EEA). The authors of every scientific article and popular work citing identical-fraternal comparisons as evidence of the role of genetics assume (explicitly or implicitly) that the EEA is correct – otherwise, they would conclude that the meaning of these comparisons is uncertain, or that they demonstrate the importance of environmental factors.

To begin this discussion, I will outline the two main ways that the equal environment assumption has been defined.

The traditional EEA definition.
The traditional definition was used by most twin researchers prior to 1972 and continues to be used in the present day. The EEA is defined as the straightforward assumption that identical and fraternal twins experience similar environments and treatment, and is exemplified by the following quotation from a leading twin researcher: 'The basic underlying assumption of the twin method is, of course, that environmental conditions of monozygotic twins do not differ from those of dizygotic twins.'[3] At times, proponents of the traditional definition acknowledge that identical twins experience more similar environments than fraternals, and that the twin method might therefore 'overestimate heritability' in these cases.

The 'equal trait-relevant environment assumption.'[4]
(Referred to here as the 'trait-relevant EEA'.) One of the first examples of this new definition is seen in a 1966 publication by Gottesman and Shields, who wrote that critics must show that the 'environments of MZ twins are systematically more alike than those of DZ twins in features which can be *shown* to be of etiological significance in schizophrenia.'[5] More recently, Kendler and associates defined the EEA as follows:

> The traditional twin method, as well as more recent biometrical models for twin analysis, are predicated on the equal-environment assumption (EEA) – that monozygotic (MZ) and dizygotic (DZ) twins are equally correlated for their exposure to environmental influences that are of etiologic relevance to the trait under study.[6]

Proponents of the trait-relevant EEA recognize that identical twins are treated more similarly and experience more similar environments than fraternals, but claim that (a) the evidence shows that greater environmental similarity does not lead to greater psychological trait correlation or to higher concordance for psychiatric diagnoses, and (b) in order to invalidate the finding of genetic factors, critics must identify the trait-relevant environmental factors for which identical and fraternal twins experience dissimilar environments.

Today, Kendler and most other twin researchers acknowledge that the environments of identical twins are more similar than fraternal twins,[7] and there is plenty of empirical evidence in support of this idea. Identical twins spend more time together than fraternals,[8] and more often dress alike, study together, have the same close friends, and attend social events together.[9] James Shields, in his 1954 study of normal twin school-children, found that 47% of the identical twins had a 'very close attachment,' which was true for only 15% of the fraternal twins.[10] In 1967 schizophrenia researcher Einar Kringlen published a survey of 117 twin pairs (75 identical, 42 same-sex fraternal; where one or both were diagnosed with schizophrenia) which stands as one of the few post-1960 attempts by a twin researcher to look at differences in association and 'ego fusion,' which in Chapter 2 we saw was central to Jackson's thesis. According to Kringlen's survey,[11] 91% of identical twins experienced 'identity confusion in childhood,' which was true for only 10% of the fraternal twins. Kringlen also found that identical twins were more likely to have been 'considered as alike as two drops of water' (76% vs. 0%), 'brought up as a unit' (72% vs. 19%), and 'inseparable as children' (73% vs. 19%). The final question made a 'global evaluation of twin closeness.' The results showed that 65% of identical twins were found to have an 'extremely strong' level of closeness, which was true for only 19% of the fraternal pairs. Kringlen's findings illustrate the striking contrast between the environments (psychological or otherwise) of identical and fraternal twins.

In the face of this type of evidence, as well as the existing widespread public belief that identical twins are treated more alike and experience more similar environments than fraternals, twin researchers began to adopt the trait-relevant EEA in the 1960s. Today, it predominates. We will look more closely at the validity of the traditional and trait-relevant definitions of the EEA a bit later, but first I would like to discuss twin researchers' explanations of why identical twins experience more similar environments.

Do identical and fraternal twins create differing environments?
We saw in Chapter 2 that early objections to the twin method by Holmes, Stocks, Wilson and others centered on the assumption that identical and fraternal twins share similar environments. These critics argued not so much that identical twins are treated more similarly because of societal traditions or, as Jackson stressed, that they share a closer psychological and emotional bond. Rather, they argued that identicals experience more

similar environments *because of* their more similar genetic makeup. Although these early critics believed in the importance of genetics *in general*, they questioned whether the twin method was a valid instrument for detecting genetic influences on *specific* traits. Schizophrenia researcher Manfred Blueler (son of Eugen Bleuler, the founder of the schizophrenia concept) made a similar argument in the 1940s, which he later summarized:

> According to the older view, it is axiomatic that identical twins are identical in their inherited predisposition (*Anlage*), while the environment in which the identical twin-partners live is on the average as different as the environment of the fraternal twin-partners. In reality, this last assumption is wrong. Identical twins, on the basis of their common inherited predisposition, create for themselves on the average a much more similar environment than is the case with fraternal twins . . . the old idea that the environment of twins varies without regard to their innate predisposition (*Anlage*) is quite erroneous; hence all further deductions which used to be based on this premise must be wrong.[12]

And in the 1970s, he wrote,

> For a long time, high concordance figures for schizophrenic twins were regarded as the 'absolute scientific truth' for the great importance of heredity . . . the line of reasoning of such rigorous precision is based on a fallacy . . . The fact that two monozygotic twins might live in a more similar environment than two dizygotic twins was forgotten . . . When monozygotics reveal high concordance figures for psychic disturbances, it could be caused just as easily by the great similarity in hereditary factors as by the great similarity in their human environment.[13]

Beginning in the 1950s, some twin researchers used the argument that identical twins create more similar environments (and that fraternals create more dissimilar environments) *in defense* of the validity of the twin method. For example, here is twin researcher James Shields in 1954:

> In so far as binovular [fraternal] twins are treated differently from one another and more differently than uniovular [identical] twins, this is likely to be due not so much to causes outside the twins as to innate differences in the needs of the binovular [fraternal] twins themselves, manifested by different patterns of behavior.[14]

Thus, Shields took the main objection of the critics and stood it on its head. The validity of drawing genetic inferences from identical-fraternal comparisons was now *supported* by the more dissimilar environments 'created' by fraternal twins. Shields was attempting to shift the debate from *whether* identical twins experience more similar environments to *why* their environments were more similar. If identical twins' greater environmental similarity was caused by their greater genetic similarity, Shields seemed to argue, the twin method is valid.

This new position was taken up by Kallmann in 1958:

> The popular notion that the behavior patterns of one-egg [identical] twins are alike chiefly because of unusual similarity in their early environments has yet to be substantiated. If confirmed, *the argument would only strengthen rather than weaken any correctly formulated genetic theory.* Psychodynamic concepts, too, are built on the premise that man is selective in respect to important aspects of his life experiences and so can be thought of as 'creating his own environment' [emphasis added].[15]

Kallmann was arguing that if identical twins' more similar environments caused them to behave more similarly, then this actually 'would only strengthen' the genetic position! As dissident psychiatrist R. D. Laing commented, 'With this two-headed penny it is not clear how Kallmann can lose.'[16] As we will soon discover, most contemporary twin researchers use the very same 'heads I win, tails you lose' argument in defense of the twin method's validity.

This argument lay dormant for several years until it was rediscovered in the 1960s as twin researchers looked for new ways to legitimize their methods. For example, in 1968 Sandra Scarr wrote, 'Differences in the parental treatment that twins receive are much more a function of the degree of the twins' genetic relatedness than of parental beliefs about "identicalness" and "fraternalness."'[17] The defense of the twin method's validity was now based on determining why—not whether—identical twins experience more similar environments.

The EEA was originally defined as the assumption of similar environments between the two types of twin pairs. When empirical studies confirmed what common sense already knew—that identical twins experience more similar environments than fraternal twins—twin researchers retreated to the position that, as Kendler put it, on the basis of an inherited 'similarity in behavior' identical twins 'might *create* for themselves more similar environments.'[18] As we shall soon see, however, the question of *why* identical twins' environments are different is irrelevant to the question of whether the twin method measures genetic influences. In addition this concept, which shall be called here the 'twins create their environment theory,' is problematic.[19] The theory is based on the belief that children are born with a genetic predisposition to manifest certain personality and behavior types. Therefore, the behavior of identical twins is more similar due to their identical genetic makeup. Because identical twins exhibit a genetically predisposed similarity of behavior, their parents and others in their social environment are induced to treat them more similarly. Fraternal twins, since they are more genetically dissimilar, inherit less similar personalities and therefore are treated more differently. The theory concludes that this inherited behavioral similarity causes more equal treatment, not the other way around. But what is conveniently forgotten is this: according to this theory, the twins' parents must also be genetically predisposed to manifest genetically determined behavior and response modes.

Proponents of the 'twins create their environment theory' argue that the genetically predisposed behavior of children is able to greatly impact the (predisposed) personalities and response modes of their parents. Identical and fraternal twins are portrayed as genetically programmed to act in rough proportion to the number of genes they share in common, but their parents are seen as readily able to change their behavior and treatment of the twins on the basis of environmentally caused factors, i.e., the twins' personalities. Children are characterized by their inborn propensity to display inherited behavior, whereas parents are characterized by their plasticity in reacting to this behavior. If parents can change their behavior on the basis of environmental influences, as this theory maintains, it should follow that children (including twins) would *also* be able to adjust their behavior in response to environmental stimuli. Genetic influences or not, we would expect adults' behavior to be far less malleable than that of five-year-old children. Yet children are portrayed as having a greater ability to change the (genetically predisposed) behavior of adults than adults have to create, through their treatment, similar behavior in identical twins.

The argument has been put forward in another form by psychiatric geneticist Edith Zerbin-Rüdin:

> Twin studies, in particular, suggest the presence of a hereditary factor. The large difference between the concordance figures for MZ and DZ twins cannot be explained exclusively by the more similar environment of MZ twins. If MZ twins create a similar environment through their greater similarity, they do so because of the greater inherited similarity in their appearance and response modes. Thus, in a roundabout way, we still come back to the importance of heredity.[20]

Zerbin-Rüdin *asserts* that response modes are genetic in order to *prove* that behavioral differences are genetic. But her argument goes beyond mere logical error; Zerbin-Rüdin's reference to the striking physical similarity of identical twins must be addressed. No one on either side of the twin method debate would claim that similarity of appearance by itself would cause a pair of twins to show high concordance rates for a particular psychiatric diagnosis. Therefore, Zerbin-Rüdin's assertion that 'the greater inherited similarity in their appearance' would lead to more similar treatment of identical twins is in accordance with the environmentalist position, which acknowledges that a greater *physical* similarity contributes to more similar treatment by the social environment and therefore plays a role in the closer bond between members of the identical twin pair.

Furthermore, the great similarity of physical appearance among identical twins is the result of the splitting of a fertilized egg; *it is not an inherited characteristic*. Rather, it is a biological phenomenon occurring in a zygote. For readers who view this claim with skepticism, a reminder is offered that what is being referred to is the similarity of identical twins' appearance *to each other*, not to their parents. A randomly selected fertilized egg, taken from a Petri dish and inserted into the uterus of a mother-to-be would, should it split, produce identical twins as identical in appearance as those produced by a split zygote conceived within the biological mother's own body. On the other hand, the degree of physical similarity between fraternal twins *is* influenced by heredity, because in this case we are dealing with two separately fertilized eggs, which require common genes in order to produce physical similarity. But Zerbin-Rüdin missed this point entirely.

Although not discussed by Kendler, Zerbin-Rüdin and others, even if the 'twins create their environment' theory were true, the twin method could still be measuring nothing more than environmental effects. Suppose for the moment that (1) the 'twins create their environment theory' is correct, and (2) that schizophrenia is caused by brain damage due to the ingestion of a dangerous environmental toxin—mercury for example. Even if it were true that identical twins' environments were more alike because of their more similar inherited behavior, identical twins—by virtue of the widely acknowledged fact that they spend more time together, eat more similar foods, etc.—would show higher concordance for exposure to mercury than fraternals, who spend less time together. It is therefore likely that identical twins would be more concordant for schizophrenia than fraternals, yet schizophrenia would still be completely environmental in origin.

Theoretically, twins' behavior could be largely under genetic control, yet the more similar behavior of identical twins could still lead to higher levels of exposure to pathological *environmental* conditions for specific traits. Therefore, whether identical twins are more alike because they are treated more alike and share a more intimate bond (conventional wisdom), or because they create more similar environments on the basis of more similar inherited behavioral tendencies (Shields/Scarr/Kendler), the simple fact that identical twins spend more time together means that the equal environment assumption is false.

As an example of the theoretical confusion exhibited by twin researchers as they attempt to explain the tortuous logic of the twin method, let us examine the following passage by Kendler:

> Twin studies are based on the assumptions that (a) monozygotic (MZ) and dizygotic (DZ) twins share their environment to approximately the same degree, but (b) MZ twins are genetically identical whereas DZ twins, like normal siblings, have on average only half of their genes identical by descent. Although the second of these assumptions is beyond question, the first, or 'equal environment' assumption, has been a focus of considerable controversy. Several studies have shown

that measures of the social environment (e.g., sharing the same friends, attitudes of parents and teachers, etc.) are more highly correlated among young MZ than among young same-sex DZ twins (Kendler, 1983). These results first appear to suggest that the equal environment assumption is false. However, reflections suggest another possible interpretation. Although the similarity in environment might make MZ twins more similar, the similarity in behavior of MZ twins might *create* for themselves more similar environments. As recently reviewed (Kendler, 1983), these two alternative hypotheses have been subject to empirical test in at least nine different studies. Consistently, these studies suggest that the environmental similarity of MZ twins is the *result* and not the cause of their behavioral similarity. Whereas most of these studies examined such traits as intelligence and personality, one specifically examined schizophrenia and found no evidence that concordance for schizophrenia was produced in MZ twins as a result of the similarity of their treatment by the environment (Kendler, 1983). Current evidence supports the general validity of the equal environment assumption of twin studies.[21]

This quotation captures the essence of the errors of the majority of Kendler's writings on the twin method and the equal environment assumption.

Kendler begins by defining the EEA in the traditional sense, writing that 'twins share their environment to approximately the same degree.' He then correctly points out that the equal environment assumption has been beset by controversy, although, as usual, he fails to discuss the names, dates, and ideas associated with this controversy. He then concedes that the traditional EEA is false, after which, upon 'reflection,' he offers a 'possible' genetically derived interpretation for these different environments. Kendler cites nine separate studies in support of his hypothesis, including one examining schizophrenia. Kendler refers to his 1983 'Overview' article as the source of the schizophrenia study. However, this article did not cite a study that 'specifically examined schizophrenia and found no evidence that concordance for schizophrenia was produced in identical twins as a result of the similarity of their treatment by the environment.' Kendler is probably referring to the summary of his and Robinette's unpublished 1982 study on the relationship between physical similarity and concordance for schizophrenia. The conclusion of this study, as stated in Kendler's 1983 article, was that 'no correlation between degree of physical similarity and concordance rate for schizophrenia' was found.[22] This study, then, only *assumed* environmental treatment differences on the basis of physical appearance.

Finally, in spite of agreeing with the idea that identical twins are treated more similarly than fraternals, Kendler concludes that the current evidence suggests that the equal environment assumption is valid *without changing his definition of the EEA*. He is claiming the following:

(a) The equal environment assumption is based on the idea that identical

and fraternal twins share similar environments.
  (b) Identical twins' environments are more similar than the environments of fraternal twins.
  (c) The equal environment assumption is valid.

Logic dictates that he would have to change his EEA definition for conclusion (c) to follow from premises (a) and (b). But Kendler does not do this. Thus we are supposed to believe something is true after he has just finished telling us that, as he defined it, it is false. If the principal defender of the twin method and the equal environment assumption cannot adequately define that which he is defending, this alone raises questions about the assumption.

Kendler's theoretical retreat

Although Kendler had staked everything on the theory that twins create their own environments, by 1994 he realized that for many parents, their differential treatment of identical and fraternal twins was an 'approach' — not simply a reaction to their twins' behavior:[23]

> The tendency for parents of MZ twins to treat their offspring more similarly than parents of DZ twins is, therefore, unlikely to result entirely from a greater similarity in behavior as children of MZ v. DZ twins.[24]

Thus, Kendler acknowledged that parents' rearing *approaches* might influence the greater behavioral similarity of identical twins. Kendler and colleagues might have closed up shop at this point, but they argued that while 'these results may seem like prima facie evidence for rejecting the twin method, such a step would be premature':

> Differential parental treatment of MZ and DZ twins would invalidate twin studies of psychiatric disorders *only if* the type of parental treatment for which MZ twins were more similarly exposed than DZ twins influenced the risk for the psychiatric disorders under examination.[25]

For Kendler, therefore, to invalidate the twin method it must be shown that the greater similarity of treatment of identical twins could influence their concordance rates for psychiatric disorders. According to Kendler, the EEA predicts that identical twins 'would develop similar phenotypes regardless of the similarity of their social environment.'[26] But a simple example casts doubt on Kendler's prediction. Suppose that a pair of identical twins is separated at birth and raised in different social environments. One twin is shifted from one foster home to another, is beaten, unloved, neglected, and abandoned several times during childhood. The other twin is raised in an exceptionally caring home where he or she is loved and treasured, respected, and is well taken care of. For

Kendler's theory to hold, he must conclude that both twins would be at equal risk for childhood or adult psychiatric diagnosis – in other words, the claim that treatment does not influence behavior. However, it is unlikely that the twins in this example would be at equal risk for psychiatric disorders, as there is considerable research evidence indicating that these experiences contribute to the causes of, or prevent the appearance of, these disorders.

A similar hypothetical scenario has been discussed by Kendler himself:

> If we selected 100 genetically diverse individuals and took them on a 'love-boat' cruise in which all of their needs (e.g. entertainment, food, and love) were met, we would find, I predict, both low levels of anxiety and depression and very little variability. Almost everybody would be happy. Suppose, however, that we then expose these same 100 individuals to a highly stressful experience, such as combat. The average level of anxiety and depression would increase but, more important, so would variability. Some individuals who are good at coping would deal effectively with the adversity and demonstrate few psychiatric symptoms. Others, with poorer coping abilities, would become very symptomatic.[27]

Here, Kendler recognizes that a given individual could be 'happy' or 'depressed' on the basis of a good or bad environment. There is no disputing this. But Kendler's EEA theory cannot allow that an identical twin sent into combat after taking the cruise would be more likely to become depressed or anxious than his or her genetically identical co-twin who remains on the 'love-boat,' since, as we have seen, Kendler insists that identical twins would develop the same phenotype regardless of the similarity of their environments. From the genetic standpoint, however, there is no difference between *an individual* placed in different environments and *identical twins* placed in different environments.

If we consider Kendler's scenario as analogous to a twin pair's rearing environment, because virtually everyone (including Kendler) agrees that identical twins share more common experiences and are treated more alike than fraternals, they would be much more likely, for example, to go on a cruise or into combat *together*. Conversely, the more variable environments of fraternal twins (or even, as Kendler would argue, their more dissimilar genetic relationship) would make it much more likely that one twin would go into combat, while the other would choose to stay on the cruise. Because Kendler would consider the twin method 'invalid' if such environmental differences 'influenced the risk for the psychiatric disorders under examination,' his position that the same individual could be depressed or happy – depending upon which environment he or she experienced – is in conflict with his views on the effects of twins' environments. Nevertheless, in 2001 Kendler continued to regard the twin method as 'a vibrant part of the field of psychiatric genetics . . .'[28]

A critique of the trait-relevant equal environment assumption

Family studies and the trait-relevant EEA

Having subtly redefined the basic theoretical assumption of the twin method by adding the trait-relevant condition, twin researchers attempted to place the burden of proof for the relevance of particular environmental factors on twin method critics. One investigator replied to a critic as follows:

> The equal environment assumption [of the twin method] is required only for trait-relevant features of the environment; features of the environment that have causal status. Causal status must be demonstrated, not assumed . . . It is absolutely mandatory that Hoffman demonstrate that the differential treatments she cites have a causal influence on the traits whose similarity she is trying to explain. This is a very difficult task.[29]

According to this investigator it is permissible to *assume* that the greater environmental similarity experienced by identical twins does not affect their greater similarity in behavior, but the assertion that the greater similarity of treatment leads to a more similar behavior must be *demonstrated*.

However, if identical twins are treated more alike, spend more time together, have more common friends, and experience greater levels of identity confusion, they are more likely to be similarly exposed to 'trait-relevant' environmental factors — known or unknown. Contrary to the wishes of the trait-relevant EEA proponents, *twin researchers* bear the burden of proof for showing that identical and fraternal twins are not differentially exposed to potentially relevant environmental factors.

The trait-relevant EEA becomes even more implausible when we realize that, as we saw in Chapter 1, most genetic researchers now acknowledge that the results of family studies are plausibly explained by environmental factors. However, genetic researchers do not make the trait-relevant argument for family studies. On the contrary, they acknowledge that the simple fact that family members share a common environment confounds the results of family studies. Similarly, the twin method is confounded by the fact that identical twins experience more similar environments than fraternal twins, even if (like family studies) the specific trait-relevant environmental factors are unknown. As an example of the failure of genetic researchers to apply their observations about family studies to the twin method, we turn to an article by prominent twin method proponents Lyons et al. These investigators recognized that family studies are 'confounded' because 'offspring are likely to share exposure to toxic or infectious agents that could lead to similar outcomes for the siblings,'[30] but they failed to recognize that *the twin method is similarly confounded by the fact that identical twins share exposure to potential toxins to a greater degree than fraternal twins*. The investigators ask for, in effect, a sort of 'special exemption' for the twin method in the face of the overwhelming evidence that, in addition to having a greater genetic similarity, identical twins experience more similar environments than fraternals.

Without realizing it, proponents of the trait-relevant EEA have redefined the assumption in a way that can lead only to the obsolescence of the twin method. If we extend their logic to family studies, it follows that the familiality of a condition demonstrates its genetic basis, unless it can be shown that the affected families were exposed to an environmental agent *proven* relevant to the condition under study. Because family studies would thereby settle the question of the existence of genetic factors, there would be little reason to conduct twin studies at all. It is therefore inevitable that the now definitive nature of the new 'trait-relevant' family studies would drive the twin method out of existence as a research method. *The trait-relevant equal environment assumption has transformed the twin method into little more than a special type of family study.*

In family studies, an affected individual shares a more similar environment with members of his or her family than with members of the general population. In twin studies, identical twins share a more common environment and a greater psychological association than fraternal twins. Most twin researchers do not dispute these claims, but argue that the correlation between common genes and common environment confounds family studies but *does not* confound the twin method. Their argument in defense of the twin method could just as easily be used in support of drawing genetic inferences from family studies. They could argue, for example, that although family members share a more common environment than shared by a group of randomly selected members of the population, critics (of drawing genetic conclusions from family studies) incorrectly infer from this finding that family members are more likely to be exposed to 'trait-relevant' environmental factors.

But twin researchers do not make this argument. Instead, they reserve the trait-relevant EEA *for the twin method* while correctly acknowledging that environmental factors confound genetic inferences from family studies. In doing so they commit the logical fallacy of 'special pleading' — that is, they apply standards to one research method but refuse to apply them to another method equally worthy of these standards. In fact, the decision to place 'the burden of proof' on critics of drawing genetic inferences from the twin method — but not on critics of drawing genetic inferences from family studies — is entirely arbitrary.

Given the similar features of family and twin studies, we should be able to test family studies for environmental confounds using logic similar to that used by researchers testing the validity of the twin method (see below). For example we could test family studies by comparing the psychological trait resemblance between biological fathers and their offspring living together in the same home, versus men who *believe* they are the biological father of children living in their home but for whom blood or DNA tests show are *not* the true biological father. We could then compare correlations between the true father-child pairs versus the perceived (but biologically unrelated) father-child pairs. The genetic hypothesis would predict that the true biological father-child pairs would correlate significantly higher than the perceived father-child pairs.

Suppose that the results showed that the true father-child pairs correlated significantly higher than the perceived father-child pairs, as the genetic theory would predict, and that the perceived father-child pairs correlated no higher than randomly selected strangers. Could we then conclude that genetic inferences from family studies are valid? Because there are many ways in which family members are exposed to a common environment, we must answer this question in the negative. A broad overview of the family study method has led twin researchers, correctly, to the understanding that genetic and environmental influences cannot be disentangled in these studies. For this reason, genetic researchers do not test for family members' 'trait-relevant' exposure to etiologically relevant factors. Twin researchers, however, have failed to explain why the trait-relevant condition is applicable to the twin method, but not to family studies.

One is tempted to propose that disinterested parties or environmentalist critics carry out a series of tests of the EEA's validity as a way of countering the inadequate measures and possible bias of twin researchers who have attempted to test the assumptions of their method. However, it is unlikely that the results of these tests could uphold the validity of the twin method. Similar to the argument I made in relation to the hypothetical investigation of environmental influences on family studies, the fact that genes and environment are correlated is reason enough to invalidate genetic inferences from identical-fraternal comparisons. The twin method is — at face value — confounded by the fact that identical twins experience more similar environments than fraternal twins. Nonetheless, the failure of the scientific community to recognize that genetic inferences from these comparisons rest on an unsupported theoretical assumption might necessitate the performance of tests looking at the effects of common environment, identity confusion, and emotional closeness on twin similarity. The design of these studies should be written up and published in advance of the collection of twin data. This would ensure that comparisons are not made after the fact on the basis of selected data favorable to the researchers' hypothesis.[31]

What leading twin researchers say about the EEA
Let us now examine some of the ways that leading twin researchers have discussed and defended the validity of the EEA, which will be followed by my commentary. I have chosen certain passages in order to make particular points, but I do not mean to imply that the views of these investigators differ greatly. They are actually quite similar.

Plomin, DeFries, McClearn, and Rutter
Plomin and colleagues are leading behavior geneticists and are the authors

of a 1997 textbook on the subject. They defined the EEA as follows:

> The *equal environments assumption* of the twin method assumes that environmental similarity is roughly the same for both types of twins reared in the same family . . . The equal environments assumption has been tested in several ways and appears reasonable for most traits.[32]

According to Plomin and associates, identical twins share a more common environment in part because 'some experiences may be driven genetically.' The authors maintained that 'Such differences between identical and fraternal twins in experience are not a violation of the equal environments assumption because the differences are not caused environmentally.'[33] This is a problematic position, as seen in the following example. Suppose a twin study is performed to investigate possible genetic influences on lung cancer. For the purpose of this example, we will assume that there is a genetic basis for the desire to smoke tobacco. Because of their more similar genetic resemblance, it is likely that identical twins would be more concordant than fraternals for cigarette smoking. According to the logic of Plomin and associates, since the propensity for *smoking* is 'driven genetically,' the higher identical twin *lung cancer* concordance rate is evidence of a genetic predisposition for lung cancer — a clearly erroneous conclusion because lung cancer could still be completely explained by exposure to tobacco smoke, which is an environmental occurrence. As another example if, as some maintain, there is a gene for 'novelty seeking,'[34] it is more likely that both members of an identical twin pair versus a fraternal pair would attempt to climb Mt. Everest. Could we then conclude that there is a genetic predisposition for frostbite?

Rowe

In his 1994 book *The Limits of Family Influence*, behavior geneticist David Rowe wrote that 'the question is not whether MZ twins receive more similar treatments (they do, and to claim otherwise would be foolish), but whether these treatments influence a particular trait.'[35] For Rowe, while the few remaining proponents of the traditional EEA are 'foolish,' the equal environment assumption remains tenable because (according to research) greater intrapair environmental similarity does not predict twin correlations for psychological traits.

Rowe addressed the environmentalist argument that societal treatment of identical twins on the basis of their greater physical similarity might make them behave more similarly, claiming that 'matching people in physical appearance should have little effect on similarity in their psychological traits . . .'[36] Rowe then cited an example:

> How alike in personality or musical talent is even the best Elvis Presley lookalike to the King of Rock and Roll? If we gathered 10 Elvis lookalikes together, would they be alike in personality at all? The remarkable similarity of MZ co-twins is attributable to genes' creating

matching neurons. Personality and temperament reside in the brain, not a face.[37]

Did Rowe wish to imply that Elvis would have been the 'King of Rock and Roll' had he looked like Winston Churchill, or if his facial pigmentation, like Chuck Berry, had been dark? One could argue that Elvis, while certainly talented, would not have been a major figure in the history of music were it not for his Caucasian ancestry plus his appeal to some for his attractiveness, and to others for his 'cool.' Presley did not write his own music and was dependent on skilled songwriters to generate hits. These writers would have likely sold their music to other performers had Presley not gained a mass following. Would an identical twin brother of Elvis have had the same chance of being the King if he had sustained serious burns and had disfiguring scars on his face? Would movie studios have featured Elvis in dozens of films had he been ordinary looking?

According to Rowe, 'personality and temperament reside in the brain, not a face.' In making this statement, Rowe seemed to discount the relevance of environmental factors in shaping psychological traits, in addition to exhibiting questionable logic. Would he argue that education is unimportant because 'learning takes place in the brain, not in a book,' or that post-traumatic stress reactions are unrelated to trauma because 'personality and temperament reside in the brain, not on a battlefield'? In fact, the logic of Rowe's argument demands that we not stop at the face and that we discount all environmental and sensory stimuli as affecting personality and temperament, since these also do not 'reside in the brain.'

Faraone, Tsuang, and Tsuang
According to psychiatric geneticists Faraone, Tsuang, and Tsuang,

> The validity of twin studies . . . requires us to assume that differences between MZ and DZ twins are due to genes; we must assume that environmental determinants of similarity are identical for the two types of twin-pairs. But if the similarity of MZ twin environments is greater than that of DZ twin environments, then the twin method will overestimate the importance of genetic influences.[38]

Faraone and colleagues noted that several studies have shown that identicals experience more similar social environments, 'Thus, twin studies may overestimate heritability if these differences in environmental similarity are etiologically relevant to the disorder under study.'[39] The claim that etiologically relevant environmental factors might 'overestimate heritability' is common, and deserves a brief comment.

The question centers on whether environmental factors influencing identical-fraternal differences point to genetic influences being overestimated, or whether they indicate that the twin method is unable to disentangle these influences. Virtually all twin researchers believe that some or all of the identical-fraternal concordance or correlational difference is explained by genetic factors, but this represents a hope or a belief and does not have solid support in the evidence. Because identical twins experience more similar environments, one could legitimately transform the twin method into a measure of the differing *environments* of the identical and fraternal twinships, based on the assumption that genetic factors are not involved. We might call this assumption the 'GEA,' or 'genes excluded assumption.' Theoretically, any twin study attempting to determine the genetic component of a trait could be transformed into a twin study utilizing the same trait as a measure of the differing environments of the identical and fraternal twinships. All that is required is exchanging one assumption for another. For too long we have allowed genetically-oriented researchers to determine which assumptions twin study conclusions could be based upon. My point is that the equal environment assumption of the twin method is implausible, and that a counterposed assumption (the GEA) is more plausible. We could pick a psychiatric disorder with an unlikely genetic component as a way of testing whether genetic or environmental differences influence identical-fraternal concordance rate differences. An example is anorexia. Twin studies have shown that identical pairs are about six times more likely to be concordant for anorexia than fraternal pairs (45% vs. 7% pairwise concordance).[40]

Bouchard

Although more well known for his studies of twins reared apart (see Chapter 4), Thomas J. Bouchard, Jr. has been a resolute defender of the twin method:

> We . . . assume that common family environment influences are the same for both types of twins (equal environment assumption). This assumption has been constantly challenged by critics of the twin method. It has, however, been studied by twin researchers for many years and the general consensus is that it is well supported.[41]

But twin researchers are hardly impartial testers of a method that is, in many cases, their life's work. The methods they use for evaluating data, the data they choose to collect, and the conclusions they reach, are influenced by their pre-existing belief in the twin method's validity. In attempts to test (or more accurately, uphold) their method, contemporary twin researchers — without exception — overlook the fact that twin studies suffer from most of the problems which they acknowledge invalidate genetic inferences from family studies.

Segal

In her 1999 book *Entwined Lives*, behavior geneticist Nancy Segal wrote that twin researchers 'assume that environmental influences (e.g., parental discipline or teacher expectations) affecting traits are the same for identical or fraternal twins. This is called the *equal environment assumption (EEA).*'[42] Although she acknowledged that critics argue that the more similar treatment of identical twins could increase their greater similarity when compared with fraternals, these critics, according to Segal,

> conclude that *similar treatment* not *similar genes* causes identical twins' behavioral resemblance. Conversely, they reason that because fraternal twins look different people treat them differently; thus fraternal twins show less similarity due to environmental effects, not genetic ones. Critics, therefore, believe that identical and fraternal twins' unequal treatment violates the EEA and invalidates findings.[43]

Segal revisited Rowe's 1994 argument about facial appearance and noted that Rowe 'reminds us that faces and brains are not one, so similar treatment will not affect biological functions underlying personality.'[44] Another way of saying this is: treatment does not affect personality, which is biological and genetic. The argument in favor of the trait-relevant EEA thus boils down to the circular claim that the twin method demonstrates genetic influences on personality and behavioral differences because . . . personality and behavioral differences are genetic.

Segal mentioned Rowe's 1994 rhetorical question asking whether we would expect ten Elvis lookalikes to be similar in personality at all, implying (like Rowe) that we wouldn't. Interestingly, we *should* expect personality similarities among these impersonators! Far from constituting ten randomly selected members of the world's population, they would have several factors in common which include: (1) they are usually Caucasian; (2) they are usually male; (3) they are willing to perform publicly, which suggests certain styles such as extroversion, narcissism, etc.; (4) they possess musical and singing ability; (5) they are more likely to have been reared in the culture of the southern United States than randomly selected people; and (6) they are more similar in age than randomly selected people, and therefore are more likely to have been raised in the same national birth cohort. Thus, we would expect these biologically unrelated lookalikes to correlate significantly above zero on personality measures, since sex and age effects alone can have a 'substantial' effect on psychological trait correlations.[45] This underscores the basic problem of Bouchard, Segal, and associates' conclusions from their reared-apart twin studies, which we will look at in Chapter 4.

Lyons

Psychiatric genetics proponent Michael Lyons has performed several twin studies since the 1980s. In a review published in 2001, he wrote that if twin pairs (combining identical and fraternal) resemble each other for a trait at a rate exceeding chance expectation, 'we can conclude that there is a family influence on the characteristic we are studying.'[46] Because most psychiatric twin studies find a much greater than chance resemblance among both identical and fraternal twins, Lyons implicitly acknowledged that 'family influences' affect concordance rates for the various disorders. However, Lyons went on write, 'Because the family environment promotes resemblance equally in MZ and DZ pairs, if we observe significantly greater similarity among MZ pairs than among DZ pairs, we can conclude that the characteristic we are observing is influenced by genetic factors.'[47] Thus, for Lyons, the validity of the twin method is dependent upon the claim that 'family environment promotes resemblance equally' among identical and fraternal twins. While it is true that Lyons is a proponent of the trait-relevant EEA (see Chapter 8), in this 2001 review he recognized that family influences are indeed 'trait-relevant' for most psychiatric conditions.

According to most twin researchers, although identical twins experience a more similar family environment than fraternal twins, no one has shown that these differing environments are trait-relevant. Conversely for Lyons, family environments are trait-relevant, but identical and fraternal twins' family environments are equal. The formulation that reflects reality, however, borrows from both of these positions: family environments are trait-relevant, and identical and fraternal twins experience far different family environments. The only valid conclusion is that the twin method is hopelessly contaminated by environmental influences.

My aim has been to illustrate the way that leading contemporary twin researchers write about the EEA. These quotations capture their general views on the subject; they were not selected because they represented isolated poor formulations. What they do show is that intellectually gifted scholars and writers, who have dedicated their adult lives to human genetic research and the twin method, become inarticulate, contradictory, and confusing when trying to uphold the validity of a thoroughly indefensible theoretical assumption.

Attempts to test the validity of the equal environment assumption

As we have seen, the broad consensus among contemporary twin researchers includes the following points:
- The observation by twin method critics, that identical twins experience more similar environments than fraternals, is correct.
- Although identical twins experience more similar environments, the evidence suggests that this does not seriously bias the results of twin studies. The twin method remains valid because identical twins' greater environmental similarity is the result, not the cause, of their greater genetic similarity.
- Identical and fraternal twin environments must be shown to differ on environmental factors etiologically relevant to the trait under study. Furthermore, the burden of proof for demonstrating

differential trait-relevant environments falls on critics.
- On the basis of the 'trait-relevant' definition of the equal environment assumption, the twin method remains a viable instrument for the detection of genetic influences on psychological trait differences and psychiatric conditions.

Investigators such as Kendler and Plomin point to twin researchers' attempts to test the twin method's validity. A common theme of Kendler's articles of the 1980s and 1990s was that the EEA has been tested 'in five different ways.'[48] Kendler and other twin researchers concluded that the results of these tests uphold the validity of the twin method. Plowing through this body of research is a tedious affair, and I will not subject readers to a critique of this literature here. Critical analyses have appeared elsewhere, and the general conclusion was that the studies comprising the five ways of testing the EEA don't provide much evidence in its favor.[49] In the words of Pam and colleagues, in their review of the 'EEA test' literature, 'much of the research supporting the EEA does not meet scientific standards, is often sophistical in the way inferences are drawn from the data, and slanted in terms of objectivity.'[50] Here, I will limit the discussion to the three most frequently cited studies in support of the EEA and the twin method.

Loehlin and Nichols, 1976
Behavior geneticists John Loehlin and Robert Nichols's 1976 twin study is one of the most frequently cited investigations in the behavior genetics literature. Its results are discussed in reference to several theoretical pillars of behavior genetic methods and findings, including the alleged importance of 'non-shared' environmental influences on human psychological trait differences.[51]

Loehlin and Nichols analyzed data from 850 pairs of reared-together twins and their parents. The sample was obtained from high school juniors who took the National Merit Scholarship Qualifying Test (NMSQT) in the Spring of 1962. Approximately 600,000 students took this examination, and each examinee was asked to indicate if he or she had a twin sibling. The researchers obtained the names of 1,507 same-sex twin pairs, who were sent questionnaires and were asked to participate in the study. The final sample of 850 twins was based on pairs who supplied 'reasonably complete' data, plus a filled-out parent questionnaire.[52]

Loehlin and Nichols found that identical twins correlated significantly higher than fraternals on personality and cognitive ability measures. This difference, of course, would be expected by 'hereditarians' and 'environmentalists' alike. The study is widely viewed by behavior geneticists as groundbreaking due to the finding of a 'weak overall relationship between differences in childhood treatment and differences in personality . . .'[53] According to Loehlin and Nichols,

Differences in the childhood treatment of twins are not very predictive of adolescent personality differences. Identical twin differences — though necessarily environmental — are not better predicted from treatment differences than are fraternal-twin differences. Stronger attempts at parental influence do not result in more similar pairs of twins. Attempts to treat twins alike do not lead to greater similarity between them.[54]

Loehlin and Nichols argued that their data showed that the similarity of identical twin pairs is not predicted by more similar parental treatment, which was measured by the treatment-related responses on the parent questionnaire. The implication was that the twin method is not biased by the more similar parental treatment experienced by identical twins.

Since the late 1970s, the results of this study have been cited repeatedly by behavior geneticists as supporting the validity of the EEA. The claims of Scarr and Carter-Saltzman are typical:

> If the usually observed behavioral differences between MZ co-twins are smaller than those between DZ co-twins, then critics assume that the experiential differences between DZs are a sufficient explanation for their greater behavioral differences. If that were true, then MZs who are treated more differently by their parents than other MZs would be less similar than MZs who are treated more alike. This was the logic of the study of Loehlin and Nichols (1976), who calculated the correlations between environmental differences of MZ co-twins on those variables that differentiated MZ and DZ twins and personality and intellectual differences between MZ co-twins. They found little relationship between differences in parental treatment and experiences and test score differences. Thus the differences in treatment between MZs who were treated similarly and those who were treated differently could not account for the magnitude of the differences between MZs and DZs.[55]

More recently, in 2001 Bouchard and Loehlin wrote, 'Loehlin and Nichols . . . long ago showed that similarity in treatment could not account for twins' similarity in personality.'[56]

In spite of its acclaim, there are problems with Loehlin and Nichols's study which include, (1) a sample that was heavily biased in the direction of more similar or 'twinlike' pairs; (2) the reliance on questionnaires and tests that were completed at home, i.e., outside of the researchers' control, and were subject to twins collaborating on their responses; and (3) the selective analysis of data.[57]

However, the most serious problem is Loehlin and Nichols's reliance on responses from retrospective questionnaires, filled out by the twins' parents, to determine treatment similarity. Several researchers have found that recall is affected by the views of respondents at the time of questioning,[58] suggesting that the NMSQT parents were influenced by a positive attitude toward their academically high-achieving twins.

In a large 1970 study assessing the accuracy of parental recall, Yarrow and associates compared the retrospective reports of 224 mothers and 190 children to ratings, observations, and reports obtained contemporaneously on the children, which included parental reports made during the child's early years. Yarrow et al. found only a modest correlation between the baseline information and mothers' later accounts, and noted that mothers 'painted a "better" or "more satisfactory" picture than yielded by the baseline.'[59] Consistent with Robbins's 1963 study, the researchers 'found mothers' reports conforming to child-care values current at the time of reporting.'[60] Yarrow and associates concluded that the use of retrospective data — especially in areas with a 'high potential for ego involvement' — is an unsatisfactory substitute for a longitudinal study and that, if used, researchers must be aware of the data's 'serious limitations.'[61] They concluded with 'a plea for caution . . . the research practitioner [should] not be too hasty in the application of the retrospective method and the consumer of the outcome of such research [should] attend to the signs saying *caveat emptor*.'[62]

Loehlin and Nichols's parent questionnaire contained 56 treatment-related items, and parents reported treating their twins alike, both identical and fraternal, on the vast majority of experiences. The possible responses to the treatment related items were: (1) 'Twin one only,' (2) 'Twin two only,' (3) 'Both twins,' and (4) 'Neither twin.' Answers #1 and #2 were counted as one point on scales measuring the frequency of reported differential treatment of the twins by their parents. According to the results, it appears that parents treated their twins remarkably alike. Based on 56 questions, parents recalled that they had treated (1) their male identical twins the same on 99.14% of the questions, (2) their female identical twins the same on 98.98% of the questions, (3) their male fraternal twins the same on 97.44% of the questions, and (4) their female fraternal twins the same on 97.7% of the questions.[63] (These figures are in conflict with the 'twins create their environment theory,' which predicts that parents would treat their fraternal twins far more differently than identicals.) Combining all responses, parents answered 'neither twin' or 'both twins' *98.3% of the time*. According to Loehlin and Nichols, 'it was exceptional for a parent to mark any treatment item as different for the twins; for the identicals it was, in fact, positively rare.'[64] Rare indeed. For identical twins ages 0–12, parents reported treating them differently only once every 175 questions. For fraternal twins in the same age range, parents answered that their twins were treated differently only once every 55 questions.[65] These questionnaire responses are therefore of doubtful validity in assessing the relationship between twin correlations and parental treatment.[66] Clearly, parents answered questions on the basis of an idealized image of their child-rearing practices, and these responses are therefore unreliable.

Ironically, although most people (as well as most contemporary twin researchers) acknowledge that identical twins are treated far more similarly than fraternals, there appears to be one exception to this near unanimous opinion: *the NMSQT twins' parents*. What better evidence could be found that these parents represented an extremely atypical population, or much more likely, that the retrospective questionnaire method is a flawed instrument for assessing parents' treatment of their children.[67] As Yarrow and associates concluded, retrospective questionnaires are an inadequate substitute for a longitudinal study.

The numerous commentators citing the Loehlin and Nichols study rarely discuss its various methodological problems, many of which were discussed by the investigators themselves. For example, Loehlin and Nichols acknowledged that their twin sample was biased because it (1) contained an excess of identical twins, (2) contained an excess of female twins, (3) consisted only of twins who were willing to respond to a mailed questionnaire, and (4) consisted of high-achieving Merit Scholarship test-takers.[68] They also cautioned that the sample was probably selected (biased) for similarity of experience,[69] and that, in general, interpretations of their data 'do not altogether exclude a completely environmentalist position.'[70] Finally, although Loehlin and Nichols were well aware of the 'fallibility of parents' retrospective accounts of the early history of their children,' they believed that 'fallible data interpreted with some caution are clearly better than no data at all . . .'[71] On the contrary, 25 years of secondary sources' caution-free and uncritical acceptance of Loehlin and Nichols's unwarranted conclusions have provided support for the EEA and the twin method. In this instance, faulty conclusions based on 'fallible data' have been far *worse* than no data at all.

Loehlin and Nichols's twin study has been hailed as a landmark achievement in the field of behavior genetics, when in fact it tells us little more than what we already knew—that identical twins correlate higher than fraternal twins on personality and IQ tests.

Scarr and Carter-Saltzman

The studies by American behavior geneticist Sandra Scarr in 1968, and Scarr and Carter-Saltzman in 1979, are the most frequently cited examples of the 'reverse zygosity' test of the twin method's validity.[72] This method looks at twins whose zygosity status has been misidentified by twins or their parents. For example, some genetically identical pairs believe themselves, or are believed to be, fraternal. Test score correlations of these mislabeled twins are compared to correctly identified twins in order to test the effects of 'true' zygosity (genetic effect) versus 'perceived' zygosity (environmental effect). Scarr considered this comparison to be an important test of the validity of the EEA:

> Critics of twins studies have assumed that differential treatment of MZ and DZ pairs constitutes *prima facie* evidence of bias. It does not. The direction of effect in the correlation between zygosity and environmental similarity is not at all clear. It is possible that the greater genetic similarity of MZ twins leads to more similarity in their environments. Parents and others may respond to the behavioral similarity of MZ pairs with more similar expectations for them, and

identical twins themselves may select and attend to more similar aspects of their environments.[73]

Thus, Scarr invoked the 'twins create their environment theory' of Shields and Kallmann in defense of the validity of the twin method.

In spite of being widely cited in support of the twin method's validity, there are problems with Scarr's studies. These include, (1) the sample sizes of misidentified twins were small. In the 1968 study there were only 11 misidentified pairs. In the 1979 study there were 41 pairs (out of a total of 342) where both members of the pair were wrong about their true zygosity status. (2) The results, as least as they pertain to personality, were mixed. In 1968 Scarr concluded, 'Data from the Vineland Social Maturity Scale and the Adjective Check List suggest that beliefs about zygosity also have an effect on MZ pairs . . .'[74] and in 1979 Scarr and Carter-Saltzman found that perceived zygosity had an effect on fraternal twin similarity for personality measures. (3) According to Kamin, who had access to the 1979 raw data, a closer look at these figures reveal environmental effects on cognitive abiliy measures not shown in the 'blunted' environmental scales reported in the published paper.[75] (4) After carefully examining Scarr and Carter-Saltzman's IQ data, developmental psychologist Ken Richardson concluded that one of the key results 'does *not* confirm the equal environments assumption in the way that is frquently claimed.'[76]

According to Scarr and Carter-Saltzman, the data showed that 'perceived similarity is not an important bias in IQ tests,' although it might have more of an effect on personality. They ended with an endorsement of the twin method:

> The critical assumption of equal environmental variance for MZ and DZ twins is tenable. Although MZ twins generally experience more similar environments, this fact seems to result from their genetic similarities and not to be a cause of exaggerated phenotypic resemblance.[77]

The problem with this argument is that — even if it were true — the EEA is not supported. The twin method is used for a wide range of traits and conditions for which the *reason* that identical twins experience more similar environments is completely irrelevant. As one critic correctly observed, 'a greater similarity in environment induced by greater similarity in genetics still means that monozygotic twins have more similar, or shared, environments than dizygotic twins.'[78]

In addition, the argument that identical twins' genetically determined similarity in personality causes their parents and others to treat them more similarly overlooks an important point which I discussed earlier: aren't parents and others, according to this theory, reacting to twins on the basis of *their own* genetically shaped behavior? How can it be that twins' behavior is genetically determined, but parents' behavior and rearing patterns are ever-changing and adaptive to their twins' needs? In the childhood struggle to create similar environments, Scarr's postulated genetically-driven twins' behavior runs up against a formidable foe — the genetically-driven behavior of heir parents.

This discussion should, one hopes, put to rest the idea that the twin method is valid or 'tenable' if twins' environments are more similar because, as Kendler and others have speculated, identical 'twins might *create* for themselves more similar environments.'[79] Because idntical twins experience more similar environments than fraternals — regardless of how genetic factors might influence this — the twin method is unable to disentangle genetic and environmental similarity.

Conclusions

Although there are several methodological flaws associated with studies using the twin method, I have focused on the assumption that identical and fraternal twins experience similar environments, which is known as the equal environment assumption. Until the early 1970s most twin researchers defined the EEA as the assumption that identical and fraternal twins experience similar environments, or if the environments were different, that the EEA remained valid because identical twins created more similar environments by virtue of their greater genetic similarity. Most twin researchers now acknowledge that identical twins experience more similar environments, but add the condition that environments must be shown to differ on 'trait-relevant' factors in order to invalidate the twin method. As Bouchard has written, 'It is certainly true that MZ twins experience more similar environments than do DZ twins,' while adding that 'no one has been able to show that such imposed similarities in treatment are trait-relevant.'[80] But it is simply not necessary for critics to identify trait-relevant factors in order to invalidate the twin method.

For several decades, twin researchers defined the EEA in the traditional sense. When confronted with the reality that the assumption was wrong, they did not discard the twin method, as they should have. Instead, they subtly redefined the EEA in a way that allowed them to go on with their work, and the new unspoken definition became the 'unequal-environments-don't-matter assumption.'[81] We have seen that genetic researchers make the 'trait-relevant' argument in support of the twin method but have failed to apply it to family studies. Their error lies in the fact that once identical-fraternal environments are acknowledged to be different, the twin method is transformed into nothing more than a special type of family study because, in both family *and* twin studies, environmental and genetic similarity are confounded. Twin researchers have led the twin method into a theoretical blind ally because they have unwittingly linked its fate to the family study method.

As three generations of critics have argued, the twin method is an environmentally confounded research technique which is unable to disentangle possible genetic and environmental factors. Thus, any conclusions in favour of genetic influences on psychiatric conditions or psychological trait differences derived from the twin method should be disregarded. Like family studies, the twin method offers only an 'initial hint' that trait differences might have a genetic component. The attempts of twin researchers to test the EEA have yielded unconvincing results, and it is debatable whether any test could establish the validity of the twin method in the face of the greater environmental similarity experienced by identical versus fraternal twins. The most reasonable (or 'parsimonious') conclusion to be drawn from twin method data is that environmental factors cause higher correlations among identical versus fraternal twins. Researchers attempting to assess possible genetic factors must find other methods to do so.

Notes

[1] Penrose, 1973, p. 90.
[2] In the foregoing discussion I draw upon the ideas expressed in articles I have published on this subject, specifically, Joseph, 1998b, 2000c, 2000e, 2001b, 2002.
[3] Kringlen, 1967a, p. 20.
[4] Carey & DiLalla, 1994.
[5] Gottesman & Shields, 1966a, pp. 4–5. Emphasis in original.
[6] Kendler et al., 1993, p. 21.
[7] e.g., Bouchard, 1993b; Gottesman & Shields, 1972; Kendler, Neale, et al., 1994; Morris-Yates et al., 1990; Rose, 1991; Scarr, 1968; Scarr & Carter-Saltzman, 1979.
[8] Wilson, 1934.
[9] Smith, 1965.
[10] Shields, 1954, p. 234.
[11] Kringlen, 1967a, p. 115.
[12] M. Bleuler, 1955, p. 13.
[13] M. Bleuler, 1978, p. 432.
[14] Shields, 1954, pp. 239–40.
[15] Kallmann, 1958, p. 543.
[16] Laing, 1981, p. 143.
[17] Scarr, 1968, p. 40.
[18] Kendler, 1987, p. 706. For a reaffirmation of this position, see Kendler, 2000.
[19] This idea is referred to by behavior geneticists as the 'reactive genotype-environment correlation.' See Plomin et al., 1977.
[20] Zerbin-Rüdin, 1972, p. 48.
[21] Kendler, 1987, p. 706.
[22] Kendler, 1983, p. 1415.
[23] Kendler, Neale, et al., 1994, p. 588.
[24] Ibid., p. 588.
[25] Ibid., p. 588.
[26] Kendler, 1983, p. 1414.
[27] Kendler, 1995, p. 896.
[28] Kendler, 2001, p. 1005.
[29] Bouchard, 1993b, p. 33.
[30] Lyons et al., 1991, p. 124.
[31] See Joseph & Baldwin, 2000.

[32] Plomin, DeFries, et al., 1997, p. 73. Earlier, Plomin and DeFries (1985, p. 17) held a position very similar to the traditional EEA definition: 'In general, identical twins do not appear to be treated much more similarly than fraternal twins. Even though identical twins are slightly more similar than fraternal twins on a few environmental measures, these experiential differences do not make a difference in their behavior.'

[33] Ibid., p. 73.

[34] Ebstein et al., 1995.

[35] Rowe, 1994, p. 45.

[36] Ibid., p. 48.

[37] Ibid., p. 48.

[38] Faraone, Tsuang, & Tsuang, 1999, p. 39.

[39] Ibid., p. 39.

[40] Treasure & Holland, 1995.

[41] Bouchard, 1997a, p. 279.

[42] Segal, 1999, p. 318.

[43] Ibid., p. 318.

[44] Ibid., p. 319.

[45] McGue & Bouchard, 1984, p. 325.

[46] Lyons & Bar, 2001, p. 318.

[47] Ibid., p. 318.

[48] For example, see Kendler, 1983, 1993.

[49] See Joseph, 1998a, 1998b; Pam, 1995; Pam et al., 1996; K. Richardson, 1998.

[50] Pam et al., 1996, p. 359.

[51] See Plomin & Daniels, 1987.

[52] Loehlin & Nichols, 1976, p. 7.

[53] Ibid., p. 60.

[54] Ibid., p. 92.

[55] Scarr & Carter-Saltzman, 1979, p. 528.

[56] Bouchard & Loehlin, 2001, p. 251.

[57] This last point is detailed by Kamin, in Eysenck vs. Kamin, 1981.

[58] e.g., Bradburn et al., 1987; Halverson, 1988; Holmberg & Holmes, 1994; Reuband, 1994; Robbins, 1963; Yarrow et al., 1970.

[59] Yarrow et al., 1970, p. 67.

[60] Ibid., p. 72. Robbins (1963) studied parental recall of occurrences in the lives of their three-year-old children. These parents had been participants in a longitudinal study since the birth of their child, and their written descriptions of experiences with their children while they were happening were known to the researchers. When parents were asked to recall events that had occurred in their child's life just two years earlier,

> The parents were quite inaccurate in their memory of details about child- rearing practices and early developmental progress, in spite of the frequent rehearsal of these data due to their participation in the longitudinal study, and in spite of the relatively young age of the children . . . Inaccuracies tended to be in the direction of the recommendations of experts in child rearing, especially on the part of the mothers (Robbins, 1963, p. 261).

We should keep in mind that some of the events recalled by the NMSQT parents went back more than 16 years. Robbins found that parents tended to remember inaccurately in the direction of the recommendations of child-rearing experts (such as Dr. Spock). This meant that parents, probably without consciously lying, had a tendency to idealize their role. They answered questions about their child-rearing practices in the direction of how they, and society, believed that they *should have* raised their children, and there is reason to believe that the NMSQT parents answered questions about the treatment of their twins in much the same way.

[61] Yarrow et al., 1970, p. 69.

[62] Ibid., p. 74.

[63] Loehlin & Nichols, 1976, p. 55, Table 5-8.

[64] Ibid., p. 56.

[65] By my calculations.

[66] Parent responses to the treatment-related questions could have been checked by looking at comparable twin questionnaire items, were it not for the fact that not one of the 1,092 questions put to the twins dealt directly with how they were treated by their parents. The closest that any twin questionnaire items came to addressing parental treatment were questions #1069 and #1070, which asked the respondent to state which twin 'Is more liked by your mother,' and 'Is more liked by your father' respectively. The possible answers were 'I am,' 'Both the same,' or 'My twin is' (Loehlin & Nichols, 1976, Student Questionnaire Appendix, p. 23). Male identical twins' answers correlated at only .21 for the question about their mothers, and .26 about their fathers (Appendix B, p. 26), even though their parents had answered that they treated both male identical twins the same 99.14% of the time.

[67] Loehlin and Nichols's parent questionnaire item #288 (Appendix A, p. 12) asked parents whether they treated their twins differently or alike, and provided five different response choices. For all twin pairs, parents answered 'We tried to treat them exactly the same' eight times more frequently than they answered 'We tried to treat them differently' (Loehlin & Nichols, 1976, p. 87). These results most likely reflected the wishful thinking that parents displayed in their responses to this question — for how many parents would be willing to admit to others (or even to themselves) that they treated their twins differently? Question #288 was also unclear about what is meant by 'different treatment.' Did this mean that twins were treated as separate and different human beings, though equally, or, as it probably appeared to most parents, that they played favorites among the twins? One could argue that the question only asked if parents 'tried' to treat their twins equally, but common sense should tell us that it is unlikely that so many parents actually did — for how can anyone treat two people 'exactly the same?'

[68] Loehlin & Nichols, 1976, pp. 8-9.

[69] Ibid., p. 87.

[70] Ibid., p. 94.

[71] Ibid., p. 47.

[72] Scarr, 1968; Scarr & Carter-Saltzman, 1979. The reverse zygosity test was proposed by Kety as early as 1959 (Kety, 1959).

[73] Scarr & Carter-Saltzman, 1979, p. 528.

[74] Scarr, 1968, p. 40.

[75] Kamin, in Eysenck vs. Kamin, 1981, pp. 130-1.

[76] K. Richardson, 1998, p. 147.

[77] Scarr & Carter-Saltzman, 1979, p. 541.

[78] Hoffman, 1991, p. 188.

[79] Kendler, 1987, p. 706.

[80] Bouchard, 1997b, p. 134.

[81] Pam et al., 1996, p. 354.

CHAPTER FOUR

Genetic Studies of Twins Reared Apart: A critical review

Many of the criticisms of behavior genetic studies that we offer may seem to place unusual demands for proof upon the researchers. But, given the potential social misuse of conclusions in this field, we believe that the utmost care is required.
Critics Paul R. Billings, Jonathan Beckwith, and Joseph S. Alper in 1992.[1]

Another area for consideration is illustrated by a case study from my own research . . . that brought out some startling similarities for two women who were both Baptists and who both considered volleyball and tennis their favorite sports. They preferred math and English in school (and listed shorthand as their least favorite subject). Both were studying nursing, and both preferred their vacations in historical places. Perhaps most astonishing — these two women were *not* identical twins and were not related in any way.
Psychologist W. Joseph Wyatt in 1993.[2]

In recent years a great deal of attention has been paid to studies of twins reared apart. Many people, unconvinced by studies of twins reared together, have been convinced by these studies. Although reared-apart identical twins might appear to be an ideal way to unlock the mysteries of the nature-nurture debate, in this chapter we shall see that while theoretically identical twins reared apart could provide interesting and important information, all studies thus far reported do not provide this information.

There are essentially two streams of evidence coming from the twins reared apart (TRA) model: the 'folklore' stream and the 'scientific' stream. The folklore stream consists of stories released by twin researchers or reported by journalists about separated twins who, upon being reunited, are said to share an amazing set of common features, traits, preferences, etc. The scientific evidence consists of data presented in books and peer-reviewed academic journals. These studies typically report on data from a group of reared-apart twins and compare them with data from reared-together twins. The researchers usually conclude that their data support the existence of important genetic influences on psychological trait differences, while adding (in recent studies) that 'shared family environment' contributes little. I use the term 'MZA' (monozygotic twins reared apart) as a shortcut for reared-apart identical twins, and 'DZA' (dizygotic twins reared apart) for reared-apart fraternals. I do so only to distinguish them from reared-together identical twins (MZTs) and reared-together fraternals (DZTs), since, as we will see, few of these twins were truly reared apart in uncorrelated environments.

Part I: Identical twins reared apart as folklore

Suppose we pick up the newspaper tomorrow morning and read about MZAs who, it is said, share remarkable similarities. It is reported that they share the same taste in food, gave their children similar names, work at the same type of job, have similar quirky habits, and so on. What should we conclude? Certainly, the story is newsworthy and interesting. But does it also tell us something important about genetic influences on human trait differences? At first glance it appears that it would; after all, these are two people who grew up apart yet share a remarkable set of similarities. What else besides the genes they share could account for these similarities?

After getting past these initial reactions, the critically-minded reader might ask some questions: How were these twins discovered? Who discovered them? Did the people who reported them have information about dissimilar pairs, which they were not reporting? Had the twins met before being studied? Did the twins have ulterior motives for exaggerating their similarities and degree of separation? Did the people reporting their story have ulterior motives for exaggerating their similarities and downplaying their differences? Are the twins available for independent verification of their stories? And most importantly, did factors other than (or instead of) genetic identity contribute to their similarities? All of these questions will be examined as we explore the meaning of the stories of reared-apart identical twins.

To begin, most of these pairs come to the attention of researchers and journalists *because of* their similarities. It's very similar to the old 'dog bites person versus person bites dog' rule in journalism. A dog biting a person is not newsworthy because it is a common occurrence, whereas a person biting a dog is news because it is an unusual occurrence. If we were to read several articles describing a person biting a dog, it would be fallacious to conclude that people bite dogs more frequently than dogs bite people. The same is true for twins. Stories of similar twins are news because they are interesting and compelling; stories about dissimilar twins are not.

The next question we want to ask is whether the pair is representative of MZAs as a whole. Because many of the pairs we have heard about since the early 1980s have been identified by the Minnesota researchers, let us start there. As of 1998, the Minnesota investigators reported having given personality tests to 71 pairs of MZAs.[3] Unfortunately, they have released information on only a few selected pairs whose stories, in most cases, support the genetically oriented views of the researchers.

Most recent popularizations of MZA studies discuss Jim Lewis and Jim Springer (commonly referred to as the 'Jim Twins'). These identical twins had been separated at birth and were reunited at age 39. The Jim Twins were said to share an uncanny set of similarities, such as the names of their wives and children, career choices, and preferences for particular brands of beer and cigarettes. They were invited to the University of Minnesota by Bouchard and became the first pair of separated twins in the Minnesota Study of Twins Reared Apart (MISTRA).

The most celebrated case has been the twin pair Oskar Stöhr and Jack Yufe. These identical twins were born in 1933 in Trinidad to a 'Jewish' father and a 'German' mother. A few months after they were born, their parents divorced. Jack remained in Trinidad with his father, while Oskar went to live in Germany with his mother. Oskar was raised a Catholic Nazi in Germany, whereas Jack was raised as a Jew in Trinidad and had spent time on an Israeli kibbutz.[4] Although the twins grew up to be what their caregivers had raised them to be, this was not the way the case was presented. The two men were described as leading 'markedly different lives,'[5] and a supposedly amazing set of similarities was discovered after they arrived at the airport:

> Both were wearing wire-rimmed glasses and mustaches, both sported two-pocket shirts with epaulets. They shared idiosyncrasies galore: they like spicy foods and sweet liqueurs, are absentminded, have a habit of falling asleep in front of the television, think it's funny to sneeze in a crowd of strangers, flush the toilet before using it, store rubber bands on their wrists . . .[6]

What usually is not mentioned is the fact that Jack and Oskar had met previously and had been in postal contact for more than 25 years.[7] After their cases were reported in the press, they sold their life stories to a Los Angeles film producer.[8]

As we see, Jack and Oskar had a personal interest in exaggerating (or inventing) their similarities and underreporting previous contact. It is possible that other pairs exaggerated their similarities in order to please the investigators, or in an effort to attain celebrity status. The Jim Twins hired an agent at one point.[9] According to a description of Daphne and Barbara (the 'Giggle Twins'), another celebrated pair of MISTRA MZAs,

> At the end of their week in Minneapolis, as they settled down to a spectacular lunch to celebrate the end of their week-long testing, one final coincidence emerged. Both had told Bouchard the same lie. 'We both said we wanted to be opera singers and neither of us can sing a note,' Barbara confessed. And they both broke into peals of laughter, yet again.[10]

As Kamin and Goldberger observed, 'This anecdote, intended to emphasize the similar behavior of MZAs, makes it clear that twins could and did lie about themselves to the investigators.'[11]

What is often overlooked is that similarities can be found between any two people, if this is what one is looking for. As Bouchard acknowledged, 'Some of those similarities are surely coincidental — complete strangers at cocktail parties routinely discover "astonishing" occurrences in their lives; imagine what they might find after fifty hours of filling out questionnaires.'[12] Wyatt and colleagues looked at the similarities of 25 biologically *unrelated* pairs of college students matched on age and sex, and compared them with a group of reared-together identical twins. One pair of unrelated individuals was found to have a lot in common: 'Both are Baptist; volleyball and tennis are their favorite sports; their favorite subjects in school were English and math (and both listed shorthand as their least favorite); both are studying nursing; and both prefer vacations at historical places.'[13]

Let us look at the way that twins are sometimes discovered. We have the case of Jerry Levey and Mark Newman, two New Jersey firefighters brought together on the basis of being identified by another firefighter:

> Capt. Jim Tedesco looked across a crowded room at a firefighters' convention and was startled at what he saw. There was Mark Newman, and that was impossible. He knew that Newman was back home in Paramus, N. J. But how many bald, 6-foot-6, 250-pound-plus New Jersey volunteer firemen are there who wear droopy mustaches, aviator-style sunglasses and a key ring on the belt on the right side?[14]

In other words, Jerry and Mark were discovered on the basis of similarity; had they been dissimilar, they might never have been reunited. It is not, as journalists often write, 'spooky' or 'eerie' that these twins were both firefighters, wore droopy mustaches, aviator glasses, and a key ring on the right side. On the contrary, they were discovered *because* they were both firefighters, wore droopy mustaches, aviator glasses, and a key ring on the right side. Had Jerry been a florist instead of a firefighter, Captain Tedesco would not have made his discovery. I should re-emphasize that Jerry and Mark's story is compelling from the human interest perspective. Personally, I'd want to read all about it. Nevertheless, it tells us nothing about genetic influences on behavioral and personality differences.

Like other publicized pairs, we are told about what Mark and Jerry have in common but little about what sets them apart. For example, in addition to the above-mentioned similarities it is said that both men are bachelors, are attracted to similar women, and enjoy hunting, fishing, going to the beach, old John Wayne movies, pro wrestling and Chinese food. However, because the descriptions by journalists and 'scientists' highlight the similarities, we could reasonably assume that they are *dissimilar* for other, unmentioned comparisons. What were Jerry and Mark's taste in music, magazines, sports, newspapers, or shoes? What were their favorite high-school subjects, planets, animals, baseball players, or musical instruments? What were their choices of toilet paper, deodorant, toothpaste, motor oil, paint, or insecticide? The list is endless. But because readers are presented with the similarities, they are led to conclude that the twins are more similar than they really are. 'When any two biographies are avidly compared, at least some overlap is likely to be found.' The bleating of a nit-picking critic, perhaps? Hardly. This sentence can be found in a 1992 article by Minnesota twin researchers Lykken, McGue, Tellegen, and Bouchard.[15]

Photographs are also used to create the impression of similarity. In Minnesota twin researcher Nancy Segal's *Entwined Lives*, she presents a photograph of Jerry and Mark each holding a can of beer in their right hand. In the photo, the twins are seen curling their little finger under the can. Segal commented, 'Note that the twins' little fingers support their can of beer.'[16] There are other photographs, however, in which only *one* twin is seen curling his little finger under his beer.[17] Why did Segal choose one photo and not the other? Like the journalists, Segal ignored non-genetic explanations for why these twins may have held their beers the same way in her photo. Some possibilities include:

- That Jerry and Mark took cues from each other before the photo was taken in order to appear more 'twinlike.'
- That many people hold cans this way, which would mean that their similarity in this regard would not be so astonishing.
- That men hold cans in this manner more often than women.
- That members of the working class are more likely to hold cans this way than randomly selected people.
- That New Jersey firefighters traditionally hold beer cans this way.
- That Jerry and Mark were asked by the photographer or the journalist to hold the cans this way.
- That people growing up in their area and culture (Jerry and Mark lived only 65 miles apart[18]) tend to hold cans this way.

For a journalist to cite this photo as evidence of a genetic basis for pinkie curling is understandable. For a 'scientist,' it is inexcusable.

Nor is this the first time that photographs of reared-apart twins have been used to create the illusion of similarity. In his classic review of the older TRA studies in *The Science and Politics of I.Q.*, Kamin commented on a photo of twins 'Edwin and Fred' found in the 1937 Newman and associates study:

> The case study includes a photograph of the twins side-by-side 'at the time of their first meeting.' The photograph was 'reproduced by permission of *The American Weekly*.' The twins are remarkably alike in appearance. They are wearing identical pin-striped suits, and identical striped ties. These, of course, might have been bought 'at the time of their first meeting.' Perhaps it is relevant to note that Fred was unemployed at the time of the [Newman, Freeman, and Holzinger] study.[19]

For the first seven or eight years of the Minnesota study, stories about twins such as I have described were the only information available to the public. A commentator observed that these stories about twins were similar to those offered in support of astrology or extrasensory perception (ESP),

> Striking coincidences are reported as supposed grounds for belief in the phenomenon itself. In literature about astrology and ESP, cases where forecasts came true or where thought of a friend was immediately followed by a phone call from that friend are offered as evidence. The cases where forecasts failed or where a thought of someone is not followed by a phone call from that person are forgotten or left unmentioned.[20]

The problem with placing faith in testimonials is described by Robert T. Carroll in *The Skeptic's Dictionary*. Carroll wrote that 'Testimonials and vivid anecdotes' are often used to support beliefs in the 'transcendent, paranormal and pseudoscientific.' 'Nevertheless,' he continued, 'testimonials and anecdotes in such matters are of near zero value in establishing the probability of the claims they are put forth to support . . .' Carroll wrote that accounts of people's encounters with angels, aliens, ghosts, miraculous dowsers, a Bigfoot, or psychic surgeons are 'of little empirical value in establishing the reasonableness of believing in such matters.' The reason is that these accounts are unreliable and biased: 'They are of no more value than the televised accounts of satisfied customers of the latest weight loss program . . .'[21] While it might sound strange to compare stories about twins to astrology and ghost sightings, there are parallels. These include the selective presentation of information by parties attempting to persuade, the frequent inability to verify information, and the desire on the part of the audience to want to believe. Carroll's reference to television commercials is relevant to this discussion. When we see people celebrating the fact that Detergent X got the stains out of their shirts, few people, if they stopped to think about it, would say that the commercial depicted a random sample of people washing their clothes with Detergent X. They would realize that the makers of Detergent X consciously chose to show only people happy with their product. Why then are people willing to believe in the power of genetics on the basis of a few pairs of selected twins? Possibly because they do not believe that scientists (and to a lesser degree journalists) would try to 'sell' them something.

Clearly, the selected information on selected pairs released by the Minnesota investigators or reported by journalists does not prove anything about the importance of genetics, but it does demonstrate the potentially misleading nature of these stories. The Minnesota strategy was used successfully by the U.S. Pentagon during the 1991 Gulf War. Photographs of missiles hitting their targets were shown to the public; those missing or hitting schools, hospitals, etc. were not.

Judith Harris concluded in *The Nurture Assumption* that 'These true stories of reared-apart identical twins are a testimony to the power of genes,'[22] but it could just as easily be said that the testimonials of people who believed they were helped by psychic surgery or astrology suggest the power of these pseudosciences.

Influence of the cohort

We have seen that there is little reason to regard individual cases of reared-apart twins as anything more than interesting stories. This brings us to a fundamental question to which I have already alluded and which bears directly on the upcoming discussion of the scientific papers coming out of the Minnesota investigation: are there important environmental factors shared by both reared-together and reared-apart identical twins that would make them more similar than two randomly selected members of the world's population? The answer is a resounding yes. These include that they are exactly the same age; that they are the same sex; that they are almost always the same ethnicity; that their appearance is strikingly similar, which probably will elicit similar treatment; that they usually are raised in the same socioeconomic class; that they usually are raised in the same culture; that they shared the same prenatal environment; and that they typically spent a certain amount of time together in the same family environment, were aware of each other's existence when studied, and often had regular contact over a long period of time. All of these factors work towards making even reared-apart twins similar, yet are rarely discussed by popularizers of the individual pairs' stories.

The mere fact that two people are born on the same day can have an important impact on their subsequent behavior and beliefs. The reason is that people with the same date of birth are members of a national birth cohort. Think of how often we make statements recognizing the influence of this cohort. Some examples include, 'she grew up during the Depression and so she doesn't trust banks,' 'he smoked marijuana because everyone was doing it in the 60s,' 'young people today aren't interested in politics,' 'their disagreements are due to the generation gap,' and so on. A 47-year-old woman loves The Beatles for the same reason that her parents love Benny Goodman and her children love Britney Spears — because it's the music she grew up with.

We also recognize that men and women in western cultures tend to think and act somewhat differently. Thus, we would expect two women or two men to be more alike than a man and a woman. In addition, people raised in the same subculture are more similar because of the cultural influence; hence statements about New Yorkers being one way, Californians acting another way, Southern hospitality, country folks, city slickers, and so on. The same could be said about cultural differences between various ethnic groups. As for physical appearance, we have ideas about how attractiveness or a lack of attractiveness can create or block opportunities, influence personality, etc. If the 'Elephant Man' had had an identical twin brother from whom he had been separated at birth, it would be reasonable to expect that *both* would have suffered anxiety in social situations. Social class also has an important influence on beliefs, preferences, tastes, and values. Had Jerry and Mark been raised by French aristocrats, we might have discussed the reasons why they curled their little fingers under a glass of Dom Perignon instead of a can of beer.

In fact, most pairs of reared-apart identical twins share all of the common factors I have just outlined and, as discussed more fully in the next section, often had frequent contact during the years they were 'separated.' Reared-apart identical twins, as one author observed, are 'not so much similar to each other as they are similar to people of their eras and SES [socioeconomic statuses].'[23] Even personality trait theorists such as Gordon Allport acknowledged that the impact of culture on personality is 'indisputable,' and that 'Everyone admits that culture is vastly important in shaping personality.'[24] In most cases, reared-apart and reared-together identical twins share at least seven different cultural influences: national, regional, ethnic, religious, economic class, birth cohort, and gender cohort.

In another example of the failure to recognize cultural and cohort effect influences on twin similarity, we have journalist Kay Cassill's description of reared-apart identical twins Keith Heitzman and Jack Hellback, who grew up in Louisiana:

> Although the mighty Mississippi divided these two physically, it could not separate their parallel lives. The welder from one side and the pump mechanic from the other found that they are both allergic to ragweed and dust. Both had done poorly in school. Both disliked sports and had cut their gym classes whenever they could. They are both addicted to candy. Their similarity of dress includes a penchant for wearing cowboy hats, which matches their parallel interest in guns and hunting.[25]

As a critic commented, 'Even if "the mighty Mississippi divided" the twins, the fact that they both wear cowboy hats and like hunting is not that unusual for two [white] working-class men in the same region of Louisiana.'[26] The same point can be made about the celebrated 'Jim Twins,' two working-class white men who grew up in the same region of Ohio at the same time.

Returning to Oskar Stöhr's childhood in Nazi Germany, Bouchard has been quoted as saying that the 'twin who was reared in Germany belonged to the Hitler Youth, but virtually every young man in this community had to belong to it.'[27] Bouchard thereby attested to the influence of the cohort. Millions of young Germans joined the Hitler Youth not only because it was compulsory but because they were caught up in the spirit of the unfortunate era they found themselves in. As an extreme example of the confounding nature of cohort effects, let us suppose that a 20-year-old pair of male German MZAs, separated at birth and brought up in different parts of the country, had been reunited in 1942. A researcher or journalist would notice several striking similarities. Both might be wearing swastika armbands, have short-cropped hair, give stiff arm salutes and rail against 'the enemies of the Reich.' The values, beliefs, behaviors, and even wardrobes of these twins might be remarkably similar, yet the conclusion that these similarities are caused by common genes would be fallacious.[28]

The stories of individual pairs of reared-apart identical twins do not tell us anything about genetic influences on human psychological trait differences. They may be, as behavior geneticist Richard Rose commented, good 'show biz,'[29] but that's about as far as it goes. Judith Harris has written that 'there are too many of these stories for them all to be coincidences,'[30] and it is true—they are not coincidences; they are selectively reported 'show biz' combined with a stunning failure to recognize the environmental factors influencing these twins' similar behaviors.

Part II: Scientific studies of twins reared apart

In this section we shall look at the systematic studies of reared-apart twins, which typically assess twins through the use of personality and IQ test score correlations. Although one might assume that scientific studies are vastly different from the single cases reported in the media, the investigators have made many of the errors made by the journalists.

Until 1979, there were only three systematic studies of separated identical twins: Newman, Freeman, and Holzinger in 1937, who studied 19 pairs (and used MZTs and DZTs as controls); Shields in 1962, who studied 44 pairs (and used MZTs and DZTs as controls); and Juel-Nielsen's 1965 study, where 12 MZA pairs were investigated (no controls were used; Juel-Nielsen updated his study in 1980). The Newman et al. study found an MZA Binet IQ correlation of .67, and an MZT correlation of .91. The MZA correlation for the Woodworth-Mathews personality test was .58; for the MZTs it was .56.[31] Shields reported IQ correlations of .77 for MZAs, and .76 for MZTs.[32] For personality, he found MZA and MZT 'Neuroticism' correlations of .53 and .38 respectively, and .61 and .42 for 'Extraversion.'[33] Juel-Nielsen gave the MZA Wechsler-Bellevue IQ test correlation as .62, but did not calculate the total sample personality test correlations.[34]

The Minnesota Study of Twins Reared Apart (MISTRA) was started in 1979 by Bouchard and his colleagues at the University of Minnesota, who have included David Lykken, Nancy Segal, Matt McGue, Auke Tellegen, Leonard Heston, and Elke Eckert. Since 1988 the MISTRA group has published papers reporting test score correlations for MZAs, DZAs, and other types of twins.[35] Two other studies saw their first publications in the 1980s—a study from Finland, and the Swedish Adoption/Twin Study on Aging (SATSA).

The controversy surrounding British psychologist Cyril Burt's claim to have given IQ tests to 53 MZA pairs has been discussed in detail elsewhere, and I will therefore touch only on the main points. Irregularities and invariant correlations were first reported by Leon Kamin in 1973 and were published in his 1974 book, *The Science and Politics of I.Q.* Two years later, London *Sunday Times* reporter Oliver Gillie wrote an article in which he raised doubts that Burt's research assistants 'Miss Conway' and 'Miss Howard' had ever existed.[36] In 1979 British psychologist Leslie Hearnshaw published a biography of Burt in which he argued convincingly that Burt had invented much of his twin data. In part because he was well respected by even Burt's strongest supporters, Hearnshaw's conclusions were widely accepted. Although some have subsequently argued that the charges of fraud leveled at Burt are unsubstantiated, it is widely agreed that, fraud or not, Burt's twin data cannot be used.[37]

In addition to pointing out irregularities in Burt's papers, Kamin produced the first in-depth critical analysis of the problems with the other TRA studies. He demonstrated that many of the twins in the Newman, Shields, and Juel-Nielsen studies had frequent contact and often were reared in correlated environments. Kamin also questioned the validity of some of the measures used to test twins, and observed that twins were often tested by the same investigator, which could lead to bias. Kamin noted further that IQ correlations were subject to the confounding influence of sex and age effects. He argued convincingly that the Newman study recruited twins on the basis of similarity and their pre-existing knowledge of each other, which meant that the sample was biased toward similarity because twins unaware of each other were excluded.

Kamin's chapter on TRA studies was written in the context of the larger debate over the importance of genetic influences on IQ scores, which had been re-ignited by Jensen's notorious 1969 article, 'How Much Can We Boost IQ and Scholastic Achievement?' There, Jensen argued that IQ was 'highly heritable' and that test-score differences between American whites and blacks were plausibly explained, at least in part, by genetic factors (that is, were due to the alleged genetic inferiority of American blacks). Jensen cited Burt's study, which in turn motivated Kamin and others to look more closely at this and other reared-apart twin studies.[38]

In 1981 American psychologist Susan Farber published her *Identical Twins Reared Apart: A Reanalysis*, which remains the most in-depth and authoritative work ever published on the subject. Farber's book contains a wealth of thoughtful discussion, theorizing, and detailed statistical analysis. Drawing from Kamin's critique and adding her own observations and analysis, Farber's book is a devastating indictment of the logic and methods of these studies. She observed that, like the Newman study, Shields's sample was biased in favor of similarity because he had recruited twins on the basis of a media appeal.[39] Looking at all pairs reported in the literature (including single-case studies), Farber pointed out that due to ascertainment bias, 'approximately 90 percent of the known cases of separated MZ twins have been studied precisely because they were so alike,' and that conclusions about their similarity therefore were based on 'circular reasoning.'[40] Only Juel-Nielsen's study was viewed as less affected by similarity bias, because he located his 12 pairs through the use of registers and other records, 'thus ensuring that their knowledge of each other was not the prerequisite for selection.'[41] According to Farber, the twins recruited to these studies, whose similarities were so eagerly generalized to the non-twin population, were not even representative of the population of reared-apart twins.

Like Kamin, Farber questioned the twins' degree of separation. Of the 121 reported pairs, in both studies and single-case reports, she found only three cases in which the twins were separated during the first year of life, were reared with no knowledge that they had a twin, and were seen at the time of their first meeting. 'Of the 121 cases reported in the last fifty years,' wrote Farber, 'only three are "twins reared apart" in the classical sense.'[42] The 'most accurate description of this sample,' she concluded, is 'MZ twins partially reared apart.'[43] Farber also observed that about 90% of MZAs were born into poor families, and that 'anywhere from 50 to 75 percent of the twins were reared in clearly deprived homes . . . Only two or three individuals were adopted into professional families.'[44] She viewed the collected data as 'not good and do not meet even elementary design requirements.'[45] A well-designed study, according to Farber, would look at randomly selected and assigned MZAs who were 'reared as single individuals with no contamination from twinning,' and having no 'organic but nongenetic features within pairs (such as prematurity, birth trauma) to artificially inflate or deflate heritability estimates.'[46] Finally, Farber called for the inclusion of subjective impressions in addition to test results, as well as the creation of a central registry to make data collected in these studies accessible to others. For Farber, a registry would act as a safeguard 'against the disreputable claims and use of data that have occurred in the past and undoubtedly will occur in the future.'[47] One objective of this chapter is to show that Farber's criticisms are valid and that the MISTRA investigators violated most of her admonitions.[48]

The myth of the separated identical twins
I have borrowed this phrase—'The myth of the separated identical twins'—from a chapter in Howard Taylor's *The IQ Game*. The reason is simple: most twins in these studies were not separated at birth and raised in separate environments. Rather, the studies suffer problems which include:

- Twin pairs sometimes were separated only after having been reared together for several years.
- Pairs were often raised by different members of the same family, and usually were placed into families correlated for socioeconomic status.
- Many pairs were aware of each other's existence and had frequent contact during much of their lives.
- Most pairs were brought to the attention of researchers on the basis of their similarity or knowledge of each other's existence.

- The material used to evaluate the similarities or correlations of twins was collected by the same researchers.
- Personality and environmental similarities or correlations were assessed by non-blinded raters.

The study of separated twins is both a twin study *and* an adoption study, because one or both twins have been removed from their biological mother and placed into an adoptive or foster family, or with a relative. A critical assumption of all adoption studies is that children are randomly placed into available adoptive homes (see Chapter 7). In the case of separated twin studies, it must be shown that trait correlations were not influenced by the possibly correlated environments of the twin pair. Genetic inferences from studies of twins reared together depend on the assumption of the rough equality of environments experienced by identical and fraternal twins. Genetic inferences from studies of *reared-apart* twins rest on the assumption that the twins' environments were not systematically more similar than a group of randomly selected biologically-unrelated paired individuals. In contrast to the study of twins reared together, they are based on an *unequal* environment assumption.

Examples of contact and late separation
In striking contrast to the newer studies, the authors of the older investigations included hundreds of pages of case history material for the twins under study. Looking through these descriptions, we have numerous examples of twins for whom the designation 'reared apart' is quite misleading.

In the Newman and colleagues 1937 study of 19 MZAs, the age at separation ranged from two weeks to six years, and most pairs had considerable contact before being tested. For example, Pair I 'had lived together for a year before tests were given,' Pair II 'had lived together seven years when tests were given,' Pair IV had 'been acquainted and have visited back and forth at intervals all their lives,' Pair VI was 'living together when examined at the age of 58 years,' Pair XIII had 'visited at long intervals during the last four years,' and Pair XIX was separated at age six and 'spent six months studying nursing together when they were 17 years old.'[49]

In Shields's 1962 study we find the following descriptions. Each depicts a different pair of 'reared-apart' identical twins.

14-YEAR-OLD MALE TWINS 'did not meet or know they were twins till they were 5 or 6. From the age of 9 they met once a week when Kenneth's family came to live in the town.'[50]

17-YEAR-OLD MALE TWINS whose 'mother died the day after the twins were born. The paternal aunts decided to take one twin each and they have brought them up amicably, living next door to one another . . . They are constantly in and out of each other's houses.'[51]

38-YEAR-OLD MALE TWINS who, at age 11, lived 'in cottages next door to one other and attended the same school.'[52]

39-YEAR-OLD MALE TWINS separated 'at 4 years . . . The twins came home again to mother at 14. Both twins say they thought a lot about one another as children.'[53]

EIGHT-YEAR-OLD FEMALE TWINS 'brought up within a few hundred yards of one another, but no contact between the families. Attracted to each other at the age of 2 . . . gravitated to one another at school at the age of 5. In order to separate them Mrs. E. removed Jessie to another school but they continued to meet in the park.'[54]

23-YEAR-OLD FEMALE TWINS 'separated at 6 months . . . They were reunited from 5 to 7 years, then parted again. They meet every weekend now.'[55]

30-YEAR-OLD FEMALE TWINS 'separated at 13 months . . . Though attending different schools, the twins met about twice a week and sometimes spent holidays together.'[56]

32-YEAR-OLD FEMALE TWINS 'separated at 5 years . . . brought up in the same town, not always same school; met regularly.'[57]

38-YEAR-OLD FEMALE TWINS 'separated at 8 years . . . Until then the twins had done everything together; thereafter they continued to attend the same school in the village where both families lived . . .'[58]

38-YEAR-OLD FEMALE TWINS 'were brought up together till the age of 7 . . . The twins meet about once a fortnight when they visit their brother in hospital.'[59]

39-YEAR-OLD FEMALE TWINS 'separated from birth till 11 years and again at 16.' Viola was taken by the grandparents at birth, 'while Olga remained with the parents, living nearby in the same industrial area . . . There were occasions when [Viola] was taken back to mother, but until 11 she was always returned to the grandmother before the day was out. At that age she went to live with the parents and was forbidden to see her grandmother.'[60]

40-YEAR-OLD FEMALE TWINS separated at birth, but 'probably together at various times before school age . . . They were reunited most of the time from 5 to 15, generally with the aunt, but sometimes with the mother.'[61]

42-YEAR-OLD FEMALE TWINS 'separated soon after birth . . .The twins generally spent summer holidays together at each other's homes.'[62]

43-YEAR-OLD FEMALE TWINS 'separated at 4 years (not met since).' Twins began corresponding at age 36. 'Though they have never met they have developed an intense affection for one another and derive great emotional satisfaction from their correspondence, which supplies something in their lives which they sometimes feel to be lacking . . . When Berta sends a letter with a lipstick kiss she presses it warmly against her lips. She longs for her twin so much that it hurts.'[63]

45-YEAR-OLD FEMALE TWINS 'separated at birth . . . From just before their ninth birthday they lived as close neighbors in a coastal town, attending the same school until they were 15. They were closely attached and went about a lot together . . .'[64]

There are many more examples in Shields's study, but from the descriptions it is clear that few (if any) of his pairs were truly 'reared apart.' As Shields himself acknowledged, his twins' environments 'have not as a rule been extremely different.'[65] In fact, 30 of the 44 MZA pairs were raised in different branches of the same family, and 33 pairs differed in family socioeconomic status 'only to a modest extent.' The defining characteristic of Shields's twins' supposed separation was only that they had been raised in different homes for at least five years during childhood. This did not prevent behavior geneticists Jinks and Fulker, in their influential 1970 paper on biometrical analysis, from writing that there is 'no evidence of correlated environments' in Shields's MZA sample.[66]

Let us turn to Juel-Nielsen's 1965 study of 12 MZAs. Although these pairs generally were more separated than those in the studies of Newman and Shields, the designation 'reared apart' is questionable for several pairs. Case III twins Maren and Jesnine were separated at six weeks because of the death of their mother, but 'were put into the care of two of their paternal aunts, who were both married to farmers and lived in the twins' home county, about 10 kilometers apart from each other.'[67] Ingegerd and Monika (Case IV) were separated at 12 months but were reunited from ages 7–14. Martha and Marie (Case VI) were reared together until age three and a half and later attended the same school. 50-year-old twins Kamma and Ella (Case VII) were separated after one day of life but grew up in rural districts 40 miles from each other: 'Although the twins never came into very close contact with each other, they had both had a feeling of strong solidarity since they met at the age of 12 . . .'[68] Signe and Hanne (Case VIII) were 54 years old when interviewed. Although separated at 3 weeks, they found out about each other at age 14 and 'corresponded with each other during the next few years . . .'[69] Astrid and Edith (Case XI) were separated at three and a half years. Viggo and Oluf (Case XII) were separated at five and three-quarter years.

In the Swedish study (SATSA), which provided only quantitative data for its subjects, pairs were considered reared apart if they had been separated by the age of eleven[70] (average age at separation = 2.8 years[71]), and about 75% had some degree of contact after separation. How much contact was not reported by the SATSA investigators apart from a complicated and inadequate mathematical formula. Nevertheless, according to the formula the SATSA twins (average age: 65.6 years) were 'separated' for an average of only 10.9 years at the time of testing.[72] The twins supplied information by mail, and many were not investigated personally. In the Finnish study, 12 of the 30 MZA pairs were separated after they turned five years old, and the degree of post-separation contact is unclear.[73]

Environmental similarity
It is common for separated twins to be raised by members of the same family, often with the biological parent(s) caring for one of the twins. Like individual pairs such as the Jim Twins and Jerry Levey and Mark

Newman, they often share many cultural influences as well as a similar appearance. For these reasons, and quite apart from possible genetic influences, we would expect even twins reared apart from birth who were unaware of each other's existence to share many similarities. The failure to recognize common environmental factors was a major theoretical blind spot in the earlier studies. For example, Juel-Nielsen noted his twins' 'outstanding similarities' despite experiencing 'markedly different upbringings and later lives,' and concluded, 'These similarities must be related to the genotypical identity of the twins, and must be taken as an expression of the importance of genetic factors for the normal development of personality.'[74]

Kamin observed that Newman and associates' reared-apart identical pairs were recruited on the basis of similarity. Many pairs showing marked behavioral differences were excluded from the study. Potential participants were excluded if they answered the following question in the negative: 'Do you yourselves believe that you are far more alike than any pair of brothers or sisters you know of?'[75] Newman and associates were attempting to screen out potential DZA pairs because of the great expense of transporting them to and lodging them in Chicago. Kamin noted that the offer of an all-expenses-paid trip to Chicago in the middle of the Depression may have induced twins to exaggerate the extent of their separation. According to Newman et al., 'Pair after pair, who had previously been unmoved by our appeals to the effect that they owed it to science and to society to permit us to study them, could not resist the offer of a free, all-expenses-paid trip to the Chicago Fair.'[76] Several pairs were excluded because they lived far apart from each other and would have created too much difficulty and expense for the researchers. According to Taylor, 'such twins might have been culturally or environmentally separated as well.'[77]

Actually, Newman and associates were quite aware of the possibility that their MZA group was biased in favor of similarity, writing, 'When it comes to the separated cases . . . It seems possible that our group is more heavily weighted with extremely similar pairs than with identical twins of less striking similarity.'[78] In an attempt to gather new cases after the study's publication, Gardner and Newman corresponded with a pair of separated male twins in Pennsylvania. Although one of the twins was 'anxious to have their case studied,' his brother 'is, we fear, somewhat of a hoodlum and refuses to submit to examination.' Gardner and Newman commented, 'This is unfortunate in view of the fact that the two brothers now seem to be so different in their personality traits.'[79] This case illustrates one aspect of the bias in reared-apart twin studies: Gardner and Newman were unable to study these twins *because of* their personality differences.

In spite of the problems of similarity bias and the questionable 'separation' of twins, the earlier investigators recognized that MZAs differed in many respects and that environmental differences have an impact. As Newman wrote three years after the publication of his study, 'if one thing is clear in our results it is that fairly large environmental differences do modify physical, mental and temperamental traits and produce proportionately large differences even between hereditarily identical individuals.'[80] Juel-Nielsen described the differences he found among his twins as follows,

> In all 12 pairs there were marked intra-pair *differences* in that part of the personality governing immediate psychological interaction and ordinary human intercourse . . . the twins behaved, on the whole, very differently, especially in their cooperation, and in their form of and need for contact. Corresponding with these observations, the twins gave, as a rule, expression to very different attitudes to life, and very divergent views on general culture, religion and social problems. Their fields of interest, too, were very different . . . Those twins who had children treated, on the whole, their children differently, and their ideas on upbringing were, as often as not, diametrically opposed. Characterologically, the twins presented differences in their ambitions and in their employment of an aggressive behavior. Emotionally, there was a deep-going dissimilarity with regard to the appearance of spontaneous emotional reactions or to the control of affective outbursts. Various traits of personality found their expression in differences in taste, mode of dress, hair style, use of cosmetics, the wearing of beard or of glasses.[81]

Thus, Juel-Nielsen found 'marked differences' in personality, emotion, and character among the genetically identical twins he studied. Clearly, if the Jim Twins were similar because of their identical genes, all reared-apart twins should be equally similar.

Contemporary genetic researchers, when they discuss these studies, tend to overlook the twins' environmental similarity and the dubious claim of separation. The evidence is usually reported by critics. After Jensen calculated an overall intraclass IQ correlation of .824 for the Newman, Juel-Nielsen, Shields, and Burt studies, while downplaying the twins' degree of contact, he wrote that this correlation 'may be interpreted as an upper-bound estimate of the heritability of IQ in the English, Danish, and North American Caucasian populations sampled in these studies.'[82] More recently, behavior geneticist David Rowe conceded only that 'MZ twins reared apart are sometimes not perfect separations.' In another gross understatement, Rowe spoke of 'the occasional social contact of "separated" twins,' which 'may not introduce strong biases.'[83]

We have come to the end of a brief discussion of the earlier reared-apart twin studies. Those interested in further analyses of these investigations should consult previous publications.[84] Focus is now placed on the Minnesota study, which has been the subject of so much attention since the early 1980s.

The 'Minnesota Study of Twins Reared Apart'

The Minnesota studies have helped tilt the nature-nurture debate in the direction of nature. Bouchard and Loehlin wrote in 2001 that only 'the most extreme skeptic' would disagree with the finding that 'virtually all human psychological traits are influenced by genetic factors to a significant degree.'[85] The MISTRA studies have been the subject of numerous articles and several books popularizing behavior genetic research, where sweeping and dramatic statements have been made about their findings. As Lawrence Wright wrote in *Twins*, 'The field of psychology has been shaken by separated-twin studies,'[86] and 'The science of behavioral genetics, largely through twin studies, has made a persuasive case that much of our identity is stamped on us from conception.'[87]

The results I now present are taken from the 1990 MISTRA article published in the prestigious journal *Science*. The article compared MZA and MZT correlations for physical variables, IQ, and personality tests. The full scale WAIS IQ scores showed an MZA intraclass correlation of .69 (48 pairs), while the MZT correlation was .88 (40 pairs). For personality measures, MZA and MZT correlations were each about .49. Because the MZA and MZT correlations were similar, the investigators concluded that psychological traits are strongly influenced by genetic factors and that 'shared family environments' have little influence on those traits. The authors of subsequent papers published in the 1990s reached similar conclusions for a host of psychological traits, interests, and tendencies. (The SATSA and Finnish studies produced lower intraclass personality correlations, ranging from .20 to .40.[88] The SATSA first principal component cognitive ability intraclass correlation was reported as .78, although the subtest correlations ranged from .15 to .65.[89])

What are we to make of these findings? Do they indeed justify the conclusions of both the investigators and the popularizers of the studies? The answer to these questions will occupy the remaining pages of this chapter.

Secret data, secret lives
In the aftermath of Kamin's and Farber's analyses of the older studies, one might expect that reared-apart twin researchers would change the way that they study twins, and in some respects they did. The Minnesota investigators were careful to test twins separately and to have different people administer the tests. This reduced the possibility of unconscious rater bias, as discussed by Kamin. They also attempted to minimize ascertainment bias by vigorously recruiting all pairs they became aware of regardless of whether they were MZA or DZA. In many ways, however, the Minnesota studies are methodologically *inferior* to the older investigations.

A major problem has been the refusal of the investigators to make life history and test score data available for independent analysis. In the older studies, hundreds of pages of biographical and test score information were published. Unfortunately, the MISTRA researchers have not published this information and have refused to make it available to others. According to Bouchard, each twin was given a Life History Interview, Clinical Interview, Sexual Life History Interview, Life Stress Interview, Child Rearing/Schooling Interview, and Briggs Life History Questionnaire.[90] Unfortunately, the investigators have reported only minimal quantitative data on rearing environment and have repeatedly denied critically minded reviewers access to the interview data.[91] Bouchard has refused to make MZA data available to a prominent critic (Leon Kamin), even under conditions where pairs are identified by code numbers and where information about age is omitted (to guarantee non-identification).[92]

The Minnesota researchers now claim that they are forbidden by federal law[93] or by the University of Minnesota Human Subjects Committee from releasing data on their twin subjects. According to the journalist William Wright, 'On agreeing to take part in the study, the twins signed an informed-consent agreement in which the Minnesota scientists promised not to divulge information that could be traced to specific twin volunteers.'[94] However, this does not explain how the investigators were able to discuss the developmental and sexual histories of six MZA pairs in a 1986 study of homosexuality.[95] The authors discussed cases while changing details 'in order to protect the twins [*sic*] identities,' and mentioned no legal or ethical restrictions on releasing or publishing this information.[96] In 1991 the investigators were challenged by Beckwith and colleagues in the pages of *Science* to publish case history information.[97] Although Wright would subsequently claim that 'because of the informed-consent agreement, which promised the twins anonymity, the [1990 *Science*] article provided no biographical information . . .,'[98] in Bouchard and colleagues' reply to Beckwith there is no mention of any such restriction.[99]

During the 1980s Bouchard and associates supplied information on their twins to journalists and others, which led to the publication of their stories in leading U.S. magazines and newspapers. In addition, the investigators released selected information on twins who signed the informed-consent agreement, and had specifically requested anonymity. In 1981 Peter Watson devoted an entire chapter of his book about twins to MISTRA-supplied stories, which included at least two case histories of anonymity-seeking twin pairs:

> This completes the number of identical twins raised separately whom Bouchard has seen and whose names he is willing to release. For a variety of reasons the other five sets of twins he has seen *do not want any personal publicity*, and two I now mention are therefore identified only by a 'case number' [emphasis added].[100]

As usual, the readers of Watson's book were presented with the cases of the two most similar pairs; the other three were not discussed. Apparently, it was acceptable to supply information to a journalist on pairs who 'did not want any personal publicity' if it helped support the genetic bias of the MISTRA researchers.

The Minnesota investigators have kept important information about the twins they have studied secret. In doing so, they ignored Farber's timely call for making this type of data available to others for independent review. They also ignored Arthur Jensen, who in many ways is Bouchard's intellectual inspiration.[101] According to Jensen, MZA data should be made available to 'anyone who wishes,'

> Especially rare data, such as those of monozygotic twins reared apart, siblings from cousin matings, double first cousins, and the offspring of two mated pairs of monozygotic twins . . . should be published in full, along with complete descriptions of the tests or measurements and procedures. Perhaps this should be a general requirement for the publication of studies based on such valuable data, so that quantitative analytical techniques other than those used by the original author can be applied to the data by anyone who wishes.[102]

Unfortunately, Bouchard and colleagues did not follow Jensen's 1974 recommendation. This has made it very difficult for independent researchers to analyze the data and offer alternative interpretations, as they had with the earlier studies. With access to the raw data, a reviewer could determine the similarity of twins' childhood environments and could assess their degree of contact, interaction, and mutual influence. A reviewer could also examine the quantitative data in order to see if there were important trends overlooked by the original investigators. Apart from an early article discussing the first 15 pairs, Bouchard and colleagues have failed to provide a table containing information such as sex, age, age of separation, and so forth for the individual pairs. Only mean figures for the various twin types have been provided. In 1978 Shields noted that one of the 'obvious limitations' of Burt's study was 'the lack of information provided about the environments and life histories of the twins.'[103] Shields continued, 'Burt provided very scanty information in his 1966 paper — he did not even give their ages.'[104] In fact, Shields could just as easily have been referring to the MISTRA studies.

How 'separated' were the MISTRA twins?
An important task for any reanalysis of the MISTRA data would be to determine the degree of the twins' separation. Like the Newman and Shields studies, MISTRA twins were recruited on the basis of similarity and their pre-existing knowledge of each other. We cannot determine how 'separated' these twins really were, however, because we are not provided life history information. Instead, the investigators developed the wholly inadequate 'contact time' formula. Time spent before and after separation is included in the total and reflects the amount of time (measured in months) that twins spent together before being studied.

Bouchard and McGue wrote that their MZAs had 'minimal contact' before entering the study, implying that a pair of separated twins could influence each other only when together.[105] However, intimate relationships are based on an ongoing association between people which is not necessarily limited to the time they are in physical proximity. According to the MISTRA contact time formula, 'Twins who met for a week at Christmas and for a week in the summer each year over a ten-year period are credited with 20 weeks of contact.'[106] Thus, a MISTRA table would show such a pair to have had as much contact as twins who had spent only the first 20 weeks of life together. Clearly, these two pairs cannot be considered to have had the same influence on each others' development. Instead of regarding twins who spent 14 days a year together over a 10-year span as having accrued 20 weeks of contact time, it is far more accurate to note that these 'reared-apart' twins *had a ten-year relationship*. And according to Bouchard and colleagues' 1990 *Science* article, the total contact time range was 1–1233 weeks.[107] This means that the members of at least one pair had over 23 years of contact with each other.[108] In addition, as Kamin noted, volunteer twins seeking inducements or scientific approval, on whose verbal accounts contact time has been calculated, might be tempted to underreport the amount of contact they had.

The MISTRA contact time formula does not capture the essence of a relationship between two people, and its usefulness rests on the assumption that environmental similarity and mutual influence increase proportionately with each hour twins spend together. Unfortunately, no evidence supporting this assumption has been presented by the investigators. In a discussion of the older studies, Farber observed that the investigators often assumed that 'a brief meeting leaves insignificant impact' because they, like Bouchard and associates, viewed contact in terms of quantity as opposed to quality. Although it may be possible to assess the contact, environmental similarity, and mutual influence of twins on the basis of an examination of their life histories, the impact of these variables cannot, as Bouchard and colleagues imagine, be measured *as a function of time*.

Farber pointed out that losing one's virginity may occupy only a moment in time, and asked, 'Does this mean that the event is forgotten or had no psychological impact?'[109] The concept of post-traumatic stress disorder (PTSD) is based on the recognition that a brief yet traumatic experience can severely impact a person's subsequent psychological functioning, yet the trauma (or 'contact') itself may consume only a few seconds of a person's life. Obviously, these examples are more relevant to behavioral than to cognitive ability (IQ) similarity. In any case, the MISTRA contact time formula is no substitute for the publication and availability of the life history information of the twins under study.

How the researchers' assumptions influenced interpretations

When large differences were found between the members of MZA pairs, the investigators often sought explanations that would minimize the importance of the social and family environment. Here, I offer three examples.

Lawrence Wright discussed a male MZA pair who had been reared in very different environments. One of the twins had been adopted by illiterate parents; the other had been raised by a 'better educated' family. The difference in their IQ scores was 29 points, with the better-educated twin scoring higher. According to Wright, Lykken believed that 'the lower-IQ twin may have suffered some kind of brain damage at birth.'[110] But why should such unsupported speculation be accepted?

The second pair, described by Leonard Heston, consisted of two 55-year-old female MZAs who were separated at four months and were adopted into different homes.[111] 'Twin A' appeared 'overweight and slovenly' and expressed 'chronic dissatisfaction with her life.' She had been treated for depression for eight years before being studied, had experienced anxiety all her life, and had several unexplained medical complaints. She had completed high school at age 17 and worked as a clerk for the next 30 years. Twin A described her marriage at age 38 as 'miserable.' Her MZ sister, 'Twin B,' weighed 60 pounds less than Twin A and was described as an 'engaging, even vivacious woman.' Twin B had gone to college and had worked as a head technician in a large hospital laboratory. She was happily married and described her life as fulfilling and happy. According to Heston, his MISTRA colleagues 'searched exhaustively through [the twins'] histories for some explanation of the striking differences between these women.' A clear bias is evident in this description, since the researchers did not exhaustively investigate the histories of *similar* twins for the purpose of discovering an environmental explanation for their alikeness. It turned out that Twin B had been taking thyroid medication since age 18, but Twin A had not. The MISTRA laboratories determined that Twin A's thyroid function was in the 'low normal range.' Heston offered Twin B's thyroid medication as a 'tenuous explanation' for the twins' differences.

In a third example, in *Entwined Lives* Segal briefly discussed two MZA pairs showing 'marked differences on the MMPI [Minnesota Multiphasic Personality Inventory].' Although she might have attributed the results to environmental influences, Segal instead concluded that the differences were 'traceable to the development of psychiatric disorder in one case, and probable brain damage due to head injury in the other case.'[112] Unfortunately, neither Segal, Heston nor Lykken indicated whether these pairs and their test scores were retained in the study, leading a reviewer to wonder whether they were removed on the basis of a suspicion of organic problems.

The researchers' assumptions have influenced this study at every step: the Jim Twins were recruited to the study because they were so similar and were then offered as evidence of the similarity of MZAs, whereas differing pairs are assumed to have become that way because of an organic impairment in one of the twins. It is quite remarkable that the Jim Twins were included in the study and had their test scores figured into all subsequent statistical formulations. Like most previous studies, the Jim Twins came to the attention of researchers *because of* their similarity. (Bouchard recruited them after reading about their reunion in the newspaper.) The Jim Twins epitomize the circular reasoning discussed by Farber.

Instead of publishing case history material, the MISTRA investigators have provided only mean scores from a questionnaire, the Family Environment Scale (FES). One might ask how this scale would be necessary in light of the fact that twins underwent extensive interviews covering most aspects of their life histories, or how it could be considered a substitute for an evaluation of family environment by blinded raters. The FES, in fact, does not purport to measure socioeconomic status but only social environment within the family.[113] Bouchard and McGue acknowledged that FES correlations show that only 'modest' selective placement occurred in their MZA sample.[114] But why must we take their word for it? Until proven otherwise, we can safely assume that the 'separation' of the MISTRA twins is as dubious as it was for the studies of Newman et al., Shields, and Juel-Nielsen. Until the data are made available to 'anyone who wishes' (or at least to any credentialed reviewer who wishes), *all conclusions about the role of genetics must be rejected.*

This position is unrelated to whether or not the researchers are legally or ethically required to keep their data secret; it is more a question of the social and political importance of the study's conclusions and implications. What Bouchard and colleagues ask of us is that we radically alter our views on the importance of heredity, with a possible dramatic shift in social policy, on the basis of how *they* interpret the data on a few hundred individuals. Essentially, we are asked to believe on faith. Fortunately, science demands that claims be subject to replication. When this is extremely difficult due to the nature of the subjects under study (such as reared-apart twins), it becomes mandatory that independent analysts be permitted to analyze the data and offer alternative interpretations.

Environmental factors contributing to MZA similarity

We have seen that there are many environmental influences shared by MZTs and even the most perfectly separated MZAs. These influences were not only uncontrolled for in the earlier investigations, they were not even thought of in most cases. It simply was not understood that because two people are the same sex, the same age, and grow up in similar cultures, they will behave more similarly and also have more similar IQ scores. Today, these influences remain misunderstood by the authors of popular works and even by some researchers. According to the SATSA investigators, 'the phenotypic resemblance of identical twins reared apart in uncorrelated environments can be attributed to genetic influences,'[115] and psychiatric geneticists Faraone and Tsuang wrote, 'Since MZ twins reared apart do not share a common environment, any phenotypic similarity must be due to genetic factors. We cannot invoke shared environment as a cause of phenotypic concordance.'[116] Even Bouchard wrote, 'When identical twins are reared apart, their personality correlations must be an effect of genetics'[117] (altough as we will see, Bouchard and associates used an age and sex correction formula).

Let us review the main environmental influences that work to make MZAs more similar than pairs of randomly selected members of the world's population:
- They are exactly the same age.
- They are the same sex.
- They are almost always the same ethnicity.
- Their appearance is strkingly similar, which probably will elicit similar treatment.
- They usually are raised in the same socioeconomic class.
- They usually are raised in the same culture or subculture.
- They frequently share the same religious beliefs.
- They are both volunteers. Studies have shown that research volunteers share several psychological and behavioral characteristics.[118]
- They shared the same prenatal environment.
- They typically spent a certain amount of time together in the same family environment, were aware of each other's existence when studied, and often had regular contact over a long period of time.

While the previous investigators failed to control for any of these factors, Bouchard and associates attempted to control for two of them. In 1984 McGue and Bouchard published a paper outlining their strategy to deal with sex and age confounds.[119] They acknowledged that age and sex effects for most psychological variables are 'substantial,'[120] and wrote that researchers can either ignore these effects or can use information on age-sex effects obtained from within the twin sample. McGue and Bouchard devised a formula, based on the twin sample itself, that adjusts twin correlations to account for these effects. This formula was used to adjust all subsequent MISTRA twin correlations and was also used by the SATSA investigators. Unfortunately, MISTRA papers do not report the original raw correlations alongside those which have been adjusted, which would allow readers to see how much impact the adjustment had on the data.[121]

It is unlikely that the confounding effects of sex and age can be determined through the use of twin data, as McGue and Bouchard attempted to do in their complicated formula. In any case, their formula purports to adjust for only two influences on MZA similarity, which is completely inadequate. There is actually a simple way to assess the effects of common age and sex that has been overlooked by all reared-apart twin researchers and which relates to the central thesis of this chapter: genetic and environmental influences can be determined only by comparing twin correlations to the correlations of pairs of biologically unrelated age- and sex-matched strangers.[122]

MZA personality and IQ test score correlations in the MISTRA studies

The Minnesota samples consist of twins whose life histories, degree of separation, quality of contact, and similarity of socioeconomic status (SES) and cultural background has been kept secret by the investigators. Apart from this important problem, MISTRA results are based on the assumption that 'intelligence' as measured by IQ tests and 'personality' as measured by personality tests can be quantified and accurately measured. I will discuss the merits of this assumption as it relates to IQ in Chapter 9. Here, the discussion is limited to personality.

There are many ideas about what constitutes 'personality.' For some, personality means a set of underlying traits that are stable over time and which can predict the behavior of a person in a given situation. Personality in this sense is a construct inferred from behavior and from the way that people fill out personality inventories. Trait theorists such as Allport began by identifying thousands of words in the English language that describe distinctive forms of personal behavior. Researchers grouped these terms into categories and claimed the existence of traits such as introversion, neuroticism, authoritarianism and so on. For others, behavior is best understood in the context of the specific situation in which a person finds himself or herself, and the existence of stable traits is rejected.[123] Personality theory is one of the most contended areas in psychology. The existence, nature, causes, and endurance of traits are still fiercely debated, and there exists no unifying personality theory in psychology. For this reason, introductory personality textbooks often consist of a series of chapters discussing the various theories and their proponents.

For traits theorists such as Allport, personality can be studied in the same way as stars and planets are studied:

> Does an astronomer who studies Arcturus think of the star as a construct tied together by a name? Hardly; to him it is a celestial body, truly existing , and possessing a composition and structure that he will try to scientifically comprehend. When a biologist dissects a plant he does not believe that the plant's structure and physiology reside only in his manipulations. Personality is even more difficult to study than stars or plants, but the situation is the same.[124]

Thus Allport argued that 'personality' has as much physical reality as a plant or a star.

For the proponents of the theory of stable underlying human traits, the next step was the claim that traits can be *measured and quantified*. The results of the Minnesota studies are dependent on the assumption that traits both exist and can be measured. People can feel happy or sad, but can we measure and quantify how much happiness or sadness we possess? Can 'extroversion' be quantified?

My purpose is to illustrate the doubtful nature of personality as a quantifiable entity, as an example of one of the many assumptions of the MISTRA work that is rarely if ever mentioned. Twin research, as we have seen, has a long history of failing to adequately discuss and provide evidence in support of its many assumptions.[125] In many cases, the conclusions in support of genetic factors depend on the validity of many assumptions, *all of which must be true in order to sustain the conclusion*. Typically, assumptions are tested only after critics point out the lack of evidence in their favor. If the MISTRA papers listed all of the assumptions that must be true in order to accept their conclusions, far fewer people would accept them. The implicit assumptions about personality in the MISTRA studies form only a small subset of the assumptions of the entire project. All of the following assumptions must be true before we can even begin to consider the possibility that the MISTRA investigators' conclusions are valid:

- Humans exhibit underlying stable personality traits which can be identified through the use of statistical methods.
- Personality traits can be identified through the use of words describing behavior.
- Personality traits can be measured and quantified on the basis of people's responses to a questionnaire.
- Personality traits are normally distributed in the population (in a bell-shaped curve).

We have seen that the MISTRA personality studies found MZA correlations of about .50. Because MZT correlations are also in this range, the investigators concluded that genes and non-shared family environments are the main influences on personality. Leaving all other problems aside, this conclusion is faulty because it fails to account for the environmental factors shared by MZAs and MZTs, and there are ways of checking this. A search through the psychology literature finds a few studies that looked at age and sex effects on test scores of unrelated people. One study was performed by Martin and colleagues in 1981.[126] The subjects (N = 179, 118 female, 61 male, mean age 28.4 years) completed the California Psychological Inventory (CPI), which was used in the MISTRA studies and which purports to measure 'normal' psychological traits or 'folk concepts.' Each CPI scale was correlated with the age of the respondents. Remarkably, the mean correlation between age and scale scores was .28 across all 18 scales, with ten scales showing a correlation of .35 or higher. The Martin et al. sample was drawn from a population of college students who were probably similar in socioeconomic and educational status, and therefore had more in common than age alone. Nevertheless, if these figures roughly reflect age effects in the population as a whole, the influence of common age, which is only one of many environmental similarities shared by MZAs, is enough to account for more than half the value found in MISTRA MZA personality correlations.

We can also compare MISTRA personality results with those of other types of studies less vulnerable to environmental confounds. In a 1998 article coming out of the Colorado Adoption Project, the investigators found that the mean personality scale correlation between birthparents and their adopted-away biological offspring (N = 245) — a relationship that the authors considered 'the most powerful adoption design for estimating genetic influence'[127] — was zero (.01 to be precise).[128] The results of this longitudinal adoption study by leading behavior geneticists are strikingly different from the MISTRA results, and cannot be attributed solely to error. One could reasonably conclude that at least one of the studies contains serious methodological flaws.

What about the DZA correlations?

The MISTRA was the first TRA study to collect a significant sample of reared-apart fraternal twins (DZAs). Although a greater MZA versus DZA correlation can be explained on environmental grounds (see below), from the genetic standpoint MZAs *must* correlate higher, because of their identical genetic makeup. Interestingly, we will see that there is little evidence that the MISTRA DZAs correlate very differently than MZAs on personality measures and IQ tests.

There is reason to believe that DZAs would correlate lower than MZAs for purely environmental reasons. Parents, families, and adoption agencies probably would attempt to place identical twins into more similar homes because of the greater physical similarity of MZs and societal expectations that they should be treated more similarly. In the cases of DZ twins separated after infancy, the difference in their physical appearance would make it more likely that they would be placed into less correlated environments than MZs, whose similar appearance might increase the desire to place them into more similar environments. Because of their less similar physical appearance, DZAs would experience more dissimilar appearance-related treatment in their social environment.

In addition, MZAs are more likely to be aware of each other's existence and to have had more contact than DZAs. Several MZA pairs located each other because people had confused them with their separated twin sibling who lived in a neighboring community (Jerry and Mark, for example). Upon discovering the existence of a separated co-twin, an MZA would be more likely to seek out a twin sibling in the belief that this person would be just like him or her. This idea is supported by the fact that the MISTRA MZA:DZA sample ratio includes far more MZAs than chance would expect, and by information demonstrating that MISTRA MZAs had about twice as much prestudy contact time than DZAs.[129] Therefore, several environmental factors indicate that separated identical twins would be more similar than separated same-sex fraternals. According to a 1998 paper by Bouchard et al., MZA/DZA mean correlations on the 20 California Personality Inventory (CPI) folk scales were .46 and .27 respectively (with several scales showing no statistically significant differences),[130] and it is likely that the environmental factors described above largely accounted for the mean scale difference. The 53 DZAs in the sample included 16 opposite-sex pairs (30%), whose correlations were not listed separately and whose scores probably reduced the total mean DZA correlation. Opposite-sex DZAs should have been excluded from this comparison because, as we have seen, McGue and Bouchard acknowledged that sex (and age) effects on psychological variables can be 'substantial.'[131]

Where are the MISTRA DZA IQ correlations?

The striking failure to adequately report DZA IQ correlations has been documented in Kamin and Goldberger's review of the MISTRA studies,[132] and I will discuss and comment further on their findings here. Kamin and Goldberger noted that, as of 2000, the MISTRA investigators had failed to publish Wechsler IQ correlations for the entire DZA sample, in direct contrast to the publication of DZA *personality* correlations. In Bouchard and colleagues' 1990 *Science* article, DZA correlations were not reported 'due to space limitations and the smaller size of the DZA sample (30),' even though MZAs and DZAs were said to be 'a fascinating experiment of nature' who 'provide the simplest and most powerful method for disentangling the influence of environmental and genetic factors on human characteristics.'[133] Space limitations? One column in Table 4 of the 1990 article would have been sufficient to supply the DZA personality and IQ correlational data. Smaller sample size? A sample of 30 DZAs was deemed sufficient to provide enough data to view this 'most powerful' comparison in a 1990 MISTRA personality paper, where the correlations of 26 DZA pairs were reported.[134] The decision not to publish the correlations of the 30 DZA pairs in the 1990 *Science* article was completely arbitrary and was made by researchers who possessed the raw data at the time the decision was made.

Let us look at some examples of the selected and incomplete reporting of the MISTRA DZA IQ data. In a 1989 paper by McGue and Bouchard, the authors reported MZA and DZA correlations on tests of 'Special Mental Abilities.' For Verbal Reasoning, the correlations were MZA = .44, DZA = .49; for Spatial Abilities, MZA = .52, DZA = .27; for Perceptual Speed MZA = .46, DZA = .30; for Visual Memory, MZA = .29, DZA = .22.[135] As we see, only one of the four variables saw a substantially larger MZA correlation. To the extent that DZA correlations are lower, they could be explained by the environmental factors I have already discussed. In addition, 16% (4/25) of the DZAs were opposite-sex and therefore should have been removed from the DZA sample. In 1998 Newman and colleagues published a MISTRA article on adult ego development which compared subsamples of MZAs and same-sex DZAs. As noted by Kamin and Goldberger, this was the first published report of MISTRA full-scale Wechsler IQ DZA correlations. Although the full-scale IQ correlations for the MZAs and DZAs in the subsample were .75 and .47 respectively, the Verbal Reasoning correlations were .46 and .53.[136] Finally, another MISTRA paper on authoritarianism by McCourt and colleagues reported correlations for 'general cognitive ability' as .74 MZA, .53 DZA.[137] The MZAs and DZAs were a subsample of twins who had been tested for 'right-wing authoritarianism,' and 26% (10/38) of the DZA sample consisted of opposite-sex pairs.

Clearly, the MISTRA investigators have failed to produce adequate DZA IQ score correlations in spite of having provided correlations for

DZA's personality scores. As an example of a more recent paper providing personality data, in 1998 Bouchard and colleagues reported the CPI correlations of 53 DZA pairs.[138] In a paper published the following year, Bouchard and Pedersen reported DZA personality correlations in both the SATSA and the MISTRA, yet failed to publish MISTRA DZA IQ correlations even though IQ was discussed.[139] As we have seen, the reported IQ data reveal instances when, contrary to the genetic hypothesis, there were no significant differences between the genetically identical MZAs, versus DZAs, who share only 50% of their genes on average. True differences which might exist can be explained on environmental grounds.

MZTs (or DZAs) as the control group: the invalidating flaw of all twins reared apart studies

In 1981 behavior geneticist Richard Rose reviewed Farber's *Identical Twins Reared Apart* and speculated that cohort effects have confounded the results of MZA studies:

> A colleague suggests that we cannot know [the importance of MZA resemblance] without necessary control data on similarities found in pairs of age-matched strangers . . . Were one to capitalize on cohort effects by sampling unrelated but age-matched pairs, born, say, over a half-century period, the observed similarities in interests, habits, and attitudes might, indeed, be 'astonishing.'[140]

Rose thereby highlighted the invalidating flaw of the studies of Newman, Shields, Juel-Nielsen, *and* Bouchard — *the investigators mistakenly compared MZA correlations to the correlations of MZTs or DZAs, when the only valid control group would consist of biologically unrelated pairs of strangers matched on the environmental variables shared by MZAs.* As its name implies, the purpose of a control group is to control for factors other than those tested by the researchers' hypothesis, but MZA-MZT comparisons fail to control for the environmental factors I have already discussed.

Readers may find it difficult to believe that prominent researchers could have made such an elementary mistake. Genetic studies, however, are full of these types of mistakes. For example, over 40 years went by before the majority of twin researchers would acknowledge that reared-together identical twins share a more similar environment than shared by reared-together fraternals. Another example was discussed by psychologist Stephen Ceci in his book, *On Intelligence*. Ceci took a second look at IQ pioneer Lewis Terman's influential *Genetic Studies of Genius*, where high IQ children (who came mostly from the upper classes) were tracked throughout their lives in order to test whether IQ correlates with 'success.' Terman compared the outcomes of his 'geniuses' to the general population, and Ceci observed, 'But surely, this is the wrong comparison group! Terman should have compared the graduation rates, college grades, and earnings of his selected students to those of others from their same family background social class.'[141] Ceci called for an alternative control group: 'What is needed is a sample of men with unremarkable IQs who came from the same family social background as Terman's high-IQ men . . . and who were subjected to the similar special treatments.'[142] Bouchard and his predecessors simply repeated the error made by Terman, that is, they used the wrong comparison group.

Behavior geneticists may object to my call for the formation of an alternative control group of biologically-unrelated matched strangers, by arguing that Bouchard and colleagues corrected their correlations for age and sex effects. They might also argue that the finding of greater IQ or personality similarity for MZAs versus same-sex DZTs is proof of the importance of genetics factors. They might present other data and other comparisons. However, these objections cannot eliminate the necessity of using the correct control group. Rather than argue against this proposal, it should be carried it out in order to see what results would be found.

What a valid MZA study would look like

Although no valid conclusions about genetic influences on personality or IQ differences can be drawn from the MISTRA data, a description of a valid MZA study seems in order. First, a systematic ascertainment of MZAs separated at or near birth would be performed in a way that did not lead to the formation of a similarity-biased sample. Twins having had post-separation contact or who knew about each other's existence could not participate in the study. After an experimental group of MZAs is collected in this manner, twins would be evaluated separately and would not be reunited until after the evaluation process is completed. The twin pair correlations of this group of MZAs would be compared with the correlations of a control group consisting of *biologically unrelated* pairs of strangers sharing all of the following characteristics: they should be the same age, they should be the same sex, they should be the same ethnicity, the correlation of their rearing environment socioeconomic status should be similar to that of the MZA group, they should be similar in appearance and attractiveness as determined by blinded raters, and the degree of similarity of their cultural backgrounds should be equal to that of the MZA twins. Finally, they should have no contact with each other until after they are evaluated and tested. These controls will constitute the Unrelated group. Of course, it is not possible for unrelated pairs to share a common prenatal environment, which twins do share.

At this point we would be able to compare the MZA and Unrelated groups for correlations of personality, IQ, and other measures. After the scoring is completed, the code concealing the individual's group and pair status will be broken, and each set of scores will be assigned to the proper MZA or Unrelated pair. The scores between groups will then be compared. The genetic hypothesis predicts that MZAs would correlate more highly on the basis of their identical genetic makeup. The environmental hypothesis predicts that there would be no differences between the MZA and Unrelated groups, other than those accounted for by prenatal factors.

Conclusions

Contrary to popular belief, media reports describing the behavioral similarities of ostensibly reared-apart twins, while interesting from an entertainment standpoint, do not tell us anything about the role of genetic influences on psychological trait differences. The same is true for the systematic investigations of reared-apart twins, which include the Minnesota studies. The problems with these studies include the dubious 'separation' of twins, the similarity bias of the samples, the failure to publish or share raw data and life history information for the twins under study (MISTRA), and the bias of the researchers in favor of the importance of genetics. Most importantly, there has been a failure to recognize and control for the many environmental similarities shared by reared-apart twins. Therefore, we can draw no valid conclusions in support of genetic influences on psychological trait differences from the studies of twins reared apart published through mid-2002.

In some respects the problems with the 'journalistic' and 'scientific' approaches to MZA study converge, in that the similarities of the twins presented were (1) based on twins who were discovered *because* they were similar, (2) reported by people who were interested in demonstrating similarity, and (3) plausibly explained by factors other than the twins' common genes. Bouchard dismissed Farber's 1981 book as a 'pseudo-analysis,' and proceeded to complete a study whose flaws had just been laid bare by Farber. In one important sense, Farber's book is a manual on how *not to* perform a study of twins reared apart. An updated edition of Farber's book could use Bouchard's MISTRA as a case study.

In the early part of 2001 the sequence of the human genome was published simultaneously in *Science* and *Nature*. One of the lead investigators, Craig Venter of Celera Genomics, concluded that although the Human Genome Project found surprisingly little evidence that genes influence behavioral differences, we know that genes are important because of the results from reared-apart twin studies. There is no doubt that Venter sincerely believed that the results of these studies are conclusive. He was simply perplexed by the contrasting results. This chapter has provided an explanation for the seemingly contradictory evidence reviewed by Venter and others.

Notes

[1] Billings et al., 1992, p. 236.
[2] Wyatt, 1993, p. 1295.
[3] Bouchard et al., 1998.
[4] Holden, 1980.

[5] Ibid., p. 1324.

[6] Ibid., p. 1324.

[7] See Horgan, 1993; L. Wright, 1997.

[8] Horgan, 1993.

[9] Donald Dale Jackson, 1980.

[10] Watson, 1981, p. 43.

[11] Kamin & Goldberger, 2002, p. 86.

[12] Bouchard, 1997c, p. 53.

[13] Wyatt et al., 1984, p. 64.

[14] Lang, 1987, p. 63.

[15] Lykken et al., 1992, p. 1566.

[16] Segal, 1999, p. 144.

[17] For example, see Lang, 1987, p. 63.

[18] Lang, 1987.

[19] Kamin, 1974, p. 54.

[20] Dusek, 1987, p. 21.

[21] Robert T. Carroll, 'The Skeptic's Dictionary' website: http://skepdic.com/testimon.html.

[22] Harris, 1998, p. 33.

[23] Farber, 1981, p. 77. Farber was speaking of twins' dental records in this quotation, but her observation applies to most aspects of twin similarity.

[24] Allport, 1961, p. 165.

[25] Cassill, 1982, p. 183.

[26] Dusek, 1987, p. 21.

[27] Quoted in Cassill, 1982, p. 193.

[28] As an experiment, I call on readers to identify a biologically-unrelated person of the same sex who is close in age and was reared in the same class and culture. Sit down with this person and list all of the things you have in common—likes, dislikes, hobbies, etc. Look hard for similarities, since journalists and genetically-oriented researchers do the same with their twins. When you are finished, you might have a list as impressive as any you have read about separated twins. Of course, it is also *possible* that you will have little in common with this person, just as there are probably several dissimilar pairs of reared-apart twins for every one reported. If you and your partner discover many similarities, consider contacting the media to see if they are interested in reporting your story. If you can't find similarities, you probably won't bother contacting the media because it is doubtful that they would be interested. Also be aware that between you, similarities can be invented. The inherent bias in the stories of reared-apart twins will become more apparent after completing this experiment. As another exercise, imagine that you have just been reunited with your reared-apart identical twin sibling who was (somehow) born 200 years before you and grew up in another part of the world. How much do you think that you would have in common with this person?

[29] Rose, 1982, p. 960.

[30] Harris, 1998, p. 293.

[31] Newman et al., 1937, p. 347.

[32] Shields, 1962, p. 61.

[33] Ibid., pp. 69–70.

[34] Juel-Nielsen, 1980, p. 107.

[35] The first MISTRA paper making these comparisons was published in 1988 by Tellegen et al.

[36] Gillie, 1976.

[37] See Hearnshaw, 1979. For the arguments of Burt's defenders, see Fletcher, 1991; Joynson, 1989. Although Fletcher and Joynson attempted to exonerate Burt of the charges of scientific fraud, they did not call for the rehabilitation of Burt's data. According to Bouchard and Pedersen (1999, p. 73), 'there is agreement that Burt's data should not be included in any summaries of kinship correlations of the IQ literature.'

[38] Contemporary popularizations of behavior genetic research usually present Kamin as, at best, a defender of ideas that research has shown to be untrue. At worst, he is presented as a discredited, politically-motivated villain seeking to impede the work of 'true scientists' (see, for example, William Wright's *Born That Way*). In fact, Kamin has provided a valuable service to society. He deserves praise for his willingness to publicly expose bad research, as well as some of the unsavory history of his own profession.

[39] This point was also made by Taylor in 1980, and by Peter Mittler in his 1971 book on twin research.

[40] Farber, 1981, p. 36.

[41] Ibid., p. 17.

[42] Ibid., p. 60.

[43] Ibid., p. 273.

[44] Ibid., p. 62.

[45] Ibid., p. 22.

[46] Ibid., p. 30.

[47] Ibid., pp. 273–4.

[48] The individual reviews by behavior geneticists Gottesman, Heston, Loehlin, and Rose recommended Farber's book as an interesting, if flawed, analysis (see Gottesman, 1982; Heston, 1981; Loehlin, 1981; Rose, 1982). On the other hand, Bouchard called it a 'pseudo-analysis' in his 1982 review. By 1999 Bouchard and Pedersen were reporting that Farber's book was 'severely criticized' by Heston, Loehlin, Rose, and Bouchard (Bouchard and Pedersen, 1999, p. 73).

[49] Newman et al., 1937, pp. 144–5.

[50] Shields, 1962, p. 163.

[51] Ibid., p. 164.

[52] Ibid., p. 176.

[53] Ibid., p. 180.

[54] Ibid., p. 189.

[55] Ibid., p. 192.

[56] Ibid., p. 192.

[57] Ibid., pp. 193–4.

[58] Ibid., p. 206.

[59] Ibid., pp. 208–9.

[60] Ibid., p. 210.

[61] Ibid., p. 214.

[62] Ibid., p. 219.

[63] Ibid., pp. 221, 223.

[64] Ibid., p. 225.

[65] Ibid., p. 20.

[66] Jinks & Fulker, 1970, p. 337.

[67] Juel-Nielsen, 1980, Part II, p. 68.

[68] Ibid., p. 176.

[69] Ibid., p. 201.

[70] Pedersen, Plomin, et al., 1992, p. 347.

[71] Pedersen, McClearn, et al., 1992.

[72] Pedersen, Plomin, et al., 1992, p. 347.

[73] Langinvainio, Koskenvuo, et al., 1984.

[74] Juel-Nielsen, 1980, p. 77.

[75] Newman et al., 1937, p. 135.

[76] Ibid., p. 134.

[77] Taylor, 1980, p. 82.

[78] Newman et al., 1937, p. 31.

[79] Gardner & Newman, 1940, p. 119.

[80] Ibid., p. 126.

[81] Juel-Nielsen, 1980, p. 75.

[82] Jensen, 1970, p. 146.

[83] Rowe, 1994, p. 39.

[84] For further analysis of the pre-MISTRA reared-apart twin studies, see Farber, 1981; Joseph, 2001d; Kamin, 1974; Taylor, 1980.

[85] Bouchard & Loehlin, 2001, p. 243.

[86] L. Wright, 1997, p. 8.

[87] Ibid., p. 143.

[88] The SATSA personality results are found in Pedersen et al., 1988; the Finnish results are from Langinvainio, Kaprio, et al., 1984.

[89] Pedersen, Plomin, et al., 1992.

[90] Bouchard, 1984.

[91] Horgan, 1993; L. Wright, 1997.

[92] Leon J. Kamin, personal communication of April 8, 2001.

[93] See W. Wright, 1998.

[94] Ibid., p. 215.

[95] The homosexuality study is by Eckert et al., 1986.

[96] Eckert et al., 1986, p. 422.

[97] Beckwith et al., 1991.

[98] W. Wright, 1998, p. 56.

[99] Bouchard et al., 1991.

[100] Watson, 1981, p. 61.

[101] L. Wright, 1997.

[102] Jensen, 1974, pp. 26–7.

[103] Shields, 1978, p. 79.

[104] Ibid., p. 80.

[105] Bouchard & McGue, 1990, p. 267.

[106] Ibid., pp. 266–7.

[107] Bouchard et al., 1990, p. 224.

[108] This point was made by K. Richardson (1998).

[109] Farber, 1981, p. 19.

[110] L. Wright, 1997, p. 71.

[111] See Heston, 1988, pp. 211–12.

[112] Segal, 1999, p. 134.

[113] See Moos & Moos, 1986.

[114] Bouchard & McGue, 1990, p. 278.

[115] Pedersen, McLearn et al., 1992, p. 256.

[116] Faraone & Tsuang, 1995, p. 91.

[117] Bouchard, 1997c, p. 54.

[118] Rosenthal & Rosnow, 1975.

[119] McGue & Bouchard, 1984.

[120] Ibid., p. 325.

[121] The McGue & Bouchard 1989 MISTRA IQ publication listed corrected and

uncorrected correlations for environmental measures, but not for the IQ correlations themselves.

[122] In 1931 Sims obtained test scores of 203 sibling-pairs raised in the same home and recorded an intraclass correlation of .40. He then tested 203 unrelated pairs who were matched on the basis of age, school attended, and 'home background.' The unrelated group correlated at .29, which was not significantly different from the sibling-pairs. Sims then formed a group of 203 unrelated pairs matched only on age and school attended. He found that this group correlated at .04. Sims concluded that 'intelligence, as far as we are today able to measure it, is greatly influenced by environment' (p. 65). In discussing this report, Kamin (1974, p. 81) noted that 'This sophisticated and elegant study has simply disappeared from contemporary reference lists.' In a subsequent paper, Bouchard (1983) wrote that Sims had obtained a correlation of .04 for the group of unrelated students matched on age (and school attended), but Bouchard neglected to cite the .29 figure for the group matched on these factors plus home background.

[123] See Mischel, 1968.

[124] Allport, 1961, p. 28.

[125] See Taylor, 1980 for a detailed discussion of this point.

[126] Martin et al., 1981.

[127] Plomin et al., 1998, p. 211.

[128] Ibid., p. 214.

[129] Bouchard & McGue, 1990, p. 267; Bouchard et al., 1998, p. 309.

[130] Bouchard et al., 1998, p. 314.

[131] McGue & Bouchard, 1984, p. 325.

[132] Kamin & Goldberger, 2002.

[133] Bouchard et al., 1990, p. 223. Four years earlier, Bouchard and colleagues wrote, 'DZA twins allow us to test the two most common competing hypotheses proposed as alternatives to the genetic hypothesis as an explanation of the similarity between MZA twins: placement bias and recruitment bias. We believe placement to be approximately comparable and we will be able to evaluate this assumption' (Bouchard et al., 1986, p. 300).

[134] Bouchard & McGue, 1990, pp. 280–1.

[135] McGue & Bouchard, 1989.

[136] Newman et al., 1998, p. 992.

[137] McCourt et al., 1999, p. 1001. In his 1996 Galton Lecture, Bouchard reported the SATSA 'first principal component' mental ability correlations (MZA = .80, DZA = .32), but failed to report the MISTRA DZA correlations (see Bouchard, 1996).

[138] Bouchard et al., 1998.

[139] Bouchard & Pedersen, 1999.

[140] Rose, 1982, p. 960. Rose identified his colleague as G. P. Frommer.

[141] Ceci, 1996, p. 60.

[142] Ibid., p. 61.

CHAPTER FIVE

The Heritability Concept:
A measure of inheritance or
inherently misleading?

Heritability analysis of human behaviour has become the dominant paradigm in academic psychology and now appears prominently in introductory texts, where it is presented to naïve students who have no understanding of the false assumptions inherent in the calculations. The pre-eminence of heritability analysis is the outcome of a power struggle, not the resolution of a debate among scientists.

Dissident behavior geneticist Douglas Wahlsten in 1994.[1]

Hereditarians *claim* to have the same interest as anti-hereditarians, to determine how much change is possible so that we may act rationally in social programs. But if that is really the hereditarian agenda, why do they keep studying heritability, which simple logic tells us cannot give the answer to this problem. Why do they not design studies to ask the questions about changeability directly? Because the answer would come out in the wrong direction.

Biologist Richard C. Lewontin in 1987.[2]

It is ironic that heritability estimates are made most often in behavior-genetic research using human subjects, where the only real application of h2 estimates — selective breeding — is prohibited.

Psychologist Scott F. Stoltenberg in 1997.[3]

The nature-nurture debate is over. The bottom line is that everything is heritable . . .

Behavior geneticist Eric Turkheimer in 2000.[4]

Heritability is perhaps the most misunderstood and misused concept in the psychological and psychiatric literature, as well as in popular works based on this literature. This admittedly difficult concept, widely used in the behavior genetics field, has helped solidify a belief in the importance of genetic influences on human psychological trait differences. The claim is frequently made that both IQ and personality 'are highly heritable,' although the meaning of this statement is often misunderstood. This chapter attempts to clarify the meaning of heritability and will look closely at two important questions relating to the concept: (1) what is the purpose of a heritability coefficient? and (2) what does heritability say about the importance of genetic influences on psychological trait differences?

What heritability is and what it is not

Heritability, according to Plomin and colleagues, is 'the proportion of phenotypic variance that can be accounted for by genetic differences among individuals.'[5] In this case 'phenotypic variance' refers to how a trait is distributed in a population. Heritability is expressed as a number ranging from 0.0 (no heritability) to 1.0 (complete heritability). Plomin and colleagues made several important observations about the concept: (1) heritability refers to genetic contributions to the phenotypic variance in a particular population; it does not describe the importance of genetic factors as they relate to an individual. (2) Heritability is applicable to a specific population, at a specific time, and in a specific environment. (3) The finding of genetic contributions in differences *within* populations does not mean that genetic factors contribute to differences *between* populations. (4) High heritability does not mean that environmental changes or interventions cannot have an important impact. (5) High heritability does not imply that traits are fixed or unchangeable: 'finding genetic influence on complex traits does not mean that the environment is unimportant.'[6]

These observations are extremely important if one is to properly understand the meaning of heritability, yet they are often unknown, misunderstood, or ignored by people using or attempting to understand the concept. Before going on, therefore, we will look at each point more closely.

(1) *Heritability refers to the genetic contribution to the phenotypic variance in a particular population; it does not describe the importance of genetic factors as they relate to an individual.* This point has been made frequently, yet many people still believe that heritability refers to the portion of an individual's phenotype attributable to genetic factors. In fact, the concept refers only to the percentage of the population variance that can be attributed to genetic factors, and there *must* be variance in order to calculate heritability. This leads to the paradoxical situation in which a trait could be 100% *inherited*, yet have a *heritability* of zero — human beings having two eyes, for example. Because virtually all people with one eye became that way on the basis of an environmental occurrence, heredity explains none of the observed *variance* of 'eyedness,' and heritability is therefore zero.[7] Thus, a genetically determined trait could have a heritability of zero. It is also important to understand that heritability is the property of a population and not of a trait itself.[8]

(2) *Heritability is applicable to a specific population at a specific time and in a specific environment.* In Kendler's words, 'a heritability estimate is a population- and time-specific "snapshot".'[9] Heritability estimates applicable to a particular population in a particular environment could

be meaningless in another environment.[10] Suppose we have a population of cows and wish to breed for increased milk production. The fact that we might be able to use a heritability coefficient to successfully select for milk production in this particular environment (characterized by geography, climate, feed, etc.) does not mean that the same result would be obtained in another environment. For this reason, Plomin and associates wrote that 'heritability describes *what is* in a particular population at a particular time rather than *what could be*.'[11]

(3) *The finding of genetic contributions in differences within populations does not mean that genetic factors contribute to differences between populations.* This point has been acknowledged by most genetic researchers, even if their conclusions do not always flow from it. In his controversial 1969 article, Jensen concluded that, due to the allegedly high IQ heritability *within* the U.S. black and white populations, the existence of mean IQ differences *between* blacks and whites is plausibly explained, in part, by genetic factors.[12] The erroneous nature of this extrapolation was demonstrated in Richard Lewontin's well-known 'sack of seeds' example.[13] We could take two handfuls of genetically heterogeneous open-pollinated corn seed, wrote Lewontin, and plant one handful in nutrient-rich soil, and the other in nutrient-poor soil. After a few weeks, we measure the height of the plants. Because no environmental variation in either field was allowed, differences in plant height *within* each field is completely explained by genetic variation (heritability = 1.0). However, the mean difference in plant height *between* the two fields is the result of differences in soil quality (heritability = 0.0). Thus, a trait could have a heritability of 1.0 within each of two populations, with the difference between populations being completely and easily explainable on environmental grounds.

(4) *High heritability does not mean that environmental changes or interventions cannot have an important impact.* This point also has been widely discussed, yet it is commonly believed that a highly heritable trait is difficult to change through treatment or intervention. A classic example of this fallacy is phenylketonuria (PKU), a genetic disorder of metabolism which, without a specific environmental intervention, can cause mental retardation. Although the population variance for susceptibility to PKU is completely explained by genetic factors (heritability = 1.0), the administration of a low phenylalanine diet to the at-risk infant prevents the disorder from appearing. PKU exemplifies Lewontin's observation that a 'trait can have a heritability of 1.0 in a population at some time, yet could be completely altered in the future by a simple environmental change.'[14] Interestingly, in a society with universal and accurate screening for PKU the heritability of the disorder would be *0.0*, because 100% of the *variance* of PKU in the population would be explained by the failure to receive the proper environmental intervention. Thus, a 'highly heritable' trait could be drastically altered or eliminated through a simple environmental intervention, and a heritability estimate can change from 1.0 to 0.0 even though the susceptibility for the trait is transmitted genetically.

In his 2001 book *The Dependent Gene*, developmental psychologist David Moore argued that while heritability addresses factors influencing a trait's variation, it says nothing about a trait's *causes*. As an analogous example he cited snowflake formation, which requires both high relative humidity and a temperature below 32 degrees Fahrenheit. Because the temperature at both poles is constantly sub-freezing, while relative humidity varies, all snowfall *variation* at the North and South Poles can be explained by differences in relative humidity. This does not mean, however, that temperature is an unimportant factor in the creation of snowflakes, or even that relative humidity is more important than temperature. Conversely, Moore observed that snowfall variation in a tropical country such as Costa Rica (which receives snow at the highest mountaintops) is accounted for by *temperature* variation alone. Thus, investigators seeking to determine the causes of snowflake formation would find 'temperature heritabilities' of 0.0 (explaining *none* of the variation) at the poles, and 1.0 (explaining *all* of the variation) in the tropics. Self evidently, these figures tell us nothing about the relative importance of humidity and temperature on snowflake formation; they merely reflect the fact that a key variable has been held constant. Moore concluded, 'since heritability estimates account for variation *and do not explain causation*, they tell us *nothing at all* about how genetic and environmental factors influence the development of our traits.'[15]

(5) *High heritability does not imply that traits are fixed or unchangeable.* This point is related to point #4 and has also been a common criticism of hereditarian interpretations of the heritability coefficient. The classic example in this case is human height, which has risen steadily in the industrialized nations (most dramatically in Japan). This increase has occurred in spite of the frequent claim that human height is a highly heritable trait.

Origins of the heritability concept

Although the word 'heritability' has been used since the nineteenth century, the concept (as it is understood today) was described by Jay Lush as early as 1936.[15] Lush, a professor of animal breeding at Iowa State College, was interested in facilitating the breeding of animals for economically desirable traits. Animal breeders needed a quantifiable method for predicting the results of a program of selective breeding for these desired traits. Lush wrote that heritability estimates could be derived from the resemblance of genetically related animals: 'All methods of estimating heritability rest on measuring how much more closely animals with similar genotypes resemble each other than less closely related animals do.'[16]

Lush's clearest statement on heritability is found in a 1949 paper.[17] Here, Lush defined heritability as 'the fraction of the observed or phenotypic variance which was caused by differences between the genes or the genotypes of the individuals.'[18] Lush discussed two types of heritability: broad and narrow. *Broad heritability* refers to the entire genotype, which includes genes which interact non-additively. *Narrow heritability* refers only to additive genetic effects, meaning those genes that are of interest to plant and animal breeders. Lush understood that highly heritable traits in animals were nonetheless influenced by the environment:

> Most of the numerical estimates for heritability of economically important characters have been surprisingly high in view of the fact that many of these characters are known to be readily and strongly influenced by changes in nutrition and by other environmental circumstances.[19]

This observation stands in direct contrast to the position taken by Jensen 20 years later, where it was argued that there was little hope of significantly 'boosting' IQ scores because intelligence was 'highly heritable' — an argument revisited in Herrnstein and Murray's *The Bell Curve*.[20] Lush understood that the phenotypic resemblance among livestock relatives was influenced by environmental factors, writing, 'The correlation between relatives is likely to have an environmental component.'[21]

Lush also viewed comparisons of identical twin *livestock* as vulnerable to environmental influences: 'Both members of a pair of identical twins would be subject to many of the same peculiarities of environment . . . Consequently this method overestimates heritability to the extent that the common environment of identical twins is important.'[22]

What is the purpose of a heritability coefficient?

We have seen that the heritability concept was created for the purpose of assisting animal and plant breeders to select for economically desirable traits. Because the statistic is now widely used in human populations for IQ and personality, and for psychiatric constructs such as schizophrenia and ADHD, one might ask what relevance heritability has for humans.

For Lush, 'To know whether heritability is high or low is important when making efficient breeding plans.'[23] A heritability estimate is important, according to Lush, because 'The possibilities of improving farm animals by suitable genetic methods are real and important.'[24] Lerner discussed heritability in some detail in his 1958 book on genetics and selection, referring to it as a 'measure of *accuracy of selection*.'[25] Lerner cautioned, however, that a heritability estimate was only valid for the particular population and generation from which data are derived.

Fuller and Thompson's 1960 *Behavior Genetics* is usually regarded as the founding document of the behavior genetics field. According to its authors,

> The heritability of a trait is of primary importance in behavior genetics, since it provides a measure of the expected rate of gain from selection . . . For the practical breeder or the psychologist trying to develop a strain characterized by particular behavioral qualities, heritability determinations provide a guide to the most efficient method of selection.[26]

Fuller and Thompson did not specify for which species they were advising psychologists to use a heritability coefficient, but we can safely assume they were referring to laboratory animals. Towards the end of their book Fuller and Thompson noted that while heritability estimates have 'no logical meaning when applied to an individual,' with regards to a population, 'one may well ask how much of the observed variation in behavior is attributable to genetic differences and how much to environment.'[27] Fuller and Thompson appear to have been the first authors to extend the definition of heritability to include the estimation of 'how much' of the observed variation is due to genetic factors, although many previous investigators had expressed interest in this question without linking it to heritability estimates.

In a 1978 work, Fuller and Thompson acknowledged that the heritability concept 'was developed by animal and plant breeders who were concerned with predicting the results of selection,' while noting that heritability 'has been taken over frequently in behavior genetics, sometimes with a less limited sense than is implied in the preceding definition.'[28] Fuller and Thompson wrote that broad heritability was useful 'for quantifying the extent of genetic influence on continuously distributed characters.' They noted that 'because the term is often misinterpreted, it has been suggested that behavior geneticists avoid the concept, but there is as yet no satisfactory alternative for quantifying the extent of genetic influence on continuously distributed characteristics.'[29]

In one of his early publications, behavior geneticist Gerald McClearn wrote about heritability in the context of animal research:

> One of the central concepts of modern genetics is that of *heritability*, which is defined as the ratio of the variance attributable to additive gene effects to the total phenotypic variance. This quantity represents the genetic contribution which is 'useful' in the sense that it provides for firm prediction of the outcome of various matings (e.g., in a selection program).[30]

Again, heritability is 'useful' as a means of predicting the outcome of matings. Although McClearn made several references to the importance of estimating the extent of genetic versus environmental influences in a population, he did not link the heritability concept to this idea.

In 1969 Jensen wrote, 'Heritability is a population statistic, describing the relative magnitude of the genetic component (or set of genetic components) in the population variance of the characteristic in question.'[31] Jensen's definition derived from Fuller and Thompson's 1960 extension of the heritability concept from an estimate for use by breeders, to an indicator of the magnitude of genetic influences on the population variance for the trait in question. I shall return to this subject later.

In the 1996 edition of his influential *Introduction to Quantitative Genetics*, geneticist D. S. Falconer wrote,

> The heritability of a metric character is one of its most important qualities. It expresses . . . the proportion of the total variance that is attributable to differences of breeding values, and this is what determines the degree of resemblance between relatives. But the most important function of the heritability in genetic study of metric characters is its predictive role, expressing the reliability of the phenotypic value as a guide to the breeding value.[32]

Like most earlier writers, Falconer discussed a heritability estimate as being important for the purposes of selective breeding.

Returning to the Plomin and colleagues 1997 behavior genetics textbook, we have seen how these authors discussed the limitations (and by implication misuse) of the concept. Given these limitations, how did Plomin and colleagues describe the purpose of a heritability coefficient? According to them, heritability is the 'statistic that estimates the genetic effect size . . . Heritability is the proportion of phenotypic variance that can be accounted for by genetic differences among individuals.'[33] The authors defined 'effect size' as meaning the 'proportion of individual differences for the trait in the population accounted for by a particular factor.'[34] Thus according to Plomin et al., 'genetic effect size' and 'heritability' are synonymous. They identified broad heritability as describing the genetic effect size of a population, whereas narrow heritability is 'interesting in the context of selective breeding studies.'[35]

Plomin and colleagues defined the heritability concept and discussed its limitations, yet offered no explanation of its practical use in human populations other than to estimate the genetic effect size. In light of their discussion of the concept's limitations, they failed to convincingly demonstrate the heritability statistic's *raison d'être* in human populations.

As we have seen, the two most frequently cited justifications for a heritability estimate are: (1) that it can predict the results of a breeding program, and (2) that it represents the percentage of phenotypic variance in a population attributable to genetic factors (that is, the magnitude of genetic influences). Regarding the first point, heritability estimates were designed for (and continues to be widely and profitably used in) agriculture. They might also be useful in experiments with laboratory animals under conditions where breeding and environment can be controlled. In human populations, a selective breeding program is called eugenics, which few people currently using heritability estimates advocate. Let us now look at the second justification.

Is it true that a heritability coefficient estimates the magnitude of genetic influences on a trait? Several critics of the concept have answered this question in the negative. According to dissident behavior geneticists Terry McGuire and Jerry Hirsch, whereas narrow heritability (h2) is an important statistic in selective breeding programs, broad heritability (H2) has been used by psychologists 'in the belief that it is their nature-nurture ratio or index to the causes of a trait.'[36] However, H2 does not

> describe the average influence of heredity in determining the level of trait expression in a population. Broad heritability estimates only the contribution of genetic factors (additive, dominance deviation, interaction and genotype-environment covariance) in the total phenotypic variance for one trait in a specified population.

McGuire and Hirsch wrote that while H2 is 'of little interest to geneticists,' it has been used and 'misused' in psychology.[37]

As noted, many critics have correctly pointed out that 'high heritability' does not say anything about the likelihood of a particular intervention's success. For example, the finding that children's reading ability is 'highly heritable' does not mean that a new teaching method could not dramatically improve reading scores.[38] Thus, a heritability estimate for reading ability has no practical purpose, and, in general, heritability 'is nearly equivalent to no information at all for any serious problem of human genetics.'[39]

Although heritability estimates have no practical purpose apart from animal and plant breeding, they are frequently misunderstood by people unfamiliar with the concept. It is often believed that the words 'heritable' and 'inherited' are synonymous, Hirsch has emphasized that 'heritability' and 'heredity' are 'two entirely different concepts that have been hopelessly conflated' in several texts.[40] He pointed out that 'because of their assonance, when we hear one of the two words, automatically we think the other.'[41] Geneticist Scott Stoltenberg proposed that the word 'heritability' be replaced by 'selectability' so as not confuse the scientific and 'folk' definitions of heritability. (Stoltenberg defined the 'folk definition' as being synonymous with 'inherited.') A change in terms would go a long way in eliminating the confusion between 'heritability' and 'inherited,' although, as Stoltenberg pointed out, many investigators would object to the eugenic implications of the word 'selectability.' 'Because of this,' wrote Stoltenberg, 'researchers would need to carefully consider the utility of calculating such a statistic, and when it is deemed necessary, careful use of language would be required.'[42]

The unfortunate conflation (or confusion) of 'heritable' and 'inherited' (or 'hereditary') is more understandable when we realize that even the creator of the heritability concept confused the terms. In a paper considered by Lush to be 'one of my best summaries on the [heritability] topic,'[43] he wrote:

> If a character can be influenced strongly by environmental variation, its *heritability* would be low in a population in which the environment varies widely. But in another population in which the physical control

is so rigid that the environmental variations are very slight, the same character would appear highly *hereditary* [emphasis added].[44]

And on the preceding page Lush spoke of 'heredity in the broad sense,' when he was clearly referring to heritability in the broad sense.[45] Of course, it is unlikely that Lush anticipated that his concept would be used in human populations. In later years he would complain that 'it does little good to warn against oversimplifying the idea of heritability. Some feel it necessary to repeat those warnings each time they write. Others ignore the warnings.' Lush believed that the time had come for 'coining a new word . . . to be presented *precisely* and in a way in which it *cannot be misunderstood.*'[46]

An exchange on the meaning of heritability

In a 1990 *Behavioral and Brain Sciences* target paper, maverick Canadian behavior geneticist Douglas Wahlsten remarked, 'The only practical application of a heritability coefficient is to predict the results of a program of selective breeding.'[47] Of the 29 authors commenting on Wahlsten's article (both favorably and unfavorably), only a few offered alternative applications. According to Cheverud, heritability analysis 'is commonly used because it can at least provide evidence that genes play a role in the development of a particular phenotype.'[48] Crow commented, 'one can be interested in the heritability of IQ without advocating a breeding program,' and claimed that low heritability 'says that changes of exiting variables within the existing range can have a substantial effect on the trait,' whereas with high heritability, 'environmental manipulations will have little effect unless they extend beyond the existing range, or bring in new factors.'[49] Crusio wrote that heritability figures had been justified in the past in order to convince psychologists and ethologists of the causal role of heredity for psychological trait differences, while 'It is by now clear that this approach is basically sterile and that these efforts should be abandoned.'[50] Detterman wrote that 'Heritabilities are nothing more than a way of representing portions of variance according to a particular model.'[51] According to Goodnight, heritability estimates 'reflect the potential for evolutionary change in a particular population at a particular time.'[52] Finally, McGuffin and Katz wrote that heritability estimates can help researchers choose the correct definition of a particular psychiatric diagnosis, such as schizophrenia, and claimed that specific numerical estimates are of little or no interest to researchers.[53]

Let us examine these claims a bit more closely. Cheverud wrote that heritability estimates provide evidence that genes play a role in the development of a phenotype. However, the findings of genetic studies — if each study is methodologically sound and if its underlying assumptions are valid — are sufficient to demonstrate this. For example, if the assumptions of the classical twin method are correct, a significantly higher identical twin correlation when compared with the fraternal twin correlation demonstrates the role of genetics for the trait in question. A heritability coefficient is merely the product of this finding rather than evidence in favor of genetic factors.

Crow claimed that environmental interventions for traits with high heritability have little effect 'unless they extend beyond the existing range, or bring in new factors.' The problem as this applies to humans is that environments are frequently capable of going 'beyond the existing range,' and 'new factors' are always possible; a new drug or teaching method, for example. This is particularly true in our current age of fast-paced technological changes.

In Detterman's response to Wahlsten, the common position was repeated: heritability estimates help gauge the magnitude of genetic effects for a particular trait using a particular model. However, there is little evidence in favor of this position. Goodnight argued that heritability estimates 'reflect the potential for evolutionary change' in a population. This is similar to the claim that high heritability implies that environmental interventions are unlikely to be successful. As we have seen, behavior geneticists such as Plomin and colleagues correctly argue against this idea.

McGuffin and Katz argued that it is not important to quantify precise estimates of the heritability of a trait. This position has been taken up by other researchers. According to behavior geneticists Rutter, Silberg, and Simonoff in 1993,

> We may well agree with the environmentalists' criticism that knowledge of the level of heritability of any given trait is of very little interest with respect to policy and practice because quite a high heritability does not rule out effective environmental intervention. Nevertheless, it is a serious mistake to suppose that behavioral genetics is mainly involved with quantifying heritability. There is indeed very little interest in calculating the precise level of heritability as such . . .[54]

In fact, behavior geneticists are very interested in calculating heritability coefficients, as seen in current research as well as in various studies published in the period when Rutter and colleagues made this claim. For example, Bouchard and McGue calculated heritability coefficients for personality variables from the Minnesota Study of Twins Reared Apart, Tsuang and Faraone estimated the heritability of several mood disorders, Goodman and Stevenson did the same for inattentiveness and hyperactivity, and in the edited book in which the Rutter et al. chapter appeared, Cardon and Fulker presented exact heritability estimates for various dimensions of cognitive ability.[55]

In summary, when confronted with Wahlsten's serious charge that a heritability estimate's only practical application is for use in a selective breeding program, several leading genetic investigators were unable to provide a viable alternative explanation. This included Plomin's response, where he did not address the issue.[56] Of note is geneticist Oscar Kempthorne's comment on Wahlsten's article, where he observed that 'most of the literature on heritability in species that cannot be experimentally manipulated, for example, in mating, should be ignored.'[57]

The heritability estimate: a product of environmentally confounded research

I have discussed problems associated with the heritability concept while only briefly discussing how these figures are arrived at. The calculation of heritability is based on the assumption that genetic makeup and environment do not interact, when it is clear that they do.[58] Obvious examples are height and skin color. In the United States and the UK, a person genetically coded to have dark pigmentation will usually experience a far different (i.e., vastly inferior) environment than someone born with light pigmentation. In the same way, adult males who are four feet tall experience different environments than adult males who are six feet tall. Kempthorne argued that because of gene-environment interaction, 'we cannot reasonably state that such and such a proportion of variance is "due to" heredity,' and concluded that '"heritability" does not even exist in the human IQ context.'[59]

Even if heritability were a valid and important concept for humans, the fact remains that heritability estimates are derived from twin, family, and adoption studies. We have already seen the problems associated with conclusions based on twin research, and as we will see more clearly in later chapters, adoption studies have their own set of problems and potential confounds. In twin research, heritability usually is estimated as $2(rMZ - rDZ)$, where rMZ is the identical twin concordance or correlation, and rDZ is the fraternal twin concordance or correlation.[60] Put more simply, heritability is calculated by doubling the reared-together identical-fraternal correlation difference. For example, if identical twins' IQ scores correlate at .70, and fraternal twins correlate at .40, twin researchers would estimate heritability at .60. However, these estimates are based on the validity of the equal environment assumption and are therefore dubious. In studies of twins reared apart, the MZA intraclass correlation is sometimes said to represent a direct estimate of heritability. These studies, as we saw in Chapter 4, contain methodological flaws and questionable assumptions which render them useless as indicators of genetic influences.

It is often said that the results from family, twin and adoption studies converge on the conclusion that genes are an important factor influencing human psychological trait differences. In fact, the results from these studies are plausibly explained on the basis of environmental factors, methodological error, and bias. Thus, even if heritability estimates were valid and useful in human populations, it is doubtful that they could be obtained from these methods.

Heritability and psychiatric disorders

Although the heritability debate has centered mainly on IQ and to a lesser extent personality, heritability estimates also have been used in psychiatry. Schizophrenia has been the most studied of all psychiatric disorders and will serve as an example. Kendler calculated heritability coefficients for schizophrenia on the basis of the results from twin studies.[61] He concluded that 'the major twin studies of schizophrenia agree in estimating the heritability of liability of schizophrenia at between 0.6 and 0.9,' while noting that these figures are based on the multifactorial threshold model 'which may or may not be appropriate for schizophrenia.'[62] According to Kendler and Diehl, the 'heritability of liability . . . ranges from 0.0 if genetic factors play no role in susceptibility of a disorder to a maximum of 1.0 if genes entirely determine disease risk.'[63] A similar formulation is found in the 1999 psychiatric genetics textbook of Faraone, Tsuang, and Tsuang.[64] This description obscures the critical point that a simple environmental intervention could completely eliminate a true genetic disorder, such as PKU. Conversely, we have seen that a heritability estimate of 0.0 for a particular trait does not necessarily mean that 'genetic factors play no role in susceptibility,' but only that such factors do not play a role in the population variance of the condition.

The questionable methods used by Kendler to calculate twin concordance rates, from which heritability has been calculated, are beyond the scope of this chapter, as are the problems with schizophrenia genetic studies in general (see Chapters 6 and 7). What is relevant here is the meaning of Kendler's heritability estimates and how they might be misunderstood by people interested in the causes of schizophrenia. A reader of Kendler who understands heritability in the 'folk' meaning of the term would conclude that schizophrenia 'is 60–90% inherited.' A more knowledgeable reader with an accurate yet incomplete understanding of heritability estimates might conclude that genetic factors are more important than environmental factors. As we have seen, both of these conclusions are erroneous. Gottesman and Shields concluded that '70% of the variance in the combined liability to developing schizophrenia is genetically transmitted and 20% is "culturally" transmitted.'[65] Again, many people would mistakenly take this to mean that 70% of schizophrenia cases are the result of genetics and that 20% are due to cultural transmission.

In fact, heritability estimates for psychiatric conditions are as misleading as they are for IQ and personality, for even if it were true that a diagnosis such as schizophrenia had an important genetic component, it could still develop that, like PKU or farsightedness, an simple environmental intervention would greatly reduce or eliminate its impact. That this has not occurred with schizophrenia is a result of the failure to discover specific environmental factors — the existence of which all sides in the debate acknowledge — rather than the alleged importance of genetic factors.

Conclusions

In this chapter I have discussed and elaborated upon the observations of several critics of the heritability concept, who have argued against its usefulness in assessing the importance of genetic factors for human trait differences. The concept is widely misunderstood by the public and by some professionals as well. Indeed, in 1985 historian Diane Paul looked at genetics textbooks' handling of the 'genetics of intelligence' topic and concluded that the discussions of heritability were 'confused in the extreme.'[66]

The heritability concept should be retained in animal and plant breeding, where it has been widely and successfully utilized. But even in agriculture it is used for the specific purpose of economic gain through increased production. Selection does not necessarily produce a 'better' cow or chicken, or one that would even survive outside of a controlled environment.[67] Although the vast majority of people utilizing the heritability concept for human traits are opponents of the eugenic program, its only practical application, as Wahlsten pointed out, is to predict the results of a selective breeding program. Ironically, heritability coefficients would have little value even in eugenic selection among humans, because a heritability calculation in the current environment might be very different in another environment.

The heritability concept has little or no value in helping us understand the origins of human trait differences, but to the extent that it is widely misunderstood and misused, it is at best irrelevant[68] and is more often *a harmful and misleading concept*. Hirsch spoke of the 'tidal wave . . . of unjustified human heritability estimates' that 'has done so much harm over two decades . . .'[69] As such, its use should be discontinued in the social sciences except when it refers specifically to selective breeding plans. Because even in this limited manner the heritability concept would continue to sow confusion about its true meaning, Stoltenberg recommended using the word 'selectability' in its place. This step would go a long way in clearing up the confusion over the term.

In conclusion, although a researcher might wish to determine whether a particular human trait is influenced by genetic factors, the claim that a trait or population has a particular heritability value is meaningless. Therefore, the heritability statistic has no practical purpose in psychology and psychiatry, where its use should be discontinued. Its disappearance would eliminate confusion over the concept, and researchers would utilize other means to show that traits or conditions have a genetic basis. What they would lose is a word (and its corresponding quantification) falsely purporting to gauge the magnitude of genetic influences on a trait since, contrary to popular belief, 'heritability is not a nature-nurture ratio measuring contributions to individual development.'[70] 'Heritability,' therefore, should become the sole property of plant and animal breeders, for whom it was originally intended.

Notes

[1] Wahlsten, 1994, p. 254.
[2] Lewontin, 1987, p. 32.

[3] Stoltenberg, 1997, p. 92.
[4] Turkheimer, 2000, p. 160.
[5] Plomin, DeFries, et al., 1997, p. 79.
[6] Ibid., p. 85.
[7] Kamin, in Eysenck vs. Kamin, 1981.
[8] Jacquard, 1983.
[9] Kendler, 2001, p. 1005.
[10] Feldman & Lewontin, 1975.
[11] Plomin, DeFries, et al., 1997, p. 84.
[12] Jensen, 1972.
[13] Lewontin, 1976, pp. 14–15.
[14] Lewontin, 1974, p. 400.
[15] Moore, 2001, p. 43. A discussion of the origins of the heritability concept can be found in Bell, 1977.
[16] Lush, 1945, p. 93.
[17] Lush, 1949.
[18] Ibid., p. 356.
[19] Ibid., p. 372.
[20] Herrnstein & Murray, 1994.
[21] Lush, 1949, p. 360.
[22] Ibid., p. 362.
[23] Ibid., p. 359.
[24] Ibid., p. 374.
[25] Lerner, 1958, p. 63.
[26] Fuller & Thompson, 1960, p. 64.
[27] Ibid., p. 317.
[28] Fuller & Thompson, 1978, p. 60.
[29] Ibid., p. 61.
[30] McClearn, 1964, p. 234.
[31] Jensen, 1972, p. 115.
[32] Falconer & Mackay, 1996, p. 160.
[33] Plomin, DeFries, et al., 1997, p. 79.
[34] Ibid., p. 312.
[35] Ibid., p. 300.
[36] McGuire & Hirsch, 1977, p. 46.
[37] Ibid., p. 46.
[38] Hirsch, 1970.
[39] Feldman & Lewontin, 1974, p. 1168.
[40] Hirsch, 1997, p. 220.
[41] Ibid., p. 220.
[42] Stoltenberg, 1997, p. 96.
[43] Quoted in Bell, 1977, p. 298.
[44] Lush, 1949, p. 358.
[45] Ibid., p. 357.
[46] Quoted in Bell, 1977, p. 299. Emphasis in original.
[47] Wahlsten, 1990, p. 119.
[48] Cheverud, 1990, p. 124.
[49] Crow, 1990, p. 127.
[50] Crusio, 1990, p. 128.
[51] Detterman, 1990, p. 132.
[52] Goodnight, 1990, p. 134.
[53] McGuffin & Katz, 1990.

[54] Rutter et al., 1993, p. 437.

[55] Bouchard & McGue, 1990, p. 277; Tsuang & Faraone, 1990, p. 74; Goodman & Stevenson, 1989, p. 697; Cardon & Fulker, 1993, p. 112.

[56] Plomin, 1990.

[57] Kempthorne, 1990, p. 139.

[58] McGuire & Hirsch, 1977; Wahlsten, 1990.

[59] Kempthorne, 1978, p. 19.

[60] Falconer, 1965; Smith, 1974.

[61] Kendler, 1988.

[62] Ibid., p. 17.

[63] Kendler & Diehl, 1995, p. 945.

[64] Faraone, Tsuang, & Tsuang, 1999, p. 32.

[65] Gottesman & Shields, 1982, p. 225.

[66] Paul, 1998, p. 37. For the original study, see Paul, 1985.

[67] McGuire & Hirsch, 1977.

[68] Lewontin, 1987.

[69] Hirsch, 1990, p. 137.

[70] Hirsch, 1997, p. 220.

CHAPTER SIX

The Genetics of Schizophrenia I: Overview

The claim that some people have a disease called schizophrenia (and that some, presumably, do not) was based not on any medical discovery but only on medical authority; that it was, in other words, the result not of empirical or scientific work, but of ethical and political decision making.

Psychiatry critic Thomas Szasz in 1976.[1]

One can divide the investigators in the [schizophrenia genetics] field into two main types: those who like to look at numbers and those who like to look at patients. The former tend to be hereditarians, the latter environmentalists.

Schizophrenia researcher David Rosenthal in 1968.[2]

Why don't they take me, too.

A young Norwegian woman upon discovering that her identical co-twin sister was being taken away to a mental hospital.[3]

Schizophrenia, the paradigmatic psychiatric diagnosis, has been studied for more years and in more depth than any other condition in psychiatry, and serves as a model for investigations into the genetics of psychiatric disorders in general. In this chapter and the next we will see that— contrary to what most people in and out of psychiatry believe—there is little evidence supporting a genetic basis for schizophrenia. Furthermore, we will see that a 'genetic predisposition,' even if it existed, is not a meaningful or helpful concept.

In 1990 psychologist Richard Marshall argued that the genetic basis of schizophrenia is best understood as a hypothesis, not as an axiom.[4] Nevertheless, the question of whether schizophrenia carries a genetic component is rarely debated anymore in leading psychiatry and psychology journals. One therefore could conclude that the evidence in favor of the genetic perspective has been so overwhelming as to permanently resolve the issue, but this is simply not the case. This question, in fact, has been 'resolved' many times since the days of Kraepelin. Before any twin, adoption, or even family study had been performed, Rosanoff and Orr could write, on the basis of the family histories of a few hospitalized patients, 'It would seem, then, that the fact of the hereditary transmission of the neuropathic constitution as a recessive trait, in accordance with the Mendelian theory, may be regarded as definitely established.'[5] This view was echoed by Charles Davenport of the Eugenics Record Office in 1920: 'The conclusion seems supported that the tendency [for dementia praecox, a term later replaced by 'schizophrenia'] is inherited as a simple Mendelian recessive.'[6] In 1958 a leading British psychiatrist would write that 'The evidence for a genetic cause of schizophrenia is now generally agreed to be overwhelming, although certain details may not yet be known.'[7] In 1970 David Rosenthal wrote that the results from schizophrenia adoption studies provided evidence 'so consistently and so strongly in favor of the genetic hypothesis that the issue must now be considered closed.'[8] More recently, a pair of twin researchers wrote that the 'substantial hereditary component in schizophrenia is surely one of the two or three best-established facts in psychiatry.'[9] The American Psychiatric Association's DSM-III-R concluded that 'genetic factors have been proven to be involved in the development of the illness,'[10] but the next edition (DSM-IV) softened this position, stating that 'much evidence suggests the importance of genetic factors in the etiology of Schizophrenia.'[11]

There are no physical or biological markers distinguishing people diagnosed with schizophrenia from 'normals.' Rather, schizophrenia is diagnosed on the basis of history, observation, behavior, and self-report. The question of whether schizophrenia is caused by a disease process of the brain is beyond the scope of this book, although the evidence in favor of the disease model, as Breggin and others have demonstrated, is weak.[12] Theoretically, schizophrenia could involve a malfunctioning of the brain and yet have no genetic basis whatsoever. Still, the alleged biological basis of schizophrenia is frequently cited in support of the genetic position, and the alleged genetic basis of schizophrenia is frequently cited in support of the biological position. As Laing observed, however, 'These two theories do not . . . reciprocally validate each other. Rubbing two phantom flints produces only the illusion of fire.'[13]

There is also the question of whether 'schizophrenia' should be viewed as a disease or illness, or whether it is better characterized as a set of socially disapproved behaviors labeled as a disease. Szasz described schizophrenia as 'a concept wonderfully vague in its content and terrifyingly awesome in its implications,'[14] which served as the model for redefining 'the criterion of disease . . . from abnormal bodily structure to abnormal personal behavior.'[15] The validity and usefulness of the schizophrenia concept has been questioned by several other critics, such as Boyle, Breggin, Cohen and Cohen, Gosden, Hill, Laing, and Sarbin and Mancuso.[16] The symptoms used to diagnose schizophrenia often reflect judgments about the appropriateness of personal conduct or beliefs. A person's beliefs, as critics frequently point out, become 'delusions' only when they are culturally unacceptable.[17] If someone says that Jesus speaks to him, he is a 'good Christian.' If another person says that Napoleon Bonaparte speaks to her, she becomes a 'schizophrenic.' How can a choice of historical figures differentiate between upstanding members of the community (or even a head of state), and those with hereditary brain diseases? Clearly, the schizophrenia diagnosis is sometimes used to label social deviance as a 'disease.'[18]

A psychiatric diagnosis, wrote Mary Boyle, gains legitimacy by 'adopting the language and assumptions of medical diagnosis.'[19] She argued that this position has two essential flaws. First, unlike in physical disease, no relationship has been established between the observed phenomena (in the case of schizophrenia, for example, there is no a priori relationship between withdrawal and hallucination). Second, she argued that it has not been demonstrated that any process, whether psychological or biological, holds together the phenomena under investigation. Psychiatric diagnoses thus fail the scientific test of validity. If psychiatric diagnoses are not valid, then they cannot be reliable, and research based on them cannot be regarded as sound. The meaningfulness of such research, according to Boyle, is sociological rather than scientific.

Another reason why we should question whether schizophrenia is caused by hereditary factors is the fact that people diagnosed with the disorder have low reproductive rates. According to the DSM-IV, approximately 60%-70% of people diagnosed with schizophrenia never marry, and most have limited social contacts.[20] The persistence of a 'genetic disorder' in which the gene carriers reproduce at a low rate does not seem possible. Supporters of the genetic theory of schizophrenia occasionally acknowledge this, yet offer few plausible explanations. In a recent statement, biological psychiatrist Nancy Andreasen could only observe that it is 'fascinating' that 'schizophrenia persists in the human population despite the fact that the majority of people who develop it do not marry or procreate.'[21] Psychiatrists Torrey and Yolken were more to the point, writing in 2000:

> Since the beginning of the [American] deinstitutionalization movement in the 1960s, the vast majority of individuals with severe schizophrenia have been released from hospitals or have been hospitalized for brief periods only. They have been allowed to live in the community and are reproducing at a much higher rate than they did prior to deinstitutionalization . . . At the same time, there are suggestions from several European studies that the incidence of schizophrenia may be decreasing . . . If schizophrenia is truly a genetic disease, therefore, it is the only known genetic disease that becomes more common with less reproduction by those affected and less common with more reproduction by those affected.[22]

Today, many schizophrenia genetic researchers have moved 'beyond' twin and adoption studies and believe that genetic linkage analysis and DNA studies should be the current focus of attention. However, the search for 'schizophrenia genes' is based on the implicit or explicit assumption that family, twin, and adoption studies have settled the question of the genetic nature of schizophrenia; the only problem left being the identification of the actual gene or genes. In Chapter 10, I discuss the failure to find genes for schizophrenia, as well as the likelihood that the search will turn up nothing.

History and prevalence of schizophrenia

Schizophrenia carries a lifetime population expectancy rate of 0.8%–1.0%.[23] Of the people currently living in the United States, it has been estimated that about 3.7 million will be diagnosed with schizophrenia at some point of their lives.[24] Although it is often reported that prevalence rates are similar in all countries, some have argued that schizophrenia is found much more frequently in the industrialized parts of the world.[25] Men and women are diagnosed in roughly equal numbers.[26]

The term *dementia praecox* (dementia of early life) was popularized by the Swiss-German psychiatrist Emil Kraepelin in 1896. Kraepelin developed his concept by combining three previously reported conditions into one syndrome. An author sympathetic towards his work described Kraepelin's method:

> The major contribution of Kraepelin was the inclusion, in the same syndrome, of catatonia, already described by Kahlbaum, hebephrenia, and 'vesania typica,' also described by Kahlbaum, characterized by auditory hallucinations and persecutory trends . . . Kraepelin was able to discern the common characteristics in these apparently dissimilar cases.[27]

In 1911 Eugen Bleuler of Switzerland published his famous *Dementia Praecox or the Group of Schizophrenias*, which was translated into English in 1950. Bleuler proposed that the condition's name be changed to *schizophrenia*, meaning 'split mind' (a term that had no common meaning at the time[28]). Although Bleuler based his conception of schizophrenia on Kraepelin's work, he saw the condition as being characterized by 'primary symptoms' such as autism, blocking, and loose associations. Bleuler regarded hallucinations and delusions as 'secondary symptoms,' which were the result of a primary disturbance in thought.

Both Kraepelin and Bleuler believed that dementia praecox/schizophrenia had a strong inherited component, in spite of the lack of empirical studies supporting these beliefs. In 1899 Kraepelin wrote, 'An inherited predisposition to mental disturbances was apparent in approximately seventy per cent of those cases in which data could be evaluated.'[29] And Bleuler wrote in his monograph that 'heredity does play its role in the etiology of schizophrenia, but the extent and kind of its influence cannot as yet be stated.'[30] These views prompted Boyle to write,

Thus, before any attempt at systematic data collection was ever made, the two most prominent users of the concepts of dementia praecox and schizophrenia were disseminating the view that whatever phenomena they included under these terms were largely inherited.[31]

Following the publication of Rüdin's 1916 schizophrenia family study, Bleuler wrote, 'No one denies that an inherited predisposition plays an important role.'[32] Bleuler also supported the compulsory eugenic sterilization of people diagnosed with (and viewed as predisposed to) schizophrenia. In a rarely-cited passage from his book on schizophrenia, Bleuler wrote,

> Lommer and von Rohe have again recommended castration which, of course, is no benefit to the patients themselves. However, it is to be hoped that sterilization will soon be employed on a larger scale in these cases as in other patients with a pathological *Anlage* [predisposition] for eugenic reasons.[33]

Interestingly, as Boyle has persuasively argued, many of the people studied by Kraepelin and Bleuler may have suffered from the viral infection *encephalitis lethargica*. The pattern of this disease, whose symptoms are remarkably similar to 'schizophrenia' and 'dementia praecox,' was not described until 1917, well after Kraepelin and Bleuler had published their major works.[34]

The genetic theory of schizophrenia
Unknown to much of the general public and even to many in the mental health field is the following empirical finding: approximately 63% of people diagnosed with schizophrenia have a 'negative family history,' meaning no schizophrenia diagnoses among the patient's first- and second-degree biological relatives, including nieces and nephews.[35] The fact that a majority of those diagnosed with schizophrenia lack similarly afflicted biological relatives constitutes powerful evidence against the operation of genetic factors.

Family, twin, and adoption studies record the expression of a postulated schizophrenia *phenotype*, which refers to a person's observable behavior. Recent studies by molecular geneticists have attempted to identify a schizophrenia *genotype*, which refers to a person's genetic makeup with reference to a particular trait. Several genetic theorists have posited the existence of a single gene, while others have argued in favor of a polygenic basis for schizophrenia. The polygenic theory holds that the interaction of many genes is necessary to produce a predisposition for a trait or disease, for which individuals are said to have a genetic vulnerability. The theory postulates the existence of a liability threshold, which has been defined as 'the point on the scale of liability above which all individuals are affected and below which none are affected.'[36]

The genetic predisposition concept

The most accepted causal framework for schizophrenia and other psychiatric disorders is the 'diathesis-stress theory,' as it was called by Rosenthal in 1963. According to this theory, 'what is inherited is a constitutional predisposition to schizophrenia.'[37] A person is viewed as inheriting a predisposition which will develop into schizophrenia in the presence of the necessary (though possibly unknown) environmental triggers, which might include psychological factors, viruses, toxins, etc. The diathesis-stress theory is acceptable to everyone who believes in the role of genetic and environmental influences on schizophrenia, excluding only those who take a purely environmental or purely genetic position. Unfortunately, members of the public as well as many professionals interpret 'genetic predisposition' as being synonymous with 'it's genetic.'

An early criticism of common interpretations of the genetic predisposition concept is found in a 1932 discussion by American psychiatrist Abraham Myerson:

> Let us suppose that a brick falls from a building and fractures a man's skull. There must be some genetic background in his make-up that gives him a skull of such thinness and such disposition of bony forces as to permit it to be fractured by a brick. Yet it is a poor analysis in every way that says it is the genetic condition of the skull that gave him the fracture and disregards the brick.[38]

Myerson argued that it is 'perfectly possible' that if environmental conditions are changed, mental disorders 'would not show up, even though there were a genetic background for them.'[39]

As Myerson's example shows, the emphasis on genes or environment is related to how one approaches the problem, and whether the environmental trigger is easily identifiable. Richard DeGrandpre draws an analogy between the theory of a gene-environment interaction and a fire that destroys houses on a city block.[40] In the fire, all of the wooden houses are destroyed, while the concrete houses remain intact. Genetic researchers would likely blame the destruction on the builders of the (predisposed) wooden houses and pay little attention to the person who set the fire. Carrying the analogy further, advocates of a psychosocial intervention would seek to prevent fires from starting in the neighborhood and thus protect all of the homes. Interestingly, the importance some would place on the material used to build houses (a genetic predisposition for schizophrenia) is proportionate to our knowledge of who started the fire (the environmental trigger). If we are able to identify the person or persons responsible for setting fires, the question of what types of materials are used to build houses becomes unimportant.

In the Tienari et al. Finnish adoption study (see Chapter 7), the investigators found that virtually all adoptees they considered genetically predisposed to schizophrenia, and who ultimately received this diagnosis, were raised in 'severely disturbed adoptive families.'[41] One could conclude that schizophrenia would be eliminated by preventing people from being raised in severely disturbed families, yet the Finnish study is more often cited in support of genetic factors. Paradoxically, the genetic predisposition concept speaks more to what we *don't* know (or are unwilling to change) about the environment than it does to what we *do* know about genetics. Had the brick in Myerson's example been *invisible*, the unfortunate person suffering a fractured skull may have been viewed as having a mysterious genetic disorder.

The reception given to the Finnish study reveals the frequently *pessimistic* nature of genetic theorizing. It is far easier to say that some people have a genetic disorder or predisposition than it is to create a society which ensures that children do not grow up in seriously disturbed families. In this sense, the *decision* to emphasize genetic endowment over environment often is based on a person's values, preconceived notions, or vested interests. It is not predestined that disturbed families will always exist or that children will be forced to endure them. I am talking about preventing psychological damage before it happens and not, as typically occurs in psychotherapy, simply trying to undo damage that has already been done. Genetic theories in psychiatry tend to blame the victim and let society off the hook, and in this way help support the status quo. As Richard Lewontin once observed, the question revolves around whether we see an individual as a problem for society, or society as a problem for the individual.[42]

My point is that knowing that a person carries a genetic predisposition is helpful only to the extent that he or she can be helped to avoid known environmental triggers. As Myerson pointed out, when the trigger is known the condition may well disappear. Quite apart from whether some people are genetically predisposed to pellagra, a proper diet protects *everyone* from pellagra. Seeing people diagnosed with schizophrenia as 'genetically predisposed' promotes the unwarranted stigmatization of diagnosed individuals and, even if true, delays the discovery of schizophrenia's environmental triggers.

Turning point

In the summer of 1967 a conference on schizophrenia was held in Dorado Beach, Puerto Rico. It was there that the preliminary findings of the Danish-American adoption series were presented. The results and conclusions from these studies carried considerable influence (see Chapter 7). In particular, the Kety et al. Adoptees' Family report was viewed by the psychiatric geneticists as a 'steamroller' which had 'obviously abolished a number of more or less fragile observations and assumptions.'[43] According to Gottesman, a major consequence of the Dorado Beach conference was that 'the entire field of schizophreniology was converted, at least in public pronouncements, to some kind of interactionist stance.'[44] The papers and speeches from this conference were published the following year in a book entitled *The Transmission of Schizophrenia*.[45]

Family studies

The genetic theory of schizophrenia found early support in the discovery of disturbed or psychotic people in the family histories (pedigrees) of people diagnosed with schizophrenia, which led to the hasty conclusions of Kraepelin, Bleuler, Rosanoff and others. Given the non-systematic nature of such evidence, attempts were made to study the families of a large group of people diagnosed with schizophrenia in order to determine whether the condition was found in greater numbers among their biological relatives than would be expected in the general population.

The first systematic family study was published by Rüdin in 1916,[46] and more than two dozen have been published since. Most of these studies were carried out by strong proponents of the genetic position, most did not diagnose blindly, and some used hearsay or sketchy information in making diagnoses. Several modern family studies have used control groups and blind diagnoses based on structured interviews. Most found lower first-degree relative rates than the older studies, and at least three found no significant difference between the first-degree relatives of people diagnosed with schizophrenia versus the expected population rate, or versus the rate among the first-degree relatives of controls.[47]

As we saw in Chapter 1, most genetic researchers now agree that familial clustering is consistent with both genetic *and environmental* explanations. As Rosenthal explained,

> Although [family] studies are well worth undertaking for their own sake, the inference of a genetic basis must be held in abeyance until it can be shown that the association between incidence and consanguinity cannot be explained on some other basis. We might conceivably find a similar association with respect to some infectious diseases or with respect to a trait like poverty, where environmental factors may be of overriding importance. As a matter of fact, just such an association would be predicted by many clinical psychologists and psychiatrists who hold that the occurrence of functional behavior disorders results from peculiar or unusual behavior that takes place in certain families.[48]

The inconclusive nature of family studies led to the study of twins and adoptees in an attempt to control for environmental confounds.

Schizophrenia twin studies

In Chapter 3 we saw that the twin method, like a family study, is unable to disentangle possible genetic and environmental influences. For this reason, I will not review the schizophrenia twin studies individually. Instead, I discuss them collectively while highlighting important trends. I also discuss other ways that twin research has been cited in support of the genetic position. Chapter 7 is devoted to an analysis of the schizo-

Table 6.1

Results of Published Schizophrenia Twin Studies

| | | Pairwise Concordance Rates | | | | | |
| | | Identical Twins | | | Same-Sex Fraternal Twins | | |
Author (Year)	Country	N	C	%	N	C	%
Luxenburger (1928) [a]	Germany	17	10	**59%**	13	0	**0**
Rosanoff et al. (1934)	USA	41	25	**61%**	53	7	**13%**
Essen-Möller (1941/1970) [b]	Sweden	7	2	**29%**	24	2	**8%**
Kallmann (1946)	USA	174	120	**69%**	296	34	**11%**
Slater (1953)	UK	41	28	**68%**	61	11	**18%**
Inouye (1961)	Japan	55	20	**36%**	17	1	**6%**
Tienari (1963/1975)	Finland	20	3	**15%**	42	3	**7%**
Gottesman and Shields (1966b)	UK	24	10	**42%**	33	3	**9%**
Kringlen (1967a) [c]	Norway	45	12	**27%**	69	3	**4%**
NAS-NRC (1970/1983) [d]	USA	164	30	**18%**	268	9	**3%**
Fischer (1973) [e]	Denmark	25	9	**36%**	45	8	**18%**
Koskenvuo et al. (1984)	Finland	73	8	**11%**	225	4	**2%**
Onstad et al. (1991)	Norway	24	8	**33%**	28	1	**4%**
Franzek and Beckmann (1998) [f]	Germany	9	6	**67%**	12	2	**17%**
Cannon et al. (1998) [g]	Finland	—	—	**—**	—	—	**—**
Pooled Rates		**719**	**291**	**40.4%**	**1186**	**88**	**7.4%**

C = Number concordant; N = Number of twin pairs studied

Concordance rates based on the authors' narrow or 'strict' definition of schizophrenia; age correction factors not included.
Unless otherwise noted, when two dates are stated, the first indicates the year results were first published, the second indicates the final report, whose figures are reported in the table.
[a] Based on figures provided by Gottesman and Shields (1966a, p. 14). Hospitalized co-twins only.
[b] Identical twin figures from Essen-Möller (1970). Fraternal twin figures were not reported in this paper. Fraternal twin concordance rate based on (1941) definite cases among co-twins, as reported by Gottesman and Shields (1966a, p. 28).
[c] Based on a strict definition of schizophrenia; hospitalized and registered cases.
[d] National Academy of Sciences-National Research Council. Original report by Hoffer and Pollin (1970); final report by Kendler and Robinette (1983).
[e] Final report of an expanded sample originally collected by Harvald and Haugue (1965).
[f] Concordance rates based on DSM-III-R schizophrenia in a twin having the same condition.
[g] Cannon et al. (1998) reported probandwise concordance rates of 46% (identical) and 9% (same-sex fraternal). The pairwise equivalents of these figures are not listed in Table 1 because the number of pairs in each group was not given by Cannon et al.

phrenia adoption studies, which are usually seen (albeit incorrectly) as conclusive proof that genetic factors operate.

The results of the 15 schizophrenia twin studies are summarized in Table 6.1. All concordance rates are based on the 'pairwise' method, which is equal to the percentage of twin pairs concordant for schizophrenia. Since the 1960s an alternative concordance rate method has gained acceptance by several schizophrenia twin researchers: the so-called 'proband' method. (In the psychiatric genetic literature, a person through whom other relatives are identified is often referred to as a 'proband.') This method (which yields the 'probandwise' concordance rate) has been described as follows: 'In the proband method a pair of twins concordant for schizophrenia is counted twice if each of them was ascertained independently in the course of identifying subjects for the study.'[49] The pairwise method asks, 'In what proportion of pairs are both called schizophrenic?'; the probandwise method asks, 'In what proportion of probands is there a schizophrenic co-twin?'[50] Because the number of concordant pairs is doubled, the proband method always produces higher concordance rates. It is difficult to understand how this method could have gained the level of acceptance that it has, considering that twin subjects and their co-twins usually are not identified independently.[51] Typically, the co-twin of a person diagnosed with schizophrenia is identified because his or her co-twin had already been located, which does not constitute independent ascertainment.

As seen in Table 6.1, the pooled pairwise concordance rates for schizophrenia are 40.4% identical, and 7.4% fraternal. We see that the older 'classical' studies, using non-blind diagnoses and resident hospital samples (e.g., Rosanoff, Kallmann, Slater), reported higher rates than the more recent studies. Kallmann's sample constitutes 24% of the identical twin total in Table 6.1, and 25% of the fraternals. If we limit the analysis to the more modern register and consecutive admissions studies beginning with Tienari, the pooled pairwise concordance rates fall to identical 22.4% (86/384), and fraternal 4.6% (33/722). Furthermore, identical twin concordance is 18% or less in three of the last nine studies, as well as in two of the three studies with the largest identical twin sample. Most of the investigators concluded that their results were consistent with a genetic contribution to schizophrenia, although some pointed to the importance of (usually unstated) environmental factors as well. In fact, most genetic researchers acknowledge that identical twin concordance of less than 100% proves that environmental factors exist.

Three low concordance rate studies

As seen in Table 6.1, the Tienari, NAS-NRC, and the Koskenvuo et al. studies reported low identical twin schizophrenia concordance rates (15%, 18%, and 11% respectively). Koskenvuo and colleagues' figures were based on the computer files of Finnish hospital discharge records from the years 1972–1979. In spite of having produced the third largest identical twin sample, this study is almost never mentioned by contemporary reviewers or by the authors of psychiatry and psychology textbooks. It is never cited or discussed in the numerous reviews by Kendler or Gottesman, for example. In fact, I am aware of only one review that has mentioned the Koskenvuo et al. study, and of no review that has included its results in a table.[52] Apart from Tienari's preliminary reports, Koskenvuo and colleagues' study produced the lowest identical twin schizophrenia concordance rates ever reported, yet strangely is missing (without explanation) from textbooks and other literature reviews.[53]

The NAS-NRC and Tienari results, on the other hand, have been *mis*reported by influential proponents of the genetic hypothesis. Mary Boyle wrote about the 'selective criticism of low-rate studies' in her 1990 book, *Schizophrenia: A Scientific Delusion?*[54] What follows is an update of Boyle's discussion of the way these studies have been reported.

Tienari

In 1963 Tienari published the results of his schizophrenia twin study, which at that time comprised 16 identical and 21 fraternal pairs. He reported *zero* identical pairs and one fraternal pair as being concordant for schizophrenia. This report took the schizophrenia twin-studying world by storm, considering that Kallmann had reported an 85.8% age-corrected identical twin concordance rate just 17 years earlier. Tienari published follow-ups in 1968 and 1971, and a final paper four years later.[55] In this 1975 final report, Tienari increased his identical twin total to 20 pairs, and his fraternal total to 42 pairs. The final results were given as follows: identical twins—3/20 concordant, pairwise concordance rate = 15%; fraternal twins—3/42 concordant, pairwise rate = 7.5%.[56]

Commenting on these findings in 1982, Gottesman and Shields wrote, 'one might have expected a low concordance rate from a small sample of male twins.'[57] Four other schizophrenia twin studies had similar identical twin pair sample sizes (including their own, N = 24), but escaped criticism from Gottesman and Shields. It is also puzzling that Gottesman and Shields would expect a sample of *male* identical twins to produce a lower concordance rate, when 12 pages later they argued that register-based studies found no significant concordance rate differences between male and female identical twins.[58] Gottesman and Shields presented a table listing the figures from Tienari's 1971 report, rejecting his 1975 update on the grounds that three new identical pairs had been added and that it 'lacked detail.'[59] They claimed that Tienari's identical twin probandwise concordance rate stood at 36% — a figure not out of line with other schizophrenia twin studies. Gottesman and Shields also suggested that an age-correction factor should have been added to this figure. According to Boyle, 'it is not clear why Tienari's research should be singled out in this way, given the haphazard and often unreported information on which other researchers based their diagnoses.'[60] She pointed out that Gottesman and Shields's suggestion of adding an age-correction factor was strange, given that Tienari's twins were all over 40 years old and were therefore past the risk period for schizophrenia. (They ranged in age from 46–55 in

1975.)

Tienari's study provides an example of how the proband method can be used to artificially inflate relatively low concordance rates. Let us take Tienari's 1975 identical twin pairwise figure of 3/20 = 15% as an example. Using the proband method this figure becomes 6/23 = 26%, which represents a 73% concordance rate increase from one method to the other. Now let us suppose that the pairwise concordance rate had been 15/20 = 75%. Utilizing the proband method, this figure becomes 30/35 = 86%. In this case, the concordance rate has risen only 14.6%.

In 1991 Gottesman wrote that 'Tienari on follow-up (1975), reported a pairwise concordance rate of 36 percent for MZ twins and 14 percent for DZ twins and is no longer the odd man out.'[61] In fact, Tienari reported no such rates. According to Tienari's 1975 paper, 'The concordance rate for the MZ group is 3:20 (15 per cent) and that for the DZ group is 3:42 (7.5 per cent).'[62] In Kendler's 1987 review of the schizophrenia twin studies, he listed Tienari's study as producing a probandwise concordance rate of 7/21 = 33%.[63] A reader unfamiliar with the nuances of the pairwise-probandwise debate would not know that Tienari reported three out of 20 pairs concordant, not seven out of 21.

The NAS-NRC study

Hoffer and Pollin's 1970 study of the NAS-NRC veterans sample was based on the records of a large sample of twins (80 identical, 145 fraternal) and possessed the virtue of being based on subjects identified by a twin register, who were not matched against (potentially biased) hospital records. Hoffer and Pollin's pairwise concordance rates were given as identical =13.8%, fraternal = 4.1%, which are among the lowest rates ever reported. We will see, however, that these figures would not be allowed to stand.

In 1972 Allen and colleagues reanalyzed the data from the original sample, plus additional pairs from the register, for a total of 220 pairs.[64] The new pairwise concordance rates were listed as identical = 27%, fraternal = 4.8%. However, as Boyle noted, these figures were arrived at by widening the diagnostic criteria and rediagnosing each twin on the basis of the original medical charts. And while the armed forces doctors/ psychiatrists may have been aware of the twins' zygosity status, they had no research hypothesis to test when making their diagnoses. Allen et al., who *were* testing a hypothesis, evaluated each pair with the knowledge of their zygosity status. Acknowledging their genetic bias, the authors wrote that while analyzing the data, 'the reviewer may have been influenced toward seeing more similarities in MZ pairs than in DZ pairs. This factor would tend to increase the MZ concordance rate.'[65] In 1982 Gottesman and Shields converted the 27% pairwise figure to a probandwise rate of 43%. Thus, in the space of 12 years the NAS-NRC sample's identical twin concordance rate rose—on paper—from 13.8% identical (pairwise, uncorrected) to 43%.[66]

In 1983 Kendler and Robinette updated and reanalyzed the results of

Hoffer and Pollin's original sample. They reported probandwise schizophrenia concordance rates as identical = 30.9%, fraternal = 6.5%. Kendler and Robinette also created an artificially wide 'broad concordance' category to include any psychiatric disorder. Employing this enlarged definition, they found that 'The broad probandwise concordance for schizophrenia for monozygotic twins is 103 of 194 (53.1% ± 1.8%); for dizygotic twins it is 66 of 227 (23.8% ± 2.6%).'[67] What was mentioned only in passing is that, even though all of the twins had passed through the schizophrenia risk period (> 45 years-old), the pairwise identical twin concordance rate for schizophrenia was still only 30/164 = 18.3%.[68]

The Tienari, NAS-NRC, and Koskenvuo et al. studies comprise three of the ten 'modern' schizophrenia twin studies, yet partisans of the genetic position have artificially inflated the concordance rates of two of these studies, and have ignored or were unaware of the other. But the fact remains that in the three studies combined, 84% of identical twin pairs were judged *discordant* for schizophrenia.

Schizophrenia twin researchers' interpretations of concordance rate differences

Several prominent schizophrenia twin researchers have recognized that — contrary to the equal environment assumption — environmental factors do influence concordance rate differences. The following quotations are listed in chronological order, and I have added all emphases.

> *Rosenthal*
> The above findings do not rule out the possibility of other psychological hypotheses which might account for higher familial incidence, higher concordance in MZ than DZ twins, and no higher frequency of the illness in twins than in nontwins. *An hypothesis like 'identification' might account for the higher concordance rate of illness in MZ twins . . .* One could imagine a co-twin being drawn toward schizophrenic behavior if his twin had become schizophrenic and if he felt himself to be so much like his twin that he was completely convinced that the fate which had befallen his twin would befall him as well. He might be unable to resist this conviction, and presumably could behave in accordance with it.[69]

> Even though the bodies of data presented have suggestive value in an accounting of the sex-concordance ratios found in studies of schizophrenia, the role of genetic factors cannot be excluded. However, if the found sex-concordance ratios are valid, *it seems reasonable to conclude that some psychological factors are influencing these ratios in good part.*[70]

> *Essen-Möller*
> Quite obviously, then, the logical evidence furnished by the classical

twin method is not unambiguous, as originally believed. *A greater concordance in monozygotics must not invariably depend on their genetic identity*, since also their environment may have been more similar.[71]

Tienari
It is doubtful, moreover, whether the difference in concordance rates between identical and fraternal groups of twins can, as such, be ascribed to hereditary factors. In all likelihood, the environment, too, is more similar in the case of identical than in the case of fraternal twins . . . Furthermore, it is obvious that the intensity of the mutual relationship of identical twins is considerably greater than that of siblings in general and, also, of fraternal twins . . . *It is apparent that differences in concordance rates between groups of identical and fraternal twins, as well as between female and male pairs, are partly attributable to environmental (psychological) effects.*[72]

Inouye
The question is what will be the possible cause of concordance of neurosis in these twins. Theoretically there are several possible causes. First is an environmental influence identically shared by two members of a twin pair. This mechanism was seen in one female monozygotic pair in dissociative reaction type. In this pair the direct etiologies of the neurosis of both twins were diverse psychogenic factors and the common primitive cultural background, which coincidentally influenced the emotion of the twin subjects. The second possible cause of concordance is a susceptibility to neurosis in both twin members resulting from a particular situation to twins. This possibility was pointed out by not a few psychoanalysts . . . We pointed out the significance of mutual relationship between two members of each pair, and concluded that the particular situation is a probable cause of ego immaturity, which in turn is the probable cause of concordance of the neurosis or neurotic personality.[73]

Gottesman and Shields
Higher female concordances in two particular studies (Rosanoff and Slater) could be an artifact of sampling; or it could be associated with more environmental variability for males or with some aspect of the process of identification.[74]

Rosenthal considers that a psychological hypothesis such as identification might be used to explain differential concordance rates in MZ and DZ twins without implying a higher incidence of illness in MZ twins. *This would appear to be so* provided that the same proportion of potential schizophrenics are held back from overt illness by identifying with a normal twin as *those who became ill by identifying with an abnormal one.*[75]

Shields

It is established that the MZ co-twins of schizophrenics are at least twice as often and, in many types of samples, 4 or 5 times as often schizophrenic as DZ co-twins of the same sex. This difference will be accounted for by influences from two sources: by the effects of the greater genetic similarity of MZ twins, and by *greater similarity in environmental factors relevant to schizophrenia shared by MZ twins and not by DZ twins.*[76]

Kringlen

There is clearly brought out by the data a trend toward higher concordance rates for females, both in monozygotic and dizygotic same-sexed twins. . . . It is difficult to explain this tendency . . . With an increasing female emancipation this sex difference in upbringing and attitudes toward boys and girls has diminished. This could offer an explanation of the disappearance of the higher female concordance rates in more recent studies. This phenomenon might also be related to national culture. *One would then expect higher female concordance to be particularly pronounced in cultures where girls and women are most restricted in their activities, whereas the phenomenon would vanish in cultures where females enjoy equal rights with males.*[77]

The total difference in concordance rate between MZ and DZ twins cannot be ascribed to genetic factors only. A series of studies of both normal and abnormal twins show that the environment of the MZ twin pair is more similar than the environment of the DZ twin pair.[78]

Slater

Concordance rates are higher in pairs of female twins than in pairs of male twins. If this is confirmed, one might well conclude that the environmental contribution to causation could be responsible . . . Anything which tends to diminish environmental variance will tend to magnify the apparent contribution made by heredity, and this will show up in twin concordance rates just as much as in other measures.[79]

Hoffer and Pollin

An hypothesis like 'identification' might account for the higher concordance rate in monozygotic twins without implying a higher incidence of schizophrenia in monozygotic twins . . . Because there are not only genetic but also environmental differences between the monozygotic group and the dizygotic group, *differences in concordance rates may be explained by environmental as well as by genetic hypotheses.*[80]

Fischer

The assumption that a number of environmental factors are similar in MZ and DZ pairs may be correct only to some degree.[81]

The result that there is a significant difference between concordance for schizophrenia in same sexed DZ pairs and same sexed sibling pairs might be an expression of a real difference, but it might be kept in mind that a statistically significant difference only expresses the likelihood of an event, in this case the low likelihood that no difference exists.[82]

Koskenvuo et al.
Greater environmental similarity between MZ and DZ twin partners can also bias conclusions . . . The intrapair relationship seems to differ significantly in MZ and DZ twins. How much of this is due to different intrapair influences or parental treatment, and how much to different genetic similarity, is unknown.[83]

These quotations show that in addition to Rosenthal, who 'provided a blueprint for improving the state of the art of schizophrenia twin studies,'[84] *the authors or co-authors of at least 11 of the 15 schizophrenia twin studies have written that environmental factors influence concordance rate differences.* These observations stand counterposed to Kendler's assertion (see Chapter 3) that the behavioral similarity of identical versus fraternal twins is caused only by genetic factors.

Same-sex and opposite-sex fraternal twin concordance rates
As we have seen, the twin method compares identical twin pairs to *same-sex* fraternal twins. However, several schizophrenia twin researchers collected concordance rate data for opposite-sex fraternal pairs as well. These data provide another method of testing the equal environment assumption. Jackson posed the question quite well:

> Obviously same-sexed and different-sexed fraternal twins have the genotypical relationship of ordinary siblings. Therefore, because it is not claimed that schizophrenia is a sex-linked disorder, one would not expect a difference in concordance for schizophrenia on a hereditary basis. On the other hand, if the hypothesis is correct that identical twins are more concordant for schizophrenia because of their 'twinness,' one would expect a higher incidence of concordance for schizophrenia in same-sexed fraternal twins because they are more alike from the identity standpoint than different-sexed fraternal twins.[85]

The pooled figures for schizophrenia twin studies reporting concordance rates for both types of fraternal twins are: same-sex fraternal = 59/523 (11.3%), and opposite-sex fraternal = 20/422 (4.7%).[86] Thus, the same-sex rate is more than double the opposite-sex rate.[87] Proponents of the twin method have not satisfactorily explained the fact that the pooled schizophrenia concordance rate for same-sex fraternal twins is more than double the rate of opposite-sex pairs. Typically, the difference is either acknowledged, denied, or ignored; rarely is an attempt made to offer a genetic explanation for the difference. Psychiatrists Keefe and Harvey, the authors of *Understanding Schizophrenia*, wrote that environmental differences between identical and fraternal twins probably do not affect concordance rate differences because 'same-sex and opposite-sex fraternal twins have exactly the same concordance rate for schizophrenia.'[88] Behavior geneticists Plomin and colleagues, on the other hand, recognized that 'opposite-sex twins have a consistently lower concordance.'[89]

Like Keefe and Harvey, in their 1982 textbook Gottesman and Shields denied that there are any differences between same-sex and opposite-sex fraternals, claiming 'that opposite-sex twins, when studied, are no *less* concordant than same-sex fraternal twin-pairs.'[90] They later qualified this statement by writing that 'opposite-sex DZ pairs are as concordant as same-sex DZ pairs in recent studies.'[91] By referring to 'recent studies' they created the impression of a trend, while obscuring the fact that significant differences were found in the older investigations. In fact, the 'recent studies' they were referring to was actually one study (Kringlen's 1967 investigation).

In 1966 Gottesman and Shields had a much different analysis of the fraternal same-sex versus opposite-sex concordance rate difference, writing, 'One significant and consistent difference that emerged from our analyses was a lower concordance rate for opposite-sex fraternal pairs than same-sex fraternal pairs for studies giving information on this point.'[92] There was only one study reporting these concordance rates in the intervening years, and it therefore is puzzling how Gottesman and Shields could have come to such different conclusions in 1966 and 1982, on the basis of very similar numbers.

At the pivotal 1967 Dorado Beach schizophrenia conference, Shields observed,

> The pooled crude concordance rates in six studies of 430 DZ pairs of opposite sex is only 5–6%, which is about half the concordance rate usually reported in DZ pairs of the same sex, and the difference is consistent in all studies except Kringlen's.[93]

In 1967 Shields saw Kringlen's finding as the exception that proved the rule; it would only be in later years that this exception would become, in the minds of Gottesman and Shields, a 'trend.' Gottesman and Shields's 1982 claim is also striking when one considers that these investigators were the leading defenders of the studies of Kallmann and Slater, whose figures showed statistically significant concordance rate differences between these two types of fraternal twin pairs.[94]

In another report, Gottesman and Shields compared Kringlen's findings to those of Kallmann:

> Kringlen's overall pattern of results is impressively like that of Kallmann's: there were no significant sibling-DZ, male-female, or same-sex DZ/opposite-sex DZ differences in concordance rates, all

three 'nulls' leading to an emphasis on autosomal genetic factors and a de-emphasis on sex role and identification factors.[95]

In fact, Kallmann's 1946 same-sex fraternal twins *were* significantly more concordant than his opposite-sex fraternal pairs. Gottesman and Shields also overlooked the statistically significant differences between Slater's fraternal twin versus sibling, and same-sex fraternal versus opposite-sex fraternal comparisons, as well as the significant differences found in Fischer's 1973 fraternal twin versus sibling comparison.

In Gottesman's 1991 review of the schizophrenia twin study literature, his only mention of the fraternal twin concordance rate differences by sex was his selective discussion of Kringlen's study, which 'used males and females and studied opposite-sex as well as same-sex fraternal twins in extensive fieldwork, reporting no difference in these fraternal concordance rates with respect to schizophrenia.'[96] About the significant differences found in the studies of Slater and Kallmann, Gottesman said nothing.

The higher schizophrenia concordance rate among same-sex versus opposite-sex fraternals cannot be explained on genetic grounds, but is consistent with Jackson's position (see Chapter 2) that, without regard to genetic similarity, we should expect those pairs more alike from the identity standpoint to be more concordant for schizophrenia.

Selected case history excerpts of identical twins concordant for schizophrenia

The case histories of identical twins judged concordant for schizophrenia provide support to Don Jackson's thesis that the close association and social isolation of identical twins plays a major role in concordance. Selected passages from relevant case histories will be examined. Whether the pairs in question should have been considered concordant for schizophrenia is not our concern at the moment. What is important is the nature of the relationship between the twins. We will look at case histories recorded both before and after Jackson's 1960 critique. Most of these histories were written by people with little interest in psychodynamic processes, and it is therefore unlikely that they adequately captured the essence of the bond between the twins.

Rosanoff and associates' 1934 study provided a sketch of identical twins George and Foster, whom they counted as concordant for schizophrenia 'but in a manner quantitatively dissimilar.'[97] The twins were born in Ohio in 1906. George was committed to a mental hospital in June of 1930 and received a diagnosis of dementia praecox (schizophrenia). According to the case history, 'The twins had always been together and had always been devoted to each other. When George was taken to the hospital Foster began to worry greatly ... He lost all ambition to work, sat around at home "just thinking," [and] paid no attention to anyone who spoke to him.' Foster was finally convinced to seek voluntarily admission to a mental hospital 'for a couple of weeks' rest and change,' but he would not remain long:

> While he was there word came that George was better and could come home on a trial visit. *'The minute that Foster heard that, he snapped right out of it, brightened up, and the two boys came home the same day.'* He resumed his work and seemed as well as ever. At the time of our observation of the case he appeared quite normal mentally [emphasis added].[98]

This is a touching story of an identical twin who felt so close to his co-twin that he was willing to follow him into a mental hospital. Upon hearing that his twin brother had been released from the hospital, Foster's 'subacute schizophrenia' symptoms vanished almost immediately as he happily rejoined his twin brother in the family home. Perhaps Galton would have interpreted this story as showing that 'the clocks of their two lives move regularly at the same rate, governed by their internal mechanism.' But would the clocks have been set to run *backwards* at the same rate? Foster 'snapped out of it' because he found out that his twin brother was leaving the hospital. The 'mechanism' was the identical twin relationship, not heredity.

The next English-language schizophrenia twin study to provide case histories was Slater's 1953 report, which was the last to be published before the appearance of Jackson's critique. Case history excerpts from some of the most interesting pairs of identical twins concordant for schizophrenia are noted below:

> PAIR #20, FEMALES, BORN 1894 (Lily and Mary). Following disagreements at home Lily and Mary left in 1914 to live together, and have done so since. The first signs of their deafness came on about then. They rather shut themselves up together, with little outside contact. . . According to a neighbor, they were well known in the neighborhood for their likeness to one another . . . Both have paranoid illnesses coming on at the time of the menopause and within 2 years of one another.[99]

> PAIR #110, FEMALES, BORN 1872 (Marjorie and Cynthia). Until they were 40 they lived at home. . . Marjorie was the first to become peculiar . . . Cynthia became ill very suddenly, about two years after Marjorie had begun to be peculiar . . . Eventually they both barricaded the flat, which led to their certification and admission to [the mental hospital] on the same day in June, 1918, aged 46.[100]

> PAIR #198/199, FEMALES, BORN 1912 (Irene and Maureen). They were always much alike, short and thin with dark eyes. And dark straight hair. They were frequently mistaken for one another, and at one time one twin deputised for the other at the factory where she worked without anyone noticing. They always remained intimate and never troubled to make separate friends. Many 'telepathic' experiences are

reported; if one fell down at one spot the other would too, and hurt the same knee. Maureen is said to have had pains when Irene had her baby. . . . At the present time both twins are attending a psychiatric clinic where observation has shown that in both symptoms are much worse just before the periods. . . . The first remark [in an interview], on being introduced by their mother, was: 'Yes, we're the twins, and we're still ill. It's a mysterious illness, ours.' They seemed to share one illness between them.[101]

PAIR #287, MALES, BORN 1915 (Leslie and Reginald). In April 1940, [Leslie] was admitted to a general hospital. . . . Reggie attended the psychiatric clinic in 1940 with his twin, and he said that he himself had bad nerves, got into arguments, had few interests. . . . When Leslie was certified in 1940, Reggie was very upset and would say: 'I've murdered my brother, I've poisoned him'. . . . He wandered about at night picking at biscuits and would call his parents names. . . . In April 1941 he became violent, attacked his parents and was taken to the observation ward and to the mental hospital.[102]

Kringlen's 1967 study also provided case histories. The following excerpts describe the social isolation and mutual association of several concordant identical pairs. Twin pair members are referred to as 'Twin A' and 'Twin B':

CASE 1, MALES, BORN 1913. As children, they were mistaken for the other, even by their parents . . . They were brought up alike, dressed in the same way, and were much together as children . . . [Twin A] was transferred to a mental hospital [in 1934] . . . [Twin] B had a short period of 'nervousness' at age 19 when his brother became ill . . . Then he seemed to have been all right until the age of 21, when his brother was admitted to the hospital. He visited his brother often, and said to his family after some time that he felt he himself had to go to the hospital . . . Both twins were seen in 1965 in the same hospital where they have stayed. Both are living in the same room.[103]

CASE 2, MALES, BORN 1921. Mistaken for each other as children by the teacher. Extremely strong attachment to each other . . . Before the age of 25 they had been separated for only very short periods of time . . . [Twin A] admitted to a mental hospital in 1947 . . . B became psychotic in 1949.[104]

CASE 3, MALES, BORN 1916. They were always together as children, and made the same few friends. They were, however, living in an isolated area and consequently had slight contact with other children. If one was home from school because of illness, the other one also stayed home, even if he was not sick . . . They were not separated until they were 20 years old . . . [Twin A at age 30] was living with his twin brother on the isolated family farm up in the hills and they had practically no contact with other people . . . [Twin] A was placed in a mental hospital 1957–1958 . . . [Also in 1957, twin B] began being more overly disturbed, knocking at the walls, scolding into the air, apparently hallucinated. [Admitted to a mental hospital, 1959].[105]

CASE 4, MALES, BORN 1930. As children the twins were mistaken for one another by their teachers, and considered as alike as two drops of water . . . They were inseparable as children, and even more so as adults . . . They were never separated from one another . . . [Twin A was admitted to a mental hospital in 1949]. At admittance he was accompanied by his parents and his twin brother . . . [Twin] B was attending a technical school when his brother got sick. Some time afterwards he lost interest in the school work, started sleeping poorly and stayed up at night . . . [Twin B hospitalized in 1951].[106]

CASE 6, MALES, BORN 1908. As children, their siblings could not tell them apart — they were like two drops of water. They had only slight contact with the other children and clung together most of the time. Their attachment was extremely strong in childhood and even after puberty. They were never separated before the age of 17.[107]

CASE 9, FEMALES, BORN 1907. They always clung together, and were extremely dependent on one another also as adults. They seemed to have 'normal' contact with other children, but they never made separate companions among their friends . . . [Twin A] started getting ill when she was 39 years old . . . When [twin] A was hospitalized in 1952, [twin] B stayed home with marginal social functioning. She was longing intensely for her sister's company . . . [In 1959, after being placed in a sanitarium, twin B] was soon transferred to the same mental hospital in which her twin sister was (1959) and has since stayed there together with [twin] A . . . Both twins were seen in the hospital in 1965. They came to my office dressed in the same way, walking and behaving in much the same way — a remarkable couple. The nurses told me that the twins were together all day long.[108]

CASE 10, FEMALES, BORN 1930. As children, they were greatly dependent on each other and virtually inseparable . . . [Twin A's] first hospitalization period was from December 1953, to April 1954. . . . When [twin] A was hospitalized, [twin] B said: 'Why don't they take me, too' . . . [Twin] B became psychotic when she was 25 years old.[109]

CASE 12, FEMALES, BORN 1923. The twins grew up on a small, lonely farm . . . Neither twin went about with other girls or boys much . . . At the age of 22–3, twin A developed a catatonic-like remittent schizophrenia . . . [Twin] B began to be 'nervous' when [twin] A was admitted to the hospital in 1944 . . . Twin B fell ill shortly after [twin] A, with approximately the same symptoms.'[110]

Like Rosanoff's 'George and Foster,' several of Kringlen's twins were apparently so distraught over the hospitalization of their co-twin that they were willing to follow him or her into a mental hospital in order to continue the relationship. In Case 10, when Twin A was being taken to a mental hospital her identical twin sister cried out, 'why don't they take me, too.' A more compelling (though tragic) human drama could hardly be imagined.

The following excerpts are from the case histories of concordant identical twins studied by Gottesman and Shields, and reported in 1972:

> PAIR 1, MALES, BORN 1893. [Twins A and B both diagnosed with schizophrenia at age 30.]
> [Mother] idolized twins, gave them most similar names she could think of; overprotected them in that she did not allow them to mix with other children . . . The twins were inseparable until 18 years of age . . . Taken in isolation this pair should *not* give more comfort to genetic enthusiasts than to environmental ones with respect to weighting the etiological importance of heredity and intrafamilial factors . . . The influence of one twin upon the other could account for the similar *content* of their symptoms.[111]
>
> PAIR 7, FEMALES, BORN 1940. [Twin A diagnosed with schizophrenia at age 18, twin B diagnosed with schizophrenia at age 19.]
> Twins very close—'couldn't make a move without the other.' They had few friends . . . [The twins] were extremely dependent on their [mother] who encouraged their dependence . . . *Psychiatric history of B.* Behavior gradually became unpredictable and irrational after transfer to another library and [twin] *A* became ill (at 17); accused [mother] of causing twin's illness . . . later increasingly withdrawn . . . This East End London family parallels those described in the psychodynamic family study literature of Lidz and others.[112]
>
> PAIR 17, MALES, BORN 1934. [Both twins were diagnosed with schizophrenia at age 22. The twins were separated at birth but lived together for one year at age five.]
> The arrival, for the first time since his adoption, of the biological mother at *A*'s home early in October preceded his manifestly odd behavior by about 2 months. *B* was taken to visit his sick twin on December 22 in the belief that this might improve *A*'s condition. Instead it appears to have precipitated *B*'s illness, for the same evening he was disturbed and on January 5 was admitted to hospital, just before *A*.[113]
>
> PAIR 19, FEMALES, BORN 1926. [Twin A diagnosed with schizophrenia at age 31, twin B diagnosed with schizophrenia at age 35.]
> Both stubborn children; later, as adolescents, backward 'but nice';

regarded as moody, suspicious, 'keeping together' . . . Both remained single . . . [Twin *B* had no symptoms until she] joined 'wild scene' when *A* became disturbed (*onset*): felt persecuted, thought food poisoned, tried to jump through window; became equally disturbed as *A* and talked about attempts to murder *A*. Together admitted to observation ward (at 31) where wept and wailed continuously . . . Socialized only with twin, upset by any suggestion of separation, once responded immediately by breaking vase . . . It is reasonable to raise questions of *folie à deux* and culture shock in this pair. Clearly their twinship was responsible for their shared delusions and simultaneous onset of psychosis.[114]

The descriptions by Rosanoff, Slater, Kringlen, and Gottesman and Shields are consistent with Jackson's observation that social isolation, mutual association, and identity confusion are common themes among concordant identical twin pairs. Throughout these case histories we encounter observations such as, 'they rather shut themselves up together,' 'never troubled to make separate friends,' 'no contact with other people,' 'they seemed to share one illness between them,' 'were never separated from one another,' 'longing intensely for her sister's company,' 'did not like to mix too much with others,' 'always clung together,' 'inseparable,' 'couldn't make a move without the other,' and so on. To be sure, there have been pairs closely associated in isolation who were discordant for schizophrenia; my point is only that these conditions likely contribute to higher concordance rates among identical twins.

Cases of twins reared apart

There have been no systematic studies of reared-apart identical twins concordant for schizophrenia, but several individual case histories of ostensibly separated pairs have been reported. In Farber's 1981 review of these cases she concluded that, according to her 'lenient criteria,' nine identical pairs warranted consideration as legitimately separated twins.[115] However, in all of these cases (six pairs were considered concordant by Farber) the twins were aware of each other's existence and had periodic contact. The frequently cited 1945 report by Craike and Slater illustrates the arguable claim of 'separation' common to most case histories (as discussed in Chapter 4). Farber considered this pair to be 'the best-separated set in the literature.'[116] 'Edith' and 'Florence' were British identical twins separated nine months after birth. Florence was adopted by a maternal aunt, while Edith stayed with her father until the age of eight, when she was placed in a children's home. Although they did not meet again until age 24, each was aware of the other's existence. Edith reported that while living with her father, 'Florence was making trouble for her by writing to her father and telling him Edith had told her that he was a drunkard.'[117] Edith also believed that Florence was watching her house and had been plotting against her. Their supposed delusional systems, which contributed to their schizophrenia diagnosis, centered on mutual distrust:

Each twin occupies for the other an over-valued position: *to each the other is supremely important,* although the circumstances of their lives touch at few points. Edith at first sight places Florence at the center of her persecutors; Florence, with her own inborn tendency to paranoia, reacts to this by coming in turn to regard Edith as her chief enemy [emphasis added].[118]

Although Craike and Slater claimed that the twins 'were brought up along entirely different lines,'[119] it is clear that this 'best-separated set' had, in the words of Craike and Slater themselves, centered their 'delusions around the other.'[120] It has been argued further that neither twin's symptoms warranted a diagnosis of schizophrenia.[121]

As seen in previous chapters, the investigators' bias often influences the way they write about (and diagnose) twins. R. D. Laing noted Craike and Slater's 'linguistic tricks used to establish concordances . . . in respect of [Edith and Florence's] "childhood neurosis." '[122] While Craike and Slater reported that 'Both were nervous as children; Edith walked in her sleep and had nightmares, Florence had fainting attacks and was nervous of the dark,'[123] Laing pointed out that they could just as easily have written, 'Edith walked in her sleep, Florence did not. Edith had nightmares, Florence did not. Florence had fainting attacks, Edith did not.'[124]

In closing, one must agree with the observations of many schizophrenia twin researchers that the case histories of 'separated' identical twins concordant for schizophrenia represent little more than 'fascinating curiosities.'[125] In 1982 Gottesman concluded, 'After a quarter century of experience with twins reared together and twins reared apart, it is my conviction that twins reared apart are a wonderful source of hypothesis generation, but not a useful source for hypothesis testing.'[126]

Studies of the offspring of identical twins discordant for schizophrenia
Fischer and the study of the offspring of discordant identical twins
Another method compares schizophrenia rates among the biological offspring of discordant identical twin pairs (that is, one member of the pair is diagnosed with schizophrenia, and the other is not). Several recent reviews and textbooks have pointed to these studies as strongly supporting the genetic position. Because identical twins share the same genes, genetic theory predicts that the schizophrenia rate among the offspring of identical twins diagnosed with schizophrenia would be the same as that among the offspring of their *non-schizophrenic* identical co-twins. Margit Fischer introduced this method into twin research in 1971. Her Danish sample included three groups:

GROUP 1 — The offspring of identical twins diagnosed with schizophrenia or schizophreniform psychosis. Age-corrected rounded-off N = 32.

GROUP 2 — The offspring of a normal identical co-twin of a person diagnosed with schizophrenia. Age-corrected rounded-off N = 24.

GROUP 3 — The offspring of a normal *fraternal* co-twin of a person diagnosed with schizophrenia. Age-corrected rounded-off N = 9.

Group 3 offspring were used as a control group to test the supposed environmental expectation that there would be no difference between their rate of schizophrenia and that of the Group 2 offspring.

Fischer reported that three cases (9.4%) of schizophrenia and 'schizophrenia-like psychosis' were found among Group 1 children, three cases (12.3%) were found among Group 2 children, and zero cases (0%) were found among Group 3 children.[127] She concluded that while these results 'could probably best be explained'[128] by genetic factors, 'the present material is of such limited size that no conclusion can be drawn.'[129] In the 1973 final report on her entire twin sample, Fischer reported similar numbers but failed to add the caveat that no conclusions could be drawn from the data.

Before looking more closely at Fischer's results, two points should be emphasized. (1) In neither Fischer's study nor the 1989 follow-up by Gottesman and Bertelsen is it stated that offspring diagnoses were made blindly. This means that a bias was introduced into the diagnostic process because these researchers were, to varying degrees, partisans of the genetic cause. One could invalidate the findings of these studies on the sole basis of non-blind diagnoses. (2) It is doubtful that Fischer's strategy adequately separates genetic and environmental influences. The model assumes that a purely environmental etiology would show that (a) Group 2 children would develop schizophrenia at a lower rate than Group 1 children, and that (b) Group 2 children would develop schizophrenia near the population rate of 1%. However, in Denmark, where this study was carried out, schizophrenia was widely viewed as a hereditary disease during the period when these offspring were growing up (see Chapter 7). The child *or niece/nephew* of a person diagnosed with schizophrenia was therefore seen as the carrier of 'tainted blood,' and was likely considered at risk for schizophrenia on that basis alone. In Denmark, a pregnant woman diagnosed with schizophrenia was encouraged to abort her pregnancy for eugenic purposes, and a child she could not keep was often kept out of the adoption process by social service agencies.[130] This restriction was based on the belief that the relatives of people diagnosed with schizophrenia were carriers of the disorder, and is therefore a good indication of the prevailing sentiments in Denmark. Thus, the biological offspring of normal identical co-twins of people diagnosed with schizophrenia would have been suspected of being schizophrenia 'taint carriers,' and would have spent much of their lives with this family history hanging over their heads.

In addition, co-twins were raised in the same family environment and likely received similar treatment. It is therefore probable that many of the *environmental* factors involved in the schizophrenia-diagnosed twin's breakdown were also experienced by their 'well' co-twin, who could have manifested deviant rearing patterns toward their children. Fischer recognized this possibility but dismissed it as seeming 'a little far-fetched.'[131] It is also unclear how much contact Group 2 children had with their schizophrenia-diagnosed aunts and uncles. When one considers that identical twins spend a considerable amount of time together, it is possible that many of these children had frequent contact with these aunts and uncles. We will never know to what extent such influences existed, in part because practically none of these offspring were investigated personally by Fischer. According to Gottesman and Bertelsen, 'For both twin types the offspring of the normal co-twins share little to none of the within-family exposure to a schizophrenic parent.'[132] However, it has been shown that deviant rearing patterns of 'normal' parents are an important predictor of schizophrenia in their children.[133] Therefore, 'exposure to a schizophrenic parent' is not a necessary component of a schizophrenia-producing environment. In fact, according to Gottesman about 89% of people diagnosed with schizophrenia were reared by two non-diagnosed parents.[134]

A closer look at the cases presented in Fischer's 1971 Table 6 reveals that there were only three diagnoses of definite schizophrenia among the offspring (Group 1 = 2 cases; Group 2 = 1 case). The other diagnoses included 'acute psychotic episode; schizoid personality disorder,' '?paranoid schizophrenia/reactive psychosis,' and 'psychosis of an uncertain type.'[135] Additionally, a Group 1 definite schizophrenia diagnosis was made on the daughter of a *concordant* identical twin.[136] Fischer had included several children of concordant pairs in Group 1 in order to increase her sample size.

Fischer's small sample and handful of offspring schizophrenia cases meant that, while Group 2 children were not significantly different from Group 1 children, *neither group had schizophrenia rates significantly higher than population expectations*. For example, the Group 1 schizophrenia rate of 3/32, when compared to a population expectation of 0/32, is not statistically significant.[137] In fact, the small sample meant that Group 2 offspring had *no possibility* of being diagnosed with schizophrenia at a significantly lower rate than Group 1 offspring. Therefore, it can hardly be concluded that a non-significant difference between these two groups supports the genetic hypothesis. Fischer cautioned, quite correctly, that no conclusions could be drawn from the data, 'but that the question raised might be evaluated in other similar studies.'[138]

The Gottesman and Bertelsen follow-up
The inconclusive results from Fischer's study would be of little value apart from historical interest, were it not for an 18-year follow-up of Fischer's sample by Gottesman and Bertelsen in 1989.[139] According to Gottesman and Bertelsen,

> If the observed risk to the offspring of schizophrenic identical twins were like that shown for offspring of probands generally and if the risk to the offspring of their genetically identical but clinically normal nonschizophrenic co-twins were close to the population base rate of 1%, it would indicate that nongenetic within-family factors have the power to produce schizophrenia . . . If, on the other hand, the risks to

> the two classes of MZ offspring were both high and equal to each other,
> it would undermine any significant etiologic role for the unshared
> factor of a schizophrenic parent . . . The corresponding risks for the
> offspring of same-sex fraternal twins concordant and discordant for
> schizophrenia will serve as control values for first- and second-degree
> relatives of schizophrenics.[140]

The rates of 'schizophrenia and schizophrenia-like psychosis' and their
accompanying age-corrected 'morbid risks' (MR) were as follows (for
reasons of clarity, these groups reflect Fischer's designations): *Group 1 —*
6/47, MR = 16.8 ± 6.6; *Group 2 — 4/24*, MR = 17.4 ± 7.7; *Group 3 — 1/52*,
MR = 2.1 ± 2.1. Based on the non-significant difference between Group 1
and Group 2 offspring, in addition to finding a significantly higher Group
2 versus Group 3 rate, Gottesman and Bertelsen concluded that their
findings 'support a strong role for genetic factors, still unspecified, in the
etiology of schizophrenia.'[141] However, as with Fischer's report, the small
sample meant that the rates for Groups 2 and 3 did not significantly exceed
population expectations.

A Group 2 diagnosis (chronic undifferentiated schizophrenia) was
given to the daughter of twin 405b ('Karl'), who was the non-schizophrenic
identical co-twin of schizophrenia-diagnosed twin 405a. The case of the
schizophrenia-diagnosed identical twin 405a is quite interesting.
According to Fischer's original description,[142] 'Morten' had been an
'unremarkable' and 'quite ordinary' 23-year-old until the day that he fell
from a height of 13–16 feet onto a stone cornice and lay unconscious for
five minutes, suffering 'severe cranial trauma.'[143] Morten grew up in an
early twentieth century 'very poverty-stricken' home in rural Denmark,
and it is unlikely that he received adequate medical attention for his cranial
trauma. Although he recovered from the injury, 'the family dated his
psychic change to this time.'[144] Diagnosed with schizophrenia, he spent
much of the rest of his life in mental hospitals. Fischer was apparently
unwilling to exclude this person on the grounds that he may have suffered
from organic brain damage caused by the fall. However, this is a plausible
explanation for his subsequent 'delusional' behavior. The doubt
surrounding Morten's diagnosis should have eliminated pair 405 from
Fischer's sample. This is another example supporting Rosenthal's 1962
observation that in order to avoid small sample sizes, twin researchers
tend to retain pairs for whom they have inadequate information.

Gottesman and Bertelsen briefly retold Morton's story and commented
that 'the cranial injury appears to have been a sufficient but not a necessary
releaser for the predisposition to schizophrenia in proband MZ 405A;
the nonschizophrenic co-twin never encountered a releaser for his
diathesis but transmitted it to his daughter.' The authors continued,
'Without this type of pedigree, the schizophrenia in the proband might
have passed as a nongenetic phenocopy of schizophrenia caused by
cranial insult.'[145] We have seen that a criticism of twin study diagnostic
procedure has been that a person has a greater chance of being diagnosed

with schizophrenia by a doctor or by a researcher on the basis of having a family history of schizophrenia, and this problem becomes apparent in Gottesman and Bertelsen's argument. It is truly begging the question to offer, as evidence in favor of the genetic hypothesis, a diagnosis made on the *assumption* that schizophrenia is genetic.[146] By removing this pair and 405b's offspring, Gottesman and Bertelsen's claim of a significantly higher Group 2 versus Group 3 rate, to the extent that it means anything, becomes statistically non-significant.[147]

Biological psychiatrist E. Fuller Torrey found serious problems with this study. He considered the case material handed down by Fischer to be 'markedly unsatisfactory,'[148] and found that none of the identical twins on whom Gottesman and Bertelsen based their findings 'met the basic criteria for twins with clearly diagnosed schizophrenia in the index twin and verifiable (i.e., based on interview) normality in the co-twin.'[149] Torrey also noted that the 'schizophrenic' twins from Fischer's sample reproduced at nearly the same rate as their 'well' co-twins—which is inconsistent with the finding that people diagnosed with schizophrenia have a low reproductive rate. Gottesman and Bertelsen disputed Torrey's claim that reproduction rates were similar,[150] but their 1989 data showed clearly that both 'schizophrenic' and 'normal' co-twins averaged roughly four offspring each. This calls into question the validity of the original hospital schizophrenia diagnoses. Torrey noted further that Gottesman and Bertelsen's decisions on which twins were retained or dropped from the study were 'entirely arbitrary,' and that their conclusion in favor of genetic factors 'is premature at best.'[151]

The results of Gottesman and Bertelsen's frequently cited 18-year follow-up, therefore, are as inconclusive as Fischer's original study.

Kringlen and Cramer

In another study using the Fischer design, but making blind diagnoses on the basis of structured interviews, Kringlen and Cramer found a higher but statistically non-significant Group 1 versus Group 2 schizophrenia rate difference.[152] The authors properly drew no important conclusions from these results, and their data further demonstrated the problems that small samples create for this design. Even if none of Kringlen and Cramer's 45 Group 2 offspring had received a schizophrenia diagnosis, the difference between the groups would have remained statistically non-significant.[153] Clearly, this result would not support the genetic hypothesis on the basis of a non-significant difference between Groups 1 and 2.[154]

The studies of Fischer, Gottesman and Bertelsen, and Kringlen and Cramer provide, at best, inconclusive evidence in support of genetic influences on schizophrenia. The design itself appears unable to adequately separate genetic and environmental influences, and small samples make it likely that no differences will be found between the offspring of discordant identical pairs. In order to guard against bias, future studies using this design must ensure that diagnoses are made by blinded raters. As Rosenthal noted, 'it is essential that the diagnostician not know whether the individual examined is related to an index or control proband . . . because it is easy to be swayed by knowledge regarding index or control status.'[155]

Interestingly, in 1959 Rosenthal studied the biological relatives of 13 discordant identical pairs (including at least 24 siblings) and concluded, 'There is almost a total absence of schizophrenia in the families of the discordant twin pairs.'[156]

Conclusion

Genetic influences on schizophrenia cannot be established by the results of family studies, twin studies, published studies of the offspring of discordant identical pairs, or individual cases of 'reared-apart' identical twins. Taken together, these methods point merely to the *possibility* that genes influence schizophrenia, and nothing more. We saw in Chapter 2 that adoption studies have superseded twin studies as the most frequently cited evidence in favor of the genetic basis of schizophrenia. Let us now turn our attention to a detailed review of these studies.

Notes

[1] Szasz, 1976, p. 3.
[2] Rosenthal, 1968, p. 414.
[3] Quoted in Kringlen, 1967b, p. 58.
[4] Marshall, 1990.
[5] Rosanoff & Orr, 1911, p. 228.
[6] Davenport, 1920, p. 305.
[7] Smythies, quoted in Jackson, 1960, p. 38.
[8] Rosenthal, 1970b, pp. 131–2.
[9] Bailey & Pillard, 1993, p. 241.
[10] APA, 1987, p. 192.
[11] APA, 1994, p. 283.
[12] Breggin, 1991. See also Boyle, 1990, 1999.
[13] Laing, 1981, p. 97.
[14] Szasz, 1976, p. xiv.
[15] Ibid., p. 12.
[16] Boyle, 1990; Breggin, 1991; Cohen & Cohen, 1986; Gosden, 2001; Hill, 1983; Laing, 1967; Sarbin & Mancuso, 1980; Szasz, 1961, 1970, 1976. Also see Sarbin, 1990.
[17] See Gosden, 2001 for a recent discussion on this point.
[18] Sarbin & Mancuso, 1980.
[19] Boyle, 1999, p. 79.
[20] APA, 1994. Also see Slater et al., 1971.
[21] Andreasen, 2000, p. 109.
[22] Torrey & Yolken, 2000, p. 114.
[23] Rosenthal, 1970b; Slater & Cowie, 1971. Gottesman (1991, p. 75) has accepted a rate of 1.0, while the DSM-IV (APA, 1994, p. 282) puts the figure at between .5 and 1%.

[24] Torrey, 1995.

[25] Torrey, 1980.

[26] Gottesman, 1991; Rosenthal, 1970b; Slater & Cowie, 1971.

[27] Arieti, 1974, p. 10. Commenting on Arieti's description, Thomas Szasz (1976, p. 10) wrote,

> The point I want to emphasize here is that each of these terms refers to a behavior, not disease; to disapproved conduct, not histopathological change; hence, they may loosely be called 'conditions,' but they are not, strictly speaking, medical conditions. If none of these three items is a disease, putting them together still does not add up to a disease.

[28] Boyle, 1990.

[29] E. Bleuler, quoted in Boyle, 1990, p. 118.

[30] E. Bleuler, 1950, p. 337.

[31] Boyle, 1990, p. 118.

[32] Quoted in Rosenthal, 1978, p. 477.

[33] E. Bleuler, 1950, p. 473.

[34] See Boyle, 1990, pp. 65–71.

[35] Gottesman, 1991, p. 103.

[36] Falconer, 1965, p. 53.

[37] Rosenthal, 1963, p. 507.

[38] Myerson, 1932, p. 491. Although Myerson was known as one of his era's foremost critics of genetic determination, he, like those he criticized, supported compulsory eugenic sterilization. In the final chapter of *The Inheritance of Mental Diseases*, Myerson claimed society's 'right to ask for sterilization of those types of feeblemindedness which we know to run in families,' and called for the sterilization of second-generation institutionalized patients who came from families with a history of 'mental disease' (Myerson, 1925, p. 320). In 1942 Myerson claimed that people with serious mental disorders 'have more intelligence than they can handle,' and that 'the reduction of intelligence is an important factor in the curative process' (quoted in Whitaker, 2002, p. 73).

[39] Myerson, 1932, p. 491.

[40] DeGrandpre, 1999.

[41] Tienari, Sorri, et al., 1987, p. 483.

[42] Lewontin, 1991.

[43] Strömgren, 1993, p. 11.

[44] Gottesman, 1991, p. 83.

[45] Rosenthal & Kety, 1968.

[46] Rüdin, 1916.

[47] For family studies finding no significant clustering of schizophrenia among first-degree biological relatives, see Abrams & Taylor, 1983; Coryell & Zimmerman, 1988; Pope et al., 1982.

[48] Rosenthal, 1970b, p. 37.

[49] Torrey et al., 1994, p. 10.

[50] Boyle, 1990, p. 131.

[51] Ibid.

[52] The review mentioning the Koskenvuo et al. study was by Murray & Reveley, 1986.

[53] Because I only recently became aware of the Koskenvuo et al. 1984 study, it is not mentioned in any of my writings published through 2001.

[54] Boyle, 1990, p. 134.

[55] Tienari, 1968, 1971, 1975.

[56] Tienari, 1975, p. 35.

[57] Gottesman & Shields, 1982, p. 103.

[58] See Gottesman & Shields, 1982, pp. 114–15. Also see Joseph, 2001b.

[59] Gottesman & Shields, 1982, p. 104.

[60] Boyle, 1990, p. 134.

[61] Gottesman, 1991, p. 111.

[62] Tienari, 1975, p. 33.

[63] Kendler, 1987, p. 707.

[64] Allen et al., 1972.

[65] Ibid., p. 944.

[66] Gottesman & Shields, 1982.

[67] Kendler & Robinette, 1983, p. 1554.

[68] Ibid., p. 1553.

[69] Rosenthal, 1960, p. 303.

[70] Rosenthal, 1962a, p. 419.

[71] Essen-Möller, 1963, p. 69.

[72] Tienari, 1963, pp. 119–21.

[73] Inouye, 1965, p. 1172.

[74] Gottesman & Shields, 1966b, p. 815.

[75] Gottesman & Shields, 1966a, p. 55.

[76] Shields, 1968, p. 100.

[77] Kringlen, 1967a, pp. 91–2.

[78] Kringlen, 1976, p. 431.

[79] Slater, 1968, pp. 23–4.

[80] Hoffer & Pollin, 1970, p. 476.

[81] Fischer, 1973, p. 10.

[82] Ibid., p. 69.

[83] Koskenvuo et al., 1984, p. 322.

[84] Gottesman, 1991, p. 109.

[85] Jackson, 1960, pp. 64–5.

[86] Joseph, 1999b.

[87] This finding is consistent with the theory that twin identification is largely responsible for the differing rates. In a study of 90 pairs of six-year-old twins from the Chicago area (including 36 same-sex fraternal and 19 opposite-sex fraternal pairs), Koch (1966, p. 234) asked individual twins to answer yes or no to the question of whether they desired to be like their sibling. Among the same-sex twins, 57% answered 'yes' whereas only 16% of the opposite-sex twins answered this way. Although these twins were quite young, Koch's study went straight to the question of how much each twin identified with his or her co-twin.

[88] Keefe & Harvey, 1994, pp. 82–3.

[89] Plomin, DeFries et al., 1990, p. 339.

[90] Gottesman & Shields, 1982, p. 114. Emphasis in original.

[91] Ibid., p. 243.

[92] Gottesman & Shields, 1966a, p. 76.

[93] Shields, 1968, p. 98.

[94] In Rosenthal's 1962a paper the statistically significant difference between Kallmann's 1946 fraternal same-sex and opposite-sex pairs was presented in a table, to emphasize the point. Gottesman and Shields (1966a) also presented Kallmann's figures in a table, without calculating a probability figure for the

difference. The figures from Kallmann's 1946 schizophrenia twin study are: same-sex fraternals = 34/296 (11%), vs. opposite-sex fraternals = 13/221 (6%), p = .019, Fisher's Exact Test, one-tailed. The figures from Slater's 1953 study are: same-sex fraternals = 11/61 (18%), vs. opposite-sex fraternals = 2/54 (4%), p = .014, Fisher's Exact Test, one-tailed.

[95] Gottesman & Shields, 1976a, p. 373.

[96] Gottesman, 1991, p. 111.

[97] Rosanoff, Handy, Plesset, & Brush, 1934, pp. 257–8.

[98] Ibid., 1934, p. 258.

[99] Slater, 1953, pp. 127–9.

[100] Ibid., p. 150.

[101] Ibid., pp. 166–8.

[102] Ibid., pp. 175–6.

[103] Kringlen, 1967b, pp. 4–7.

[104] Ibid., pp. 10–11.

[105] Ibid., pp. 15–19.

[106] Ibid., pp. 22–5.

[107] Ibid., p. 34.

[108] Ibid., pp. 48–53.

[109] Ibid., pp. 56–8.

[110] Ibid., pp. 65–9.

[111] Gottesman & Shields, 1972, pp. 72–5.

[112] Ibid., pp. 88–91.

[113] Ibid., pp. 114–18.

[114] Ibid., pp. 122–4.

[115] Farber, 1981, p. 165.

[116] Ibid., p. 156.

[117] Craike & Slater, 1945, pp. 214–15.

[118] Ibid., p. 220.

[119] Ibid., p. 219.

[120] Ibid., p. 221.

[121] See Pam, 1995; Ratner, 1982.

[122] Laing, 1981, p. 124.

[123] Craike & Slater, 1945, p. 219.

[124] Laing, 1981, p. 124.

[125] Gottesman, 1991, p. 121.

[126] Gottesman, 1982, p. 351.

[127] Fischer, 1971, p. 45.

[128] Ibid., p. 46.

[129] Ibid., p. 51.

[130] Kety et al., 1994.

[131] Fischer, 1973, p. 65.

[132] Gottesman & Bertelsen, 1989, p. 868.

[133] See Wynne et al., 1976.

[134] Gottesman, 1991.

[135] Fischer, 1971, pp. 47–9.

[136] See case #18; see Fischer, 1973, p. 148.

[137] p = .12, Fisher's Exact Test, one-tailed.

[138] Fischer, 1971, p. 50.

[139] Fischer died in 1983.

[140] Gottesman & Bertelsen, 1989, p. 868.

[141] Ibid., p. 871.

[142] Fischer, 1973, pp. 124–5.

[143] Ibid., p. 46.

[144] Ibid., p. 124.

[145] Gottesman & Bertelsen, 1989, p. 871.

[146] It might be added that on the basis of the case history material, the schizophrenia diagnosis of 405b's daughter is by no means clear-cut, which is reflected by the 'provisional diagnosis' given by Gottesman & Bertelsen.

[147] According to Fischer (1973, p. 124), Group 2 twin 405b (Karl) had three offspring. If we remove these offspring from the figures shown in Gottesman & Bertelsen's Table 5 (1989, p. 869), the comparison becomes, Group 2: 3/21 vs. Group 3: 1/52, $p = .069$, Fisher's Exact Test, one-tailed.

[148] Torrey, 1990, p. 976.

[149] Ibid., p. 977.

[150] Gottesman & Bertelsen, 1990.

[151] Torrey, 1990, p. 977.

[152] Kringlen & Cramer, 1989. Group 1: $3/28 = 10.7\%$ vs. Group 2: $1/45 = 2.2\%$, $p = > .05$.

[153] 3/28 vs. 0/45, $p = .053$, Fisher's Exact Test, one-tailed.

[154] Rosenthal, 1975b, p. 20.

[155] Rosenthal, 1959, p. 9.

CHAPTER SEVEN

The Genetics of Schizophrenia II: Adoption studies

Someday in the twenty-first century, after the human genome and the human brain have been mapped, someone may need to organize a reverse Marshall plan so that the Europeans can save American science by helping us figure out who really has schizophrenia or what schizophrenia really is.

Biological psychiatrist Nancy Andreasen in 1998.[1]

Once the presence of a mental disease has been established, the specific diagnosis of schizophrenia offers further difficulties. Only a few isolated psychotic symptoms can be utilized in recognizing the disease, and these, too, have a very high diagnostic threshold value. Manic and depressive moods may occur in all psychoses; flight of ideas, inhibition and — as far as they have not assumed specific characteristics — hallucinations and delusions, are partial phenomena of the most varied diseases. Their presence is often helpful in making a diagnosis of a psychosis, but not in diagnosing the presence of schizophrenia.

Eugen Bleuler, originator of the schizophrenia concept, in 1911.[2]

It should be apparent now that if we had included in our comparisons of index and control relatives only those who clearly had process [chronic] schizophrenia, we would have found no difference between the two groups of relatives.

Danish-American schizophrenia adoption researcher David Rosenthal in 1972.[3]

If schizophrenia is a myth, it is a myth with a strong genetic component!
Schizophrenia researcher Seymour Kety in 1974.[4]

Adoption studies have played a crucial role in establishing schizophrenia as a genetic disorder. Moreover, the early studies performed in Oregon and Denmark helped pave the way for the acceptance of an important role for genetic factors in shaping human psychological and behavioral trait differences in general. Behavior geneticists Rowe and Jacobson recently summarized the impact of the Oregon and Danish-American studies, writing, 'When a single theory [environmentalism] is monolithic in a field, contrary findings can break paradigms . . . It is just this role, we believe, that the first adoption studies of schizophrenia played in the 1960s.' They added that these studies 'stirred the Zeitgeist and gave impetus to the formation of the Behavior Genetics Association at the Institute for Behavior Genetics in Boulder, Colorado.'[5] Thus, the results from the American and Danish investigations not only strengthened the genetic theory of schizophrenia, but helped establish a field (behavior genetics) whose theories are currently gaining widespread acceptance. They also helped to revive psychiatric genetics, which was still reeling from its founders' collaboration with German National Socialism.

Until the 1960s, adoption studies had been used primarily to assess genetic influences on IQ. We have seen that Galton discussed studying adoptees as early as 1865, and the issue was taken up again in 1913 by L. F. Richardson in *Eugenics Review*. Richardson proposed a study to investigate whether the 'offspring of poor parents, adopted when newly born into well-to-do and well-educated families, turn out markedly different from the birthright members of those families . . .' Richardson continued,

> A thorough study of a hundred such cases of adopted children would do more to reveal the nature of the poorer classes than statistics of 100,000 poor persons brought up in poverty. The author hopes that a co-operative study of this kind can be arranged among the adherents of the Eugenics Education Society.[6]

I will now describe the three main adoption designs used in schizophrenia genetic research.

The Adoptees' Family method

The Adoptees' Family method begins with children given up for adoption who are later diagnosed with schizophrenia or a 'schizophrenia spectrum disorder.' A control group of non-diagnosed adoptees is also established. The investigators then attempt to identify and diagnose the biological and adoptive relatives in each group. A statistically significant difference between index and control biological relatives is seen as evidence in favor of the genetic hypothesis, although we will see that there are questions about whether this was the original design of the Danish-American Adoptees' Family study.

The Danish-American Adoptees' Family study results were published by Seymour Kety and colleagues in 1968 and 1975. These studies were limited to the population of the greater Copenhagen area. The investigation was then extended to the rest of Denmark. These 'Provincial study' final results were published in 1994. The 1968 study looked at the biological and adoptive relatives of 34 adopted-away children who were diagnosed with a schizophrenia spectrum disorder, and 33 matched controls. All diagnoses were based on a blind evaluation of the subjects' institutional records. The 1975 report utilized the same group of index and control adoptees (plus one additional control), but many of the relatives had now been interviewed and this information was supplied to the raters as part of the diagnostic process.

The Adoptees method

The Adoptees method determines the rate of schizophrenia among the adopted-away offspring of parents (usually mothers) diagnosed with schizophrenia. These adoptees are compared to a control group consisting of the adopted-away offspring of biological parents who had not received a diagnosis of schizophrenia.

The first published schizophrenia adoption study (by Heston) utilized the Adoptees method. The first publication from the Danish sample, by David Rosenthal and colleagues, appeared in 1968. The Rosenthal et al. 1971 final paper reported a statistically significant difference between the index and control adoptee groups.

A third locus of investigation has been Finland. The research team is headed by Pekka Tienari, who holds the unique distinction of having performed both a schizophrenia twin and adoption study. Tienari and colleagues are distinguished from Heston and the Danish-American group by virtue of their interest in family interaction effects as well as genetics. While presenting evidence which they believe supports the genetic hypothesis, Tienari and associates have noted that a high level of adoptive family disturbance also predicts which of their adoptees will be diagnosed with schizophrenia.

The Crossfostering method

The Crossfostering method looks at the adopted-away children of non-schizophrenic biological parents, who are raised by an adoptive parent eventually diagnosed with schizophrenia. The Danish-American Crossfostering study was published by Wender and colleagues in 1974. According to the investigators, the purpose of their study was to 'explore the hypothesis that rearing by or with schizophrenic parents will produce schizophrenic psychopathology among persons who carry a normal genetic load.'[7] This 'crossfostered' group was compared with the adopted-away children of normal biological parents (who were reared by normal adoptive parents), and with a group of adopted-away offspring of schizophrenic biological parents reared by normal adoptive parents.

I will not review Wender's Crossfostering study in this chapter. Its methodological problems are numerous and are similar to those I will discuss in detail as they relate to the Kety et al. and Rosenthal et al. studies. Wender and colleagues themselves acknowledged that the results of the Crossfostering study were 'modest,'[8] and critics such as Lidz, Lewontin et al. and Boyle documented the glaring problems with the study.[9] These include (1) the use of global mental health ratings in place of diagnosing schizophrenia, (2) the use of post hoc comparisons which were used to support the genetic position, (3) the failure to find significant differences between important comparison groups, (4) the investigators' failure to consider alternative explanations for their results, and (5) that the mean age of the crossfostered adoptees at the time of their parent's diagnosis was 12 years old.[10]

By the 1980s, Wender would admit that 'The study was particularly difficult' because,

> adoption agencies do their best to place their infant charges with normal parents; only an unusual kind of schizophrenic — whom the adoption agency cannot detect — would receive a child for adoption. Such schizophrenic parents might become ill only later in life, long after having adopted a child, or they might be able to conceal their illness from prying adoption-agency eyes. So the group of adopting schizophrenic parents is different from the group of schizophrenic parents who have their own children. Thus, the question of what would happen if children born of normal parents were placed in the homes of typical schizophrenics *cannot be answered* [emphasis added].[11]

Wender's belated acknowledgment that his investigation was unable to test the investigators' main hypothesis is reason enough to refrain from burdening the reader with a detailed analysis of the study.

This chapter focuses on two broad themes relating to the schizophrenia adoption studies: (1) That genetic inferences from these studies' results are confounded by environmental factors; and (2) that to varying degrees, most are methodologically unsound and were subject to the investigators' genetic biases.

The French team of Cassou, Schiff, and Stewart, in their 1980 critique, repeatedly referred to adoption studies as 'studies of abandoned children' ('les études d'enfants abandonnés'). As Alvin Pam noted, this phrase indicates 'the irony of using throwaway kids as proof that schizophrenia is genetically transmitted.'[12] Cassou and associates posted a reminder that the decisive 'evidence' in favor of the genetic theory of schizophrenia is based the families of 150 or so abandoned Danish children.

Like the twin method, adoption studies are based on a critical theoretical assumption: that factors relating to the adoption process (including the policies of adoption agencies) did not lead to the placement of experimental (index) adoptees into environments contributing to a higher rate of the condition or trait in question. Adoption researchers assume that selective placement did not occur in their sample, meaning that children were not placed into homes correlated with the status of their biological family. In the various studies we will examine evidence supporting the position that, from a psychological standpoint, index (experimental) adoptees experienced less healthy rearing environments than those experienced by control adoptees. We will discover that most adoptees in these studies were born into societies where eugenic ideas were strong and compulsory eugenic sterilization was permitted by law. This suggests that children whose biological family had a history of 'mental illness' were considered inferior potential adoptees, and were likely placed into more psychologically harmful family environments.

Problems relating to methodology and bias will become apparent in a discussion of topics which include:

1. The questionable use of a schizophrenia spectrum of disorders.
2. Inconsistent and biased methods of counting relative diagnoses.
3. The frequent failure to adequately describe the basis upon which a schizophrenia diagnosis was arrived at.
4. The counting of second-degree biological relatives the same as first-degree biological relatives.
5. The lack of case history material, which would allow reviewers to assess the environmental conditions experienced by adoptees and relatives.
6. The bias introduced by counting relatives individually, which violates the assumptions of the statistical measures used.
7. The use of late-separated adoptees.
8. The evidence that the investigators' bias in favor of genetic explanations had an important influence on their methods and conclusions.

Table 7.1 (overleaf) lists the presence or absence of important methodological problems in these studies. A detailed exploration of these issues is the subject of the remaining part of this chapter.

The Danish-American investigators began their work in the early 1960s, but Heston was the first investigator to publish a schizophrenia adoption study. Thus, we begin with Heston's Oregon study.

Heston's 1966 Oregon schizophrenia adoption study

Leonard Heston's 1966 study has been referred to by behavior geneticists as a 'remarkable report,'[13] and a 'classic' study which 'turned the tide towards acceptance of genetic influence on schizophrenia.'[14] Although Heston's Oregon study was important in shifting the prevailing opinion for explaining the causes of psychiatric disorders in the genetic direction, there are many reasons to disagree with the conclusions of Heston and most subsequent commentators discussing his study.

Heston (who, like Gottesman and Shields, was mentored by Munich-trained Eliot Slater at London's Psychiatric Genetics Research Unit, Maudsley Hospital), identified 47 adopted-away offspring (mean age, 30.3 years) of institutionalized women diagnosed with schizophrenia. These children were separated from their mothers in the first days of life and were placed with family members or in foundling homes. These experimental group adoptees, who were born between 1915 and 1945, were matched against a control group of 50 adoptees (mean age, 35.8

Table 7.1

Important Methodological Problems of

Author (Year)	Country	Significant chronic schizo-phrenia first-degree bio *relative* rate vs. controls? [a]	Significant chronic schizo-phrenia first-degree bio *family* rate vs. controls? [b]	Used late separated adoptees?	Probable selective place-ment bias?
Heston (1966)	USA (Oregon)	Yes	N/A	No	Yes
Kety et al. (1968)	Denmark	No	No	N/A	Yes
Kety et al. (1975)	Denmark	No	No	N/A	Yes
Kety et al. (1994)	Denmark	Yes	No	N/A	Yes
Rosenthal et al. (1968, 1971)	Denmark	No	N/A	Yes	Yes
Wender et al. (1974)	Denmark	No	N/A	Yes	Yes
Tienari et al. (1987, 2000)	Finland	No [d]	N/A	Yes	Yes

[a] Using the traditional one-tailed .05 level of statistical significance utilized by all schizophrenia adoption researchers. Based on a comparison between either (1) the first-degree biological relatives of chronic schizophrenia index adoptees vs. the first-degree biological relatives of control adoptees (Kety); (2) the chronic schizophrenia rate among the adopted-away biological offspring of people diagnosed with chronic schizophrenia vs. the biological offspring of controls (Heston, Rosenthal, Tienari); or (3) the chronic schizophrenia rate of crossfostered adoptees vs. controls (Wender).

[b] Based on a comparison between the biological families of index adoptees diagnosed with chronic schizophrenia, with at least one first-degree biological relative diagnosed with chronic schizophrenia, vs. the biological families of controls with at least one first-degree biological relative diagnosed with chronic schizophrenia.

. . . contd

the Schizophrenia Adoption Studies

Blind diagnoses?	Adequate definition of schizophrenia?	Used a schizophrenia spectrum?	Studied adoptive family rearing environments?	Adequate case history information provided?	Used fabricated interviews to make diagnoses?	Counted first- and second-degree relatives equally?
No [c]	No	No	No	No	No	N/A
Yes	No	Yes	No	No	N/A	Yes
Yes	No	Yes	No	No	Yes [f]	Yes
Yes	No	Yes	No	No	?	Yes
Yes	No	Yes	No	No	No	N/A
Yes	No	Yes	No	No	No	N/A
Yes	Yes [e]	Yes	Yes	No	No	N/A

[c] Although two of the three raters made diagnoses while unaware of the adoptees' group status, the third rater, Heston, was not blind to their status.

[d] See the discussion of Tienari's results later in this chapter.

[e] Although the validity of DSM-III-R criteria (used by Tienari) is questionable, his is the only study to use clear and accepted diagnostic guidelines.

[f] For documentation of Kety et al's use of fabricated 'pseudo-interviews,' see Lewontin et al., 1984, p. 224; Kendler and Gruenberg, 1984, p. 556. It is not clear whether Kety et al. used pseudo-interviews in the 1994 study.

years) whose parents had no record of being admitted to an Oregon state psychiatric hospital. Controls were selected from the records of foundling homes where some of the experimental group children had been placed. They were matched with the experimental group on the basis of sex, type of eventual placement (adoptive, foster family, or institutional), and the length of time adoptees spent in child-care institutions. Dossiers containing psychiatric hospital records, personal interviews, and the results from psychological tests, were compiled for each adoptee. The dossiers were evaluated 'blindly and independently' by two psychiatrists, in addition to Heston.[15] The results showed that five experimental and zero control subjects received a diagnosis of schizophrenia, which reached the .024 level of significance. Heston concluded, 'The results of this study support a genetic aetiology of Schizophrenia.'[16]

Methodological problems

According to Heston, 'No attempt was made to assess the psychiatric status of the father.'[17] This meant that one-half of the experimental group gene pool was unaccounted for. The problem this creates was described by Rosenthal, who pointed out that from the standpoint of genetics,

> it is equivalent to Gregor Mendel beginning his investigation with one selected type of sweet pea plant in the F1 generation, crossing it with other plants whose characteristics he knew nothing about, and then trying to relate all the characteristics he finds in the F2 generation to those of the F1 parent whose special characteristics were known to him.[18]

Rosenthal noted the 'confusion and folly of this procedure.'[19] In addition, the control group biological parents' mental health status was checked only by searching the records of Oregon state mental hospitals. Thus it is fair to say that Heston knew little of the people who contributed over 75% of the genes to his study.

Heston personally interviewed 72 of the 97 adoptees. He described the process as follows:

> The interview was standardized, although all promising leads were followed, and was structured as a general medical and environmental questionnaire which explored all important psychosocial dimensions in considerable depth. Nearly all of the interviews were conducted in the homes of subjects, which added to the range of possible observations.[20]

Heston was familiar with the histories and psychiatric records of his 97 experimental and control adoptees and was aware of the status (experimental or control) of each of the 72 people he interviewed. The interviews may have been standardized, but it is difficult to accept that the decision to follow 'all possible leads' (and in what direction he was looking for 'leads') was not influenced by Heston's knowledge of the interviewee's status. The dossiers of the 97 adoptees were then rated 'blindly and independently by two psychiatrists. A third evaluation was made by the author.'[21] In other words, *Heston was one of the three people making diagnoses.* This meant that the diagnostic procedure was contaminated because Heston *did not* make diagnoses blindly.[22]

Although the other raters initially evaluated each case independently, 'several differences arose in the assignment of specific diagnoses. In disputed cases, a fourth psychiatrist was asked for an opinion and differences were discussed in conference.'[23] A potential bias is evident in this description, since Heston, as the lead investigator, knew whether each subject was an experimental or control adoptee. He also was aware that five of the experimental adoptees had previously received a diagnosis of schizophrenia and had spent time in a mental hospital. It is possible that the other raters were swayed by Heston's superior knowledge of these subjects. Had they been prepared to diagnose a control adoptee with schizophrenia, Heston might have been tempted to convince them to change their diagnosis.

An elementary measure against bias would have required Heston to have had no part in the evaluation process. Heston, then a resident in Psychiatry at the University of Oregon, had spent two years conducting interviews and gathering information in 14 U.S. states and Canada without the benefit of a research grant.[24] The future course of his career and stature rested on the results of this study — one decision on one case meant the difference between worldwide recognition or the anonymity of his filing cabinet.[25] No researcher could be expected to make objective diagnoses under such circumstances. As Rosenthal once commented, 'it is essential that the diagnostician not know whether the individual examined is related to an index or control proband.'[26]

As previously noted, 25 of the 97 adoptees were not interviewed. Of these 25, six refused to be interviewed, eight were dead, seven were inaccessible, and four were not approached because Heston felt that their status as adoptees might have been revealed in the process. Rather than being dropped from the study, these 25 cases were retained on the grounds that 'considerable information was available for most of them.'[27] This information often consisted of second-hand reports, prison or military records, etc. In the Danish-American studies, the name of a non-participant could be checked against a national psychiatric register. Given the lack of such a register in the United States, the best Heston could do was to check the records of public psychiatric hospitals in three West Coast American states plus those in areas where adoptees were thought to be living. It is possible that some of these 25 adoptees (we are not told how many were in each group) were diagnosed with schizophrenia without anyone in the study knowing about it. Clearly, these 25 cases should have been dropped, but 'In a conference the raters agreed that it would be misleading to discard any cases, and that all subjects should be rated by forced choice.'[28] One might argue that it was 'misleading' to retain adoptees for whom insufficient information was available. In order to guard against bias resulting from Heston's knowledge of an adoptee's history and group status, the decision of whether to count non-interviewed adoptees should have been made at the beginning of the study.

How schizophrenia was defined

Another problem is the way that schizophrenia was defined in the study. Heston wrote only that diagnoses were based on 'generally accepted standards,'[29] and that the diagnosis of schizophrenia 'was used conservatively.'[30] Thus, the dependent variable was vaguely defined, making it impossible for independent reviewers to determine whether the five adoptees diagnosed with schizophrenia meet common diagnostic standards (DSM or otherwise). A rare case history presented by Heston describes a man who would not be diagnosed with chronic schizophrenia by the criteria of any edition of the Diagnostic and Statistical Manual (DSM-I through IV).[31] This individual received a hospital diagnosis of 'schizophrenic reaction,' but it is not clear which type of 'reaction' it was. DSM-I lists nine different types of schizophrenia reactions, and the one fitting this man's case history is 'schizophrenia reaction, acute undifferentiated type.'[32] Using DSM-IV classifications, he probably would have been diagnosed with schizophreniform disorder or brief psychotic disorder. The change in status of this one case is enough to render the experimental-control comparison statistically non-significant (with alpha set at .05). One can only wonder, as did Cassou and colleagues in 1980, what else could have been learned had the scientific community been presented with more case histories to examine.

Evidence of selective placement

As we will see, the social and political conditions in *all* regions where schizophrenia adoption studies have been performed make it probable that their results were confounded by selective placement factors. Almost all of the adoptees in Heston's study were born and placed at a time when eugenic ideas were dominant and the compulsory sterilization of people diagnosed with schizophrenia was sanctioned by law.

In 1909 the Oregon legislature passed a eugenic sterilization law by a wide margin, but the bill was vetoed by the governor. A new bill was passed in 1917 and was signed into law. This statute established the nation's first official eugenical organization, the State Board of Eugenics, which was comprised mainly of superintendents of mental and penal institutions.[33] The duty of the Board was to authorize, in the words of the law, the compulsory sterilization of:

> all feeble-minded, insane, epileptic, habitual criminals, moral degenerates and sexual perverts, who are persons potential to producing [sic] offspring who, because of inheritance of inferior or antisocial traits, would probably become a social menace, or a ward of the state.[34]

Clause 2889 of the 1917 law gave the Board of Eugenics power to examine the 'family traits and histories' of such persons,

> and if in the judgment of the majority of the said board procreation by any such person would produce children with an inherited tendency to feeble-mindedness, insanity, epilepsy, criminality or degeneracy . . . then it shall be the duty of said board to make an order directing the superintendent of the institution in which the inmate is confined to perform or cause to be performed upon such inmate such a type of sterilization as may be deemed best by said board.[35]

An additional law passed in 1919 stipulated that the mere fact that a person had been admitted to a mental hospital constituted 'prima facie evidence that procreation by any such person would produce children with an inherited tendency to feeble-mindedness, insanity, epilepsy, criminality or degeneracy.'[36] These eugenic sterilization laws were on the books until 1983, almost 40 years after the last of Heston's adoptees was born. Approximately 3,000 Oregonians were sterilized for eugenic purposes between 1909 and 1983.[37] The author of a 1925 article in *Eugenical News* noted that, while sterilization laws in many states were limited to 'inmates of institutions,' in Oregon 'there is a Eugenics Commissioner, who has authority to comb the state for degenerates and enforce sterilization.'[38] And far from being a little known or discussed statute, leading newspapers such as *The Oregonian* and *The Oregon Journal* championed its cause and published summaries of eugenic lectures and statements by pro-sterilization authorities.[39] In a 1935 *Oregon Journal* article, the author wrote, 'Taking a tip from Nazi Germany, Oregon today considered embarking on a far-reaching program of sterilization of its unfit citizens.'[40]

The social conditions in Oregon suggest that the offspring of institutionalized women diagnosed with schizophrenia were viewed as the carriers of an inherited predisposition for 'insanity' and 'degeneracy.' As Kringlen commented, 'Because the adoptive parents evidently received information about the child's biological parents, one might wonder who would adopt such a child.'[41] In Oregon circa 1915–1945 it is unlikely that such children would have been allowed to be placed into, or would have been accepted by, qualified adoptive homes. And little help could have been expected from Oregon State Hospital physicians, since they were one of the groups pushing hardest for the sterilization laws.[42] It is therefore unlikely that Heston's experimental and control group adoptees were randomly placed into available adoptive homes, which violates the no selective placement assumption. Leaving aside all other methodological problems, Heston's results are plausibly explained by the likelihood that experimental group adoptees were reared in more psychologically harmful environments than control adoptees.

According to Heston's figures, significantly more index than control adoptees had spent over a year in a penal institution or were diagnosed with 'sociopathic personality.'[43] In addition, alcoholism ('problem drinking') was concentrated 'almost exclusively' in the experimental group.[44] These findings can be interpreted as evidence of the inferior rearing environments experienced by the offspring of mothers diagnosed with schizophrenia. For Heston, genetic factors provided the most plausible explanation for the higher rate of 'psychosocial disability' in the experimental group.

Abandoned children

In a discussion of the early environmental deprivations of adoptees who had spent many years in foundling homes, Heston recognized that the conditions in these homes and the lack of adult nurture experienced by the children had a harmful impact on their psychological development:

> The actual extent to which these children were deprived of maternal or emotional nurture beyond that implicit in group care *must be inferred largely from indirect evidence . . . the later history of these children strongly supports the hypothesis that significant deprivation did occur.* Random observations recorded in school or nursery records and recollections of foster parents and subjects describe several children as shy, withdrawn, demanding excessive attention, or sad; few as happy, spontaneous, or normal. Some were requiring sedation at night . . . It is certain that most of these children were unhappy and probable that there was significant deprivation of emotional nurture [emphasis added].[45]

From a psychosocial perspective we would expect that many such children, who spent an average of two years in a foundling home, would grow up to be neurotics, felons, 'sociopaths' . . . or the recipients of schizophrenia diagnosis. Most psychodynamic theories stress the importance of the first few years of a child's life in his or her subsequent psychological development, and Heston recognized that many of his adoptees had been emotionally damaged because of their experiences in the foundling home. In fact, three of the five diagnoses of schizophrenia were made on experimental group adoptees who had spent months or years in such homes.[46]

By observing the 'later history' of children raised in foundling homes, Heston was able to infer a lack of emotional nurture in their earliest years. He astutely saw their sadness and withdrawal as symptomatic of their deprivation, and not as the result of a genetic predisposition for childhood depression. Unfortunately, Heston refused to infer a lack of emotional nurture from the significantly more troubled lives of his experimental group *adults* – seeing instead the suggestion of a genetic relationship between psychosocial disability and schizophrenia. In Heston's view, the emotional damage caused by early deprivation associated with institutional care 'can be spontaneously reversed by the time adulthood is achieved.'[47] What a convenient theory for reconciling two radically different approaches to the understanding of human suffering.

Heston's Oregon study contrasts sharply to later behavioral and IQ adoption studies which were able to minimize deprivation and attachment disturbances. In the Colorado Adoption Project, for example, all adoptees were separated from their biological mothers at or shortly after birth and were placed into qualified adoption homes within a few weeks.[48] In contrast, almost half of Heston's adoptees spent a good deal of time in foundling homes, which, as Heston acknowledged, could not provide a nurturing environment and which contained sad and withdrawn children. Heston also admitted, and I will add the emphasis, that *'none of the subjects were reared in typical or 'normal' circumstances.'*[49] This was especially true for the experimental adoptees, whose adoptive parents may have treated them differently because they were aware of their adoptive child's biological background. One therefore must question whether the results of this study (already biased by methodological problems and selective placement) can be generalized to the population of non-adoptees. Although Heston claimed that there were no important differences in adult psychological disturbance between adoptees placed in foundling homes versus those placed with families,[50] it remains likely that Heston's adoptees experienced more psychological damage than non-adopted children. Unfortunately, the psychological status of these 97 'abandoned children' has been the basis (sometimes combined with the results of other studies) of sweeping yet unwarranted generalizations about the importance of genetic factors in schizophrenia.

Conclusion

There are five major problems with Heston's study. Each one by itself is reason enough to seriously question the validity of his conclusions. (1) The evidence that selective placement occurred among the experimental adoptees. (2) A diagnositic process that was contaminated because one of the raters (Heston) was aware of the group status and personal history of the adoptees. (3) The lack of case history material, which would allow independent analysis of the history and mental status of the subjects. (4) About 26% of the subjects were not interviewed, yet were retained in the study. (5) Schizophrenia was not adequately defined.

The evidence suggests that Heston's 1966 Oregon schizophrenia adoption study was methodologically unsound and was subject to the confounding influence of placement issues.

The Danish-American Adoptees' Family study: 1968

Although the Kety and associates Danish-American schizophrenia adoption studies are perhaps the most frequently cited evidence in support of a genetic basis for schizophrenia, they have been the subject of several critical analyses. My objective is to take a closer look at the key issues raised in these analyses in order to determine whether the conclusions reached by Kety and associates are consistent with the evidence.[51]

The first phase of the Danish-American Adoptees' Family study was conducted in the greater Copenhagen area. The most important papers coming out of this work were published in 1968 and 1975. The 1968 study was based entirely on institutional records. The 1975 study was based in part on interviews conducted with the biological and adoptive relatives of index and control adoptees. Following the publication of the 1975 Copenhagen study, the work was extended to the rest of Denmark. Preliminary reports on this 'Provincial' sample were based on institutional records only. The final paper, published in 1994, presented the results of the Provincial interview study. Results also were reported for the Danish National sample, meaning the combined diagnoses of the Copenhagen and Provincial studies.

Kety, Rosenthal, Wender and their Danish associates presented their first paper on the Danish-American adoption work at the Dorado Beach schizophrenia conference in the Summer of 1967, which was published the following year. The investigators established an index group of 34 adoptees who were diagnosed with 'chronic schizophrenia' (designated 'B1,' N = 16), 'acute schizophrenia' (designated 'B2,' N = 7), or 'borderline schizophrenia' (designated 'B3,' N = 10).[52] Index cases were selected from the records of all adoptions granted in the City and County of Copenhagen between 1924 and 1947. In all, 507 adoptees from this group were recorded as having been admitted to a psychiatric facility. The investigators were granted access to the Danish Adoption Register of the State Department of Justice (established in 1937[53]), the *Folkeregister* (Population Register), the Psychiatric Register of the Institute of Human Genetics, records of the Mother's Aid Organization, and police, court, and military records. The records of these adoptees were screened, and when a consensus B1, B2, or B3 diagnosis was made by Kety, Rosenthal, and Wender, that person became an index case. A control group of 33 adoptees having no record of being admitted to a psychiatric facility was also established.

Index and control adoptees were matched on the basis of age, sex, age at transfer to the adoptive parents, and the socioeconomic status of the adoptive family. After the index and control groups were established, the researchers attempted to locate the records of the adoptees' biological and adoptive relatives. A total of 463 such relatives were identified. Kety and colleagues then searched for any existing psychiatric records for these relatives, and when such records were found, a blind consensus diagnosis was made by Kety, Rosenthal, Wender, and Schulsinger. The code concealing the identity and group status of the relatives was then broken, and each case was assigned to the biological or adoptive relative category of the appropriate group (index or control). All diagnoses were based on information obtained from institutional records. There was no personal contact between the investigators and any of the adoptees or their relatives.

Diagnoses falling into what was called the 'schizophrenia spectrum of disorders' were counted as schizophrenia in comparisons between index and control relatives (see the discussion below). Diagnoses were made according to a global (or consensus) diagnostic system. This method, which was used in the entire series spanning more than 25 years, was described as follows:

> Four copies of the edited summary were prepared and distributed to the four authors who served as raters and who independently characterized each subject according to the classification described below. The individual ratings were then tabulated and those cases in which there was disagreement among the raters were discussed at a conference of all four authors where an effort was made to review additional edited information which it was possible to obtain and to arrive at a consensus diagnosis acceptable to all. In 4 cases there remained an evenly split opinion regarding the presence of schizophrenia or doubtful schizophrenia, and these were not included in those categories.[54]

Although high inter-rater reliability was claimed,[55] each investigator approached the diagnostic process somewhat differently. According to Kety, each rater's 'individual definitions of schizophrenia varied by virtue of [their] training and experience, from a substantial reliance on Kraepelin and Bleuler to the broader psychodynamic concepts which were taught in the 50s.'[56]

The distribution of spectrum diagnoses among index and control adoptive and biological relatives can be seen in tables 4a and 4b of the 1968 paper. Schizophrenia was defined as diagnoses B1, B2, B3, D1, D2, D3, and C. In addition, several 'outside the spectrum' psychiatric diagnoses were made by the authors. The 1968 schizophrenia spectrum is seen in Table 7.2.

Table 7.2

The Kety and Associates 1968 Schizophrenia Spectrum

B1	Chronic Schizophrenia
B2	Acute Schizophrenia Reaction
B3	Borderline Schizophrenia
D1	Uncertain Chronic Schizophrenia
D2	Uncertain Acute Schizophrenia Reaction
D3	Uncertain Borderline Schizophrenia
C	Inadequate Personality

Source: Adapted from Kety et al., 1968, p. 352

Kety and colleagues found a higher prevalence of spectrum disorders among index biological relatives versus control biological relatives:

Of 150 biological relatives of index cases 13, or 8.7%, had a diagnosis of schizophrenia, uncertain schizophrenia or inadequate personality compared to 3 of 156, or 1.9%, with such diagnoses among the biological relatives of the controls. The difference is highly significant (p, one sided probability from exact distribution = 0.0072).[57]

They concluded that 'genetic factors are important in the transmission of schizophrenia.'[58]

In the following sections I discuss six important topics relating to the 1968 study and all subsequent Danish-American Adoptees' Family reports: (1) the schizophrenia spectrum concept, (2) questions relating to the design of the study, (3) the failure to study environmental variables, (4) the counting of half-siblings in statistical calculations, (5) counting individual relatives versus counting families, and (6) the evidence of selective placement in the Danish adoption process.

The schizophrenia spectrum concept

Although only chronic B1 'process' cases were traditionally counted as schizophrenia by Danish psychiatrists,[59] the Danish-American researchers broadened the definition of schizophrenia considerably to include what they considered related diagnoses. I have argued against the validity of the Kety et al. schizophrenia spectrum in detail elsewhere,[60] and here I would like to touch on the main points.

In 1963 Kety, Rosenthal, and Wender began a collaboration with Fini Schulsinger, a Danish psychiatric geneticist whose expertise and language abilities would enable them to obtain the records of a large number of adoptees through the extensive population and psychiatric registers then existing in Denmark. Schulsinger has been credited with convincing the Danish Ministry of Justice to open up the national adoption records for the purpose of scientific research. His persuasiveness finally convinced the authorities to allow access to these records, in spite of the fact that earlier Danish researchers had made similar, unsuccessful requests.[61]

The researchers obtained information on 5,483 adoptees from the greater Copenhagen area. Based on rough population expectations, it probably was felt that such a large sample would yield a sufficient number of adoptees diagnosed with chronic schizophrenia to compare to a control group of adoptees with no record of a schizophrenia diagnosis. The results of prevalence studies conducted in Western countries have placed the lifetime expectancy rate for schizophrenia at between 0.8 and 1%.[62] Thus, Kety and associates were probably expecting to find 50–55 schizophrenia-diagnosed adoptees in their sample (5,483 x .01= 54.8). Epidemiological studies conducted in Denmark and summated in Zerbin-Rüdin's 1967 study produce a lower age-corrected Danish population prevalence of 0.69%.[63] Based on this relatively low rate, the expected number of schizophrenic adoptees from the greater Copenhagen sample would have been 38 (5,483 x 0.0069 = 37.8). However, contrary to genetic expectations Kety and colleagues diagnosed only 16 cases of chronic schizophrenia

(B1) among their adoptees. This rate of about three per thousand is less than one-half of the expected rate in the general population.

Thus, even before they located the hospital records of the biological and adoptive relatives of adoptees diagnosed with schizophrenia, the investigators were confronted with a powerful piece of evidence in favor of the environmentalist position—for the numbers suggested that merely being raised by parents screened for mental health by adoption agencies had reduced the chance of a person being diagnosed with schizophrenia by over 50%. The status of being a Danish adoptee, therefore, placed one in a distinct group when compared to the general population. According to Sarbin and Mancuso, the results 'show, quite simply, that the social services agencies have suitably done their job.'[64]

One can imagine the researchers' disappointment in finding only 16 cases of chronic schizophrenia among their large sample of adoptees. This number was simply too small to be able to conduct their study, as acknowledged by Rosenthal in 1972:

> The fact that the number of 16 hard-core, process [chronic] schizophrenic, index cases *is too small to give the heritability of such disorders the proper opportunity to express itself,* and the possibility that by relying only on psychiatric records rather than on personal examinations we were missing a number of cases [emphasis added].[65]

In this passage Rosenthal acknowledged that the investigators did not have enough cases with which to conduct the study. A year earlier, Rosenthal had made the same point:

> The second [important] feature [of the research] has to do with the fact that we have included a broad spectrum of disorders in the ones I am calling schizophrenic. These include not only the classical chronic, process types of cases, but patients called doubtful schizophrenic, reactive, schizo-affective, borderline or pseudoneurotic schizophrenic, or schizoid or paranoid. *If we dealt only with hardcore schizophrenia, our ns would be too small to make any of these studies meaningful* [emphasis added].[66]

He continued,

> However, a more positive reason for including the spectrum of disorders is that, in the process, we hope to be able to determine whether any or all of these disorders, which phenotypically have strong resemblances to hardcore schizophrenia, are genetically related to it as well.[67]

A careful reading of this paragraph reveals that, whereas the testing of whether the other diagnoses were related to chronic schizophrenia was a 'positive reason' for using the spectrum, the 'negative reason' was the *necessity of widening the definition of schizophrenia in order to obtain enough subjects to be able to conduct the study.*[68]

Clearly, had Kety and colleagues restricted their definition of schizophrenia to B1 only, they would have had no possibility of achieving statistically significant results in the genetic direction. Rosenthal was clear on this point also: 'It should be apparent now that if we had included in our comparisons of index and control relatives only those who clearly had process schizophrenia, we would have found no difference between the two groups of relatives.'[69]

The evidence suggests that the schizophrenia spectrum was created because — contrary to genetic expectations — Danish adoptees were diagnosed with schizophrenia 50% less often than members of the non-adoptee population. Thus, the investigators could have concluded that their results were not generalizable to non-adoptees. In any case, Kety, Rosenthal, and Wender were confronted with the choice of (literally and figuratively) packing their bags and giving up on several years of hard work, or broadening the definition of schizophrenia in order to obtain more cases. I do not mean to imply that Kety and associates, who were diagnosing blindly, knew into which group (index or control) these cases fell. But wherever they might have fallen, there were not enough to allow even the *possibility* of significant findings.

The evidence points to the conclusion that the schizophrenia spectrum was created by selecting the disorders which seemed closest to chronic schizophrenia in order to have enough cases to conduct the study, and not, as its architects maintained in their major publications, in order to test the hypothesis that the spectrum disorders were related to chronic schizophrenia. This idea receives support in a remarkable description of the origins of the spectrum by Rosenthal:

> It seems somewhat ironic that while representing the U.S. Field Center in the WHO International Pilot Study of Schizophrenia, John Strauss and Will Carpenter were working upstairs at the National Institute of Health Clinical Center, trying to hone a definition of schizophrenia as sharply as they could possibly make it, while Paul Wender and I were working downstairs, in concert with Seymour Kety, in effect *broadening the concept of schizophrenic disorder as widely as it may have ever been reasonably conceived before* [emphasis added].

> While Carpenter and Strauss emphasized the limits or boundaries of the process [B1] schizophrenia concept, *our group strained* to encompass all disorders that shared salient clinical and behavioral manifestations with process schizophrenia and to group these as a spectrum of schizophrenic disorder [emphasis added].

> Of course, we did not know which disorders, if any, should be included in such a spectrum to meet the criterion of genetic or familial commonality. Nevertheless, we selected the ones that we thought had

the highest probability of meeting this criterion, and introduced them into our research studies as a hypothesis to be tested. It was easy to read through the *APA Diagnostic and Statistical Manual*, second edition, to make such selections.[70]

These passages are relevant to the present discussion for several reasons. Because the spectrum was based on DSM-II definitions, Rosenthal seemed to date its creation to no earlier than 1967, since it was in February of that year that a draft version of the second edition was distributed to psychiatrists.[71] Assuming that Rosenthal, Kety, and associates did not have access to the draft before it was distributed to psychiatrists for commentary, we discover that as late as 1967 the Danish-American team was 'straining' to broaden the concept of schizophrenia 'as widely as it may have ever been reasonably conceived before' — well after Kety and colleagues began the collection of adoptees in the Copenhagen study. It therefore must be asked: What circumstances compelled these researchers to widen the concept of schizophrenia so many years after they began their study?

I will now briefly discuss several other problems with Kety and colleagues' schizophrenia spectrum. (1) The investigators' rationale for the inclusion of a particular diagnostic category in the spectrum was based on finding a significantly greater number of diagnosed people among index versus control biological relatives. However, with the exception of the vaguely defined 'uncertain borderline schizophrenia,' no spectrum category in either the 1968 or 1975 studies was found in significantly greater numbers among index biological relatives versus controls.[72] (2) The researchers' conclusion that the spectrum diagnoses were related to chronic schizophrenia was the result of invalid methods of counting diagnoses.[73] (3) In spite of Kety's reliance on E. Bleuler's description of latent schizophrenia,[74] Bleuler did not believe that it was possible to distinguish 'milder cases of schizophrenia' from people who were merely 'whimsical.' Bleuler therefore called on clinicians to use a 'very high diagnostic threshold value' in making a schizophrenia diagnosis.[75] (4) As the researchers themselves admitted, it was often difficult to distinguish spectrum disorders from similar non-spectrum disorders.[76] (5) The spectrum concept was based on the fallacy of believing that conditions are genetically related simply because they are (allegedly) found together. As Stephen J. Gould noted, 'the invalid assumption that correlation implies cause is probably among the two or three most serious and common errors of human reasoning.'[77] (6) Although the diagnostic formulations for the BI, B2, and B3 categories included references to homosexuality, a 'disorder' subsequently dropped by the American psychiatric establishment, there is no mention of homosexuality in the diagnostic criteria for schizophrenia in any edition of the DSM. (7) It is questionable whether diagnoses labeled 'uncertain' should have been counted by the investigators, and there is no evidence that the decision to count these cases was made before diagnoses were assigned to their respective categories. According to Kety, 'in the case of the relatives, questionable or uncertain schizophrenia had to be added if relatives with less certain diagnoses were not to be lost.'[78] One would think that these relatives could have been prevented from becoming 'lost' without being counted in Danish-American statistical calculations.

Because of the doubtful validity of Kety and colleagues' schizophrenia spectrum, chronic schizophrenia is the only recognized diagnosis in comparisons between the various relative groups made here.

Questions concerning the design of the study

Beginning with the 1968 study, the conclusions of all Adoptees' Family reports were based on a diagnostic comparison between index and control biological relatives. However, questions remain about whether this comparison reflected the original research design of the Danish-American investigators. The four groups of relatives under study are represented in Figure 7.3.

Figure 7.3

Groups of Relatives Receiving Diagnoses in the Danish-American Adoptees' Family Studies

	Biological Relatives	Adoptive Relatives
Index Adoptees	**IB** (Index Biological)	**IA** (Index Adoptive)
Control Adoptees	**CB** (Control Biological)	**CA** (Control Adoptive)

The investigators' conclusions were based on finding a statistically significant difference in spectrum diagnoses between groups IB and CB, and the finding of no significant differences between groups IA and CA. According to Rosenthal's 1970 description,

> A higher incidence among the biological relatives of index cases than of controls indicates that heredity is contributing significantly to the disorder. A higher incidence among the adoptive relatives of index cases than of controls supports the view that rearing by, of, or with schizophrenics contributes to the development of the disorder[79]

The design of the study was called into question by Lidz and Blatt in 1983, who made the unsupported claim that Kety and colleagues' original intent had been to compare the schizophrenia rates of index biological versus *index adoptive* families (IB vs. IA): 'The investigators sought to differentiate genetic from intrafamilial environmental factors by comparing the occurrence of such disorders in the biological and adoptive relatives of schizophrenic patients who had been adopted at a very early age.'[80] According to Lidz and Blatt, the fact that Danish adoptive parents were screened for psychiatric disorders compelled Kety and associates to create a control group of non-schizophrenic adoptees and their relatives. Lidz and Blatt concluded that 'the major interest of the study became the comparison of the biological relatives of the schizophrenic and control adoptees rather than the original purpose of the project.'[81]

Kety replied that Lidz and Blatt incorrectly described the aim of study and misunderstood the logic of its design. Furthermore, according to Kety,

> We anticipated that there would be differences between adoptive and biological relatives in age, socioeconomic status, life style, and other variables. For that reason we planned not to make comparisons between these two groups of relatives but, instead, as described fully in the original publications and outlined above, to compare each group with their respective controls in evaluating separately the significance of genetic or family-related environmental factors.[82]

It is true that the 1968 study discussed the importance of making direct comparisons between index and control relatives from each group. However, a paper delivered by Rosenthal in 1967, three months before the first public presentation of the study, suggests that Lidz and Blatt were correct that the intent of the researchers had been to compare the biological and adoptive relatives of index adoptees, versus the same comparison among controls:

> In Denmark, with the collaboration of Dr. Fini Schulsinger and others, we began with adoptees who are now schizophrenic. *We compare the incidence of schizophrenic disorders in their biological and adoptive families.* The same procedure is carried out for a matched group of normal adoptees, who serve as controls [emphasis added].[83]

Rosenthal is therefore on record as stating, in March of 1967, that the investigators intended to compare the difference between groups IB and IA *versus* the difference between groups CB and CA — but it was not until April of 1967 that all of the data had been collected and diagnoses were made.[84] When comparisons are made according to Rosenthal's description, the difference is not statistically significant in either group. Kety et al. found 13 spectrum diagnoses out of 150 *index biological relatives* (8.7%), and two such diagnoses out of 74 *index adoptive relatives* (2.7%) — but the difference is not statistically significant.[85] Thus, had the Danish-

American investigators analyzed their data using Rosenthal's stated approach, they would have concluded that there were no significant differences between the groups and that their study had found no evidence in support of the genetic transmission of schizophrenia. Interestingly, neither Rosenthal nor Kety cited Rosenthal's 1967 paper in any of their subsequent publications on the genetics of schizophrenia.[86]

The failure to study environmental variables

Two years before the publication of the 1968 study, Kety wrote, 'although genetic factors undoubtedly operate in schizophrenia, they do not constitute a sufficient explanation for the genesis of this disorder.'[87] If Kety had little doubt that genetic factors were operating, he might have concentrated on discovering the *environmental* component of schizophrenia—for why should so much time and money be spent in order to investigate something that is 'undoubtedly' true? Kety followed these lines by writing, 'we must continue to look for environmental factors that operate to produce what we call schizophrenia in the genetically vulnerable individual. Fortunately, this seems to be the attitude of most investigators in the field.'[88] *Unfortunately*, Kety abandoned any idea of discovering environmental factors. His studies looked at only one environmental variable, for the likely purpose of answering anticipated objections to his research design. In 1970, Kety wrote:

> We are really examining only one environmental factor and that is the presence of a person in the adoptive family with a mental illness. There are thousands of environmental factors which we have not examined: the personality of the family, their child-rearing practices, the diet which the individual has had, the lead in the drinking water and many perhaps undreamed of. Therefore, these data by no means rule out the operation of environmental factors. They simply indicate that at least one of these factors, namely, having a mentally ill person with a schizophrenic form of illness in the immediate environment, is not an important or significantly operating variable.[89]

Kety wrote of the family-rearing environment as being just another possible environmental variable, on par with lead in the drinking water, diet, etc. But it also happens to be a major psychosocial explanation of why people are eventually diagnosed with schizophrenia, and deserved to be looked at more closely. In a subsequent schizophrenia adoption study by Tienari and associates in Finland (reviewed later in this chapter), the investigators found that 'all adoptees who had been diagnosed either as schizophrenic or paranoid had been reared in seriously disturbed adoptive families.'[90] The Danish-American Adoptees' Family series would have been substantially more interesting and important had Kety followed his own advice from 1966 and, like Tienari and colleagues, taken a closer look at the psychosocial environments of the people under study.

The validity of counting half-siblings

The Danish-American investigators counted diagnoses among first- and second-degree relatives (biological half-siblings) equally. Most critics of the 1968 study pointed out that only one diagnosis of chronic schizophrenia was made among the 150 index biological relatives — and that person was a half-sibling.[91] The statistically significant index spectrum rate was dependent on non-B1 diagnoses among index biological half-siblings. Nine spectrum diagnoses were found in this group, whereas none was found among the control half-siblings. In the 1968 report, half-siblings constituted 57% of all biological relatives (173/306) and 69% of index biological relative diagnoses (9/13). As we will see, the 1975 study rested its case for the 'compelling evidence' of genetic factors on the distribution of spectrum disorders among index and control paternal half-siblings.

In the mid-1970s a debate arose among several prominent schizophrenia researchers over the importance of half-sibling relative cases in the study. Gottesman and Shields noted that half-siblings had a higher spectrum diagnosis rate than full-siblings, leading them to comment that 'genetic theory predicts a much higher risk for full siblings.'[92] Kringlen argued that the high rate among half-siblings 'is, from a genetic point of view, meaningless.'[93] Kety et al. admitted that they had 'toyed with the idea of giving them half-weight as soon as we realized how many were being identified but rejected it as being too pretentious.'[94] Boyle would later comment, 'it is difficult not to wonder whether the idea was also rejected because it leads to a quite different set of conclusions.'[95] In 1976 Benjamin wrote, 'this finding is peculiar and contradictory. It shows, in effect, that the *less* consanguinity, the *greater* the "genetic" effect. Differences should be weakest, not strongest, in the half-sibling category.'[96] Kety responded by arguing that Benjamin was incorrect because the difference was not statistically significant. He explained further that a significant difference, had it been found, could be explained by 'socioenvironmental' and other factors.[97]

Lidz and Blatt argued in 1983 that Kety and colleagues should have counted half-siblings weighted on the basis of the common parent's status, whether the other parent was known, and the type of genetic model used. They pointed out that no information was provided about the half-siblings' other parent or about their rearing environments. Kety replied that the findings from the 1975 report would have been significant even if half-siblings had been counted one-half, and that in his opinion there was just as much reason to give them a weighting of two. He accused Lidz and Blatt of giving half-siblings no weight at all, because they focused their attention on rates among first-degree relatives.[98]

Here, I will argue that only first-degree relatives should have been counted. From a *genetic* standpoint, knowledge of only one-half of the genetic background makes the determination of genetic factors difficult. In a discussion of other schizophrenia adoption studies, Rosenthal noted that an investigation that looks at only one-half of the genetic equation is dealing in 'confusion and folly,' and that 'A proper genetic study must be based on who mates with whom. In fact, although genetics has been defined in various ways, the simplest and perhaps best definition of genetics is: *the science of matings.*'[99] In the case of the Danish-American half-siblings, we know little about 'who' and *nothing* about 'whom,' yet diagnoses among the products of such 'matings' were decisive in the Kety-led studies.[100]

From an *environmental* standpoint, it is unacceptable to count people for whom little or nothing of their family and social environment is known. In another comparison, Kety and colleagues stressed that there were no significant differences in spectrum or 'outside the spectrum' psychiatric diagnoses between index and control *adoptive* families, thereby implying that both sets of adoptees grew up in comparable rearing environments (although as we will soon see, in 1984 Lewontin, Rose, and Kamin found other results). Let us suppose, for example, that most index adoptees had been raised by adoptive parents diagnosed with schizophrenia or another psychiatric disorder, while most or all of the control adoptees had been raised by 'normal' parents. Given this scenario, it is likely that Kety and associates would have concluded that the significantly higher level of index adoptive family distress was a factor in the higher rate of index adoptee schizophrenia. This type of comparison cannot be made for the Danish-American half-siblings, however, because no information was provided about their family environments. Yet we have seen Kety and associates acknowledge that this information is necessary in order to identify potentially pathological rearing environments. Furthermore, no information was provided on how many biological parents of half-siblings actually lived with their children. An index father, for example, could have sired the half-sibling in question without having been the child's rearing father. Quite literally, a spectrum half-sibling could have grown up eating out of garbage cans on the streets of Copenhagen, whereas non-diagnosed half-siblings could have been raised in exceptionally nurturing and loving environments.

Even Gottesman and Shields noted the problems of assessing the environments of the Danish half-siblings:

> The high degree of psychopathology seen in the half sibs could have a major environmental component and could reflect cultural transmission. That is, the half sibs might have stayed with a disturbed biological parent or been subjected to various kinds of institutional care. Although such happenings may not be sufficient to produce schizophrenia, they may have produced 'inadequate personalities' of the kind that might have been diagnosed as definite or uncertain latent or borderline schizophrenia (B3 or D3).[101]

Gottesman and Shields brought up an important point: how many of these half-siblings were raised in institutions? This has never been discussed in any Danish-American paper.

According to Rosenthal, in order to 'demonstrate that genes have anything to do with schizophrenia' the investigator must show that 'the frequency of schizophrenia in relatives of schizophrenics [is] positively correlated with the degree of blood relationship to the schizophrenic index cases.'[102] The fact that the Adoptees' Family series failed this test speaks to the operation of environmental factors causing the high rate of index half-sibling spectrum diagnoses.

In conclusion, the counting of half-siblings, for whom no biographical information is provided, is untenable on both genetic and environmental grounds. This suggests that there is no basis upon which they can be counted in the Danish-American adoption studies. Following Lidz and Blatt's 1983 critique, the focus here is on diagnoses found only among first-degree biological relatives.

Who gets counted, families or individuals?
In 1976 Lorna Benjamin observed that counting each biological relative separately and combining them into index and control totals is a violation of the assumption of independent observations:

> The procedure of counting up all the possible relatives of each index case and pooling them as if they were independent samples . . . would allow some families to disproportionately affect the results . . . This group of people, who were likely to have been reared together and to have had the same health care system supplying records used in the study, could falsely inflate the 'significance' of the difference.[103]

According to Benjamin, the 1975 study should have reported an N of 67 (the number of index + control biological relative *families*) instead of Kety and colleagues' N of 347 (the number of biological *relatives*). Kety replied that his study did indeed recognize the importance of counting families in addition to individuals.[104] By Kety and colleagues' calculations, spectrum disorders were found in 17 of 33 biological families, but in only five of 34 biological control families.[105] However, they had to redefine the spectrum in order to find statistical significance in this comparison.[106]

According to Loren Mosher, then editor of *Schizophrenia Bulletin* and the Director of Schizophrenia Research at the National Institute of Mental Health:

> The actual sample size of all the adoption studies is, at least for genetic purposes, the number of index probands [adoptees], *not* the number of relatives identified. The power of the adoption methodology is its separation of heredity and environment for genetic analysis; therefore, when *N*'s are reported as the number of biological relatives seen, it leaves the misleading impression that the genetic/environmental separation is applicable in this group, whereas, in point of fact, it is not. Basically, studying either biological or adoptive relatives is just a special family study.[107]

Benjamin's and Mosher's observations have merit because the clustering of schizophrenia in particular families could be explained by exposure to common environmental influences. It could also be explained by siblings receiving similar hospital diagnoses because they were viewed as sharing the same 'familial taint.' For Mosher in particular, counting all biological relatives individually in statistical comparisons transformed the Adoptees' Family method into little more than a 'special family study.' Thus, an N of 347 in the 1975 interview study (and an N of 306 in the 1968 study) violates the independence assumption of the statistical procedures used.

Evidence of the selective placement of adoptees

There is evidence that the selective placement of adoptees was a major factor influencing the Danish adoption process, and it is therefore unlikely that placements were made randomly in Denmark during the period when adoptees were placed. We begin with the findings of criminality adoption researchers Hutchings and Mednick:

> The most important limit of the adoption method is the possibility that the adoption procedure results in selective placement, promoting correspondence between the adoptive home and the characteristics of the biological parents . . . The Danish organization which arranged many of the adoptions examined in this study states clearly that they do aim at matching in certain respects.[108]

Hutchings and Mednick found a statistically significant correlation between the social class of the biological and adoptive fathers. Teasdale studied all 14,427 National sample adoptees identified by Kety and associates and found a significant correlation between the socioeconomic status of the adoptees' biological mothers and fathers and the adoptees' adoptive fathers. Teasdale concluded that 'some selective placement has occurred in the sample; i.e., the adoption agencies tended to place children into adoptive homes as a function of their biological background.'[109]

Kety and colleagues believed that placement policies in Denmark did not create a major problem for their study:

> Since the etological role of environmental variables remains obscure at present, it is not likely that a social agency, even if it set about doing so deliberately, could find sufficient of the unknown variables in the prospective adoptive parents to materially affect the risk of [schizophrenic] illness in the adoptee.[110]

According to Kety and associates, it is therefore unlikely that agency policies led to the placement of index adoptees into more schizophrenia-producing environments. But is this really the case?

The potentially confounding influence of selective placement can be seen in the example of pellagra. As we saw in Chapter 1, because of its tendency to cluster in certain families, pellagra was once thought to have an important genetic component. It later was shown that the disease is caused by a niacin deficiency. Pellagra was found mainly in poor families, whose diet did not provide enough niacin.[111] According to Kety's logic, an early twentieth century adoption agency, unaware of the true causes of pellagra, could not have systematically placed certain classes of adoptees into more 'pellagragenic' environments. However, if an agency had placed adoptees into homes corresponding to the socioeconomic status of the adoptee's biological family, then adoptees born into poor families would have been placed into poorer adoptive homes, where they would have been more likely to develop pellagra. All of this wouldhave occurred even though the adoption agency was not aware of the 'unknown variables' of pellagra. And Kety and associates have acknowledged that schizophrenia, like pellagra, is correlated with lower socioeconomic status.[112]

In the opinion of Kety and associates, knowledge of the parents' psychiatric status did not influence the results of their studies because 'in practically every case' index adoptees were born to *parents* with no record of mental disorder at the time of adoption.[113] But as we are about to see, a prospective adoptee's biological parents were not the only relatives checked for a history of mental disorder. As seen in the 1946-47 annual report of the Mother's Aid Organization of Copenhagen, which was the largest adoption service in the country, the genetic background of a potential adoptee was of great concern to the Danish authorities. We should recall that adoptees in the Copenhagen and Provincial studies were placed between 1924 and 1947:

> Before a child is cleared for adoption, it is investigated with respect to health, and an attempt is made to obtain detailed information on the child's family background and to form an impression of its developmental potential. Not only for the adoptive parents, but also for the child itself, these investigations are of great importance for its correct placement. Information is obtained on the child's mother and father; *on whether or not there are serious physical or mental illness in the family background*; criminal records are obtained for the biological parents; and in many cases school reports are obtained. By means of personal interview with the mother an impression of her is formed. Where information is uncovered on convicted criminality or on mental retardation, *mental illness, etc. in the family background*, the case is referred to the Institute of Human Genetics of Copenhagen University, with whom there exists a valuable cooperation for advice on the advisability of adoption [emphasis added]. [114]

This passage demonstrates that selective placement was operating in Denmark during the period when the Kety et al. adoptees were being placed for adoption. Potential adoptees were carefully screened for a family history of mental disorders, which went well beyond the biological parents. Where suspicions were formed of a family history of mental disorders the case was turned over to the Institute of Human Genetics, which was the keeper of the National Psychiatric Register. It is likely that a more thorough check of family psychiatric records was then undertaken. Therefore, the fact that an adoptee's biological *parents* had no record of mental disorder at the time of adoption does not diminish the likelihood that placements were influenced by the prevalence of mental disorders *among other family members*. Because mental disorders were viewed as having an important genetic component — thus branding a potential adoptee as 'tainted' — the family background of an adoptee (who often was only an infant) was considered 'of great importance for its correct placement.' Potential adoptees with a family history of mental disorders were seen as having poor 'developmental potential' — that is, poor 'genetic potential' — and were likely placed with less qualified adoptive families.

It is critically important to understand that the country in which these adoptions took place had a long history of governmental and social support for eugenic practices. As early as 1925, Denmark passed a law forbidding 'insane and highly feeble-minded persons' from marrying.[115] In 1929 Denmark became the first European nation to pass national legislation for the purpose of promoting eugenic sterilization, predating the Nazi sterilization law of 1933 by four years. The 1929 Danish legislation legalized sterilization in cases of mental retardation or mental disorder. Although the word 'eugenic' did not appear in the law, it allowed sterilization 'where suppression of reproduction must be regarded of being of great importance to society.'[116] The law passed easily in the Danish Parliament and according to historian Bent Hansen, 'The Danish version of eugenics seemed to command agreement among all political parties.'[117]

In 1935 a new law was passed allowing compulsory sterilization in certain cases — typically those labeled 'mentally abnormal.'[118] Compulsory eugenic sterilization was now widely accepted in Denmark: 'While everybody up to and during the passing of the 1929 law had recommended caution, they now spoke of eugenics legislation as something that was urgently needed.'[119] According to Hansen, there were nearly 6,000 eugenic sterilizations performed in Denmark between 1929 and 1950, and it wasn't until the 1960s that compulsory eugenic sterilization was legally abolished in Denmark.

Manfred Bleuler, who was well acquainted with European attitudes toward the 'hereditary taint' of schizophrenia, gave the following description of the effect of these attitudes on the afflicted families:

> If one knows schizophrenics and their families well, it is sometimes a matter for despair to see how much they suffer under the terrible concept of 'familial tainting.' Like a sinister shadow it darkens the lives of many people and of entire families. The stifling, uncertain fear of coming from an 'inferior breed,' of carrying within one's self the seeds of something pathological, morbid, and evil (I am speaking in

the jargon the afflicted apply to themselves), like a curse that you must pass on to someone else, causes oppressive feeling of inferiority.[120]

And there is good reason to believe that attitudes in early- to mid-twentieth century Denmark were similar to those described by M. Bleuler.

There were many more available Danish children than there were families wanting to adopt them during much of the period when adoptees were placed. As described by Mednick, Gabrielli, and Hutchings,

> Many of these adoptions took place during the Great Depression and World War II. It was more difficult to find willing adoptive homes in these periods owing partly to the relative unavailability of adoptive parents and to the additional number of adoptees available.[121]

Apparently, the adoption process was a 'buyers' market' during this period of Danish history, and in the words of Mednick and Hutchings, 'serious deviance in the biological parents was routinely reported to the prospective adoptive parents unless they refused the information.'[122] Mednick discussed the case reports of children who were put up for adoption but were never placed: 'Every weekend (at least in the 1930s), Danish people who wished to adopt would visit the orphanages and pick children . . . Children whose selection by an adoptive parent is delayed may be less attractive physically and behaviorally.'[123] It is likely that many children were also less attractive *genetically*. Crudely put, each weekend Danish prospective adoptive parents arrived at the orphanage and picked children as they might have selected apples at the market. Healthy, attractive, and non- 'hereditarily tainted' children were picked by the most qualified parents, as they would have chosen the most attractive apples. The 'best' families would have gotten the most desirable children, leaving the less desirable either unadopted or placed with families providing less favorable environments.

Continuing with the analogy, we might expect that all buyers would have an equal chance to select the best apples. It is unlikely that this was the case in the Danish adoption process, however, because placement officials had the ultimate authority to determine where children would be placed. Most probably, children with the best 'developmental potential' were placed into more stable and nurturing family environments, leaving the rest to less stable families.

An important indication of the impact of selective placement was reported by Lewontin, Rose, and Kamin in 1984, who were granted access to the Danish-American raw data. It turned out that a notation had been made in each file if a relative had been placed in a mental institution. Lewontin et al. found that in eight of the 33 index *adoptive* (i.e., rearing) families a parent had been admitted to a mental hospital, which was true for none of the control families. Thus, the higher rate of spectrum diagnoses among index versus control biological relatives might reflect little more than the agencies' placement of children with 'tainted' biological relatives into more psychologically harmful adoptive homes. Lewontin and colleagues concluded,

> A credible interpretation of the Kety et al. results [is] that the schizophrenic adoptees, who indeed had been born into shattered and disreputable families, acquired their schizophrenia as a result of the poor adoptive environments into which they were placed. The fact that one's adoptive parent goes into a mental hospital clearly does not bode well for the psychological health of the environment in which one is raised.[124]

It appears that the single environmental variable assessed by Kety and associates — the prevalence of mental disorders among adoptive family members — showed a significantly higher psychiatric hospitalization rate among index versus control adoptive parents. This finding is consistent with the previously discussed evidence of selective placement in the Danish adoption process. As Mednick, who had intimate knowledge of the Danish adoption process, acknowledged, 'In Denmark, the national adoption agency had a policy of selective placement,' although he believed that 'the agency was not successful.'[125] However, the evidence suggests that the agency was 'successful' in many respects.

To summarize, it is likely that the placement of Danish adoptees was influenced by the socioeconomic and psychiatric background of their biological families, leading to the placement of 'tainted' children into more psychologically harmful rearing environments. Therefore, before any conclusions about the role of genetic factors can be drawn from the Danish Adoptees' Family studies, it must be shown that index adoptees were not systematically placed into rearing environments inferior to those experienced by control adoptees.

Reanalysis of the 1968 study

Methodological flaws and environmental confounds aside, it has been argued here that (according to Kety and associates' post-1968 design) the prevalence of chronic B1 schizophrenia among the first-degree biological relatives of B1 index versus control adoptees is the only valid measure of the genetics of schizophrenia. Utilizing these criteria, the index/control difference for affected individuals or families is not statistically significant. Among the 34 first-degree biological relatives of the 16 B1 index adoptees, there are zero diagnoses of chronic B1 schizophrenia, versus one diagnosis among the 68 first-degree biological relatives of controls. A comparison of families with at least one first-degree biological relative diagnosis yields similar results. Thus, on the basis of these criteria the 1968 study provides no evidence in favor of the genetic theory of schizophrenia. Kety and colleagues found significant differences for both comparisons by greatly expanding the definition of schizophrenia and by counting second-degree relatives.

The Danish-American Adoptees' Family study: 1975

The next major publication of the ongoing study appeared in 1975. Using the same index adoptees from the 1968 study and adding one more control adoptee, the researchers had the difficult task of attempting to interview as many adoptive and biological relatives as possible. They believed that there were many 'schizophrenia-related disorders' among these relatives which had not been identified through the use of institutional records.

The interviews were conducted by Danish members of the research team. English language transcripts were sent to Kety, Wender, and Rosenthal in the United States, who made blind global diagnoses. According to the authors, the interviews

> were extremely exhaustive, 35 pages in length, including many check lists and much narrative material, and covered the major aspects of the life experience: sociological, educational, marital, occupational, and peer relationship history from birth, medical background, and a careful mental status examination. These interviews were transcribed in English . . . and the transcripts were edited to remove any clues which a sophisticated reader might use to guess that this was a biological or adoptive relative of an index case or of a control.[126]

There were now 347 identified biological relatives in the study (173 index and 174 control). One record-based B3 index adoptee (S3) was re-diagnosed as B1, meaning that index adoptee diagnoses now consisted of 17 B1, 9 B3, and 7 B2. A 'screened control' group of 23 control adoptees (out of the original 34) was identified. This group consisted of control adoptees who were interviewed and were judged free 'from the suggestion of schizophrenic disorder.'[127]

The investigators found a significantly greater number of spectrum disorders in the index biological relative group as compared to the control biological relative group (37/173 vs. 19/174, p = .006). According to a table provided by Kety and associates,[128] the index/control differences for the various spectrum diagnoses were as follows: B1 index versus control = 5 to 0; B3 index versus control = 6 to 3; D1+ D2 + D3 index versus control = 13 to 3; and C index versus control = 13 to 13. A full 65% of non-category C spectrum diagnoses were made on half-siblings, and nearly one-half were 'uncertain.' Of the five index relative chronic (B1) schizophrenia diagnoses, only one was a first-degree relative. No significant spectrum differences were found between index and control adoptive relatives.

Although the investigators' calculations showed the combined spectrum diagnoses to be significantly concentrated among the index biological relatives, this finding was, according to Kety et al., 'compatible with a genetic transmission for schizophrenia,' but not 'entirely conclusive.'[129] Because of factors such as birth trauma, *in utero* influences, early mothering experiences, etc. 'one cannot, therefore, conclude that the high prevalence of schizophrenia illness found in these biological relatives of schizophrenics is genetic in origin.'[130] Unfortunately, this statement did not prevent a generation of textbook authors and commentators from concluding that the higher prevalence of 'schizophrenia illness' among index biological relatives showed that schizophrenia is genetic in origin.[131]

Kety and colleagues did point to another comparison which, they argued, provided 'compelling' evidence in favor of genetic factors:

> The largest group of relatives which we have is, understandably, the group of biological paternal half-siblings. Now, a biological paternal half-sibling of an index case has some interesting characteristics. He did not share the same uterus or the neonatal mothering experience, or an increased risk in birth trauma with the index case. The only thing they share is the same father and a certain amount of genetic overlap. Therefore, the distribution of schizophrenic illness in the biological paternal half-siblings is of great interest.[132]

The investigators noted that there were 16 record- and interview-based spectrum diagnoses among these paternal half-siblings, but that the distribution was 'highly unbalanced' (14 index vs. 2 control). Kety and colleagues concluded, 'We regard this as compelling evidence that genetic factors operate significantly in the transmission of schizophrenia.'[133]

As discussed previously, the investigators realized that counting individual relatives might not be 'entirely appropriate' due to the clustering of cases 'into a limited number of biological and adoptive families.'[134] A table was presented which showed, on the basis of their criteria, that index families were more affected at the .001 level of probability.[135] Kety and associates concluded that the weight of the evidence provided strong evidence in favor of the genetic basis of schizophrenia.

The inclusion and non-inclusion of category C

A critique of the 1975 study begins with a discussion of the way the spectrum was defined. As pointed out by Lidz and Blatt in 1983, category C (schizoid and inadequate personality) *was not* dropped from the schizophrenia spectrum; rather, it simply was not counted in the most important statistical comparisons.

After breaking the code and assigning interview diagnoses to their respective categories, Kety et al. discovered that category C cases were equally distributed among index and control biological relatives (index 13, control 13). That this diagnosis was nevertheless still considered part of the schizophrenia spectrum can be seen in Kety and colleagues' Table 3.[136] This table lists category C under the 'Schizophrenia spectrum' heading, and its totals are included in the column entitled 'Total in schizophrenia spectrum.' It therefore was established that category C remained a component of the spectrum. Kety and associates noted that this diagnosis did not differentiate index from control biological relatives, thereby casting doubt on its relationship to B1 schizophrenia. However, they 'were not prepared to dismiss the possibility that there is a schizoid or inadequate personality which is genetically related to schizophrenia.'[137]

As we have seen, Kety and colleagues found 'compelling evidence' for the genetic transmission of schizophrenia in the paternal half-sibling spectrum diagnosis difference. However, as seen in a comparison published in a table (but not discussed) by Kety and colleagues,[138] they had to *exclude* spectrum category C in order to obtain a statistically significant difference. In this table, the index paternal half-sibling versus control paternal half-sibling spectrum rate comparison — including category C — was listed as index 18/63 (28.5%) versus control 11/64 (17.2%), yielding a non-significant p value of .094. Thus, Kety and associates omitted category C from their calculation *even though they still considered it a schizophrenia spectrum disorder*. Lidz and Blatt called the removal of category C from this comparison a 'post hoc change in criteria' and correctly called the procedure invalid.[139] The significant difference cited by Kety and associates as 'compelling evidence' of the genetic transmission of schizophrenia turned out to be statistically *non-significant* — a fact which leading psychiatry and psychology textbooks published since the mid-1970s have failed to mention.[140]

As a co-author of the Kety et al. 1975 study, David Rosenthal's 'compelling evidence' for genetic influences on schizophrenia was based on the 'confusion and folly' of looking at subjects for whom he knew (at best) only half of the genetic picture, and by omitting the schizoid (category C) diagnosis, which in the same year he had written was 'indeed genetically related to process schizophrenia.'[141]

Problems of the interview method

Although typically referred to as the interview-based study, only 72% of the 347 identified biological relatives and 48% of the 165 identified adoptive relatives were interviewed in the 1975 study.[142] Of the biological relatives, 79% were alive and accessible; the rest had either died, disappeared, or had emigrated outside of Scandinavia.

Although such occurrences are common in studies of this type, a problem arises when we consider that all 347 identified biological relatives were counted in the study's statistical calculations. The investigators decided that 347 was 'the most unbiased and conservative denominator.'[143] Counting all relatives created a problem of what to do with the 1968 diagnoses for people who had either died or were unavailable for interview, and there are clear inconsistencies in the way these dead or unavailable relatives were diagnosed and counted.[144] For example, by comparing the tables presented in pages 154 and 158 of the 1975 study, it is apparent that one of the three 1968 record-based D (uncertain) relatives who had either died, emigrated, or had refused an interview was given a 1975 spectrum diagnosis, whereas the other two were not. However, *all* record-based B and D diagnoses were counted in the paternal half-sibling comparison.

The case of control adoptee C9's biological father highlights the inconsistent and often arbitrary manner of counting family members. This individual had received a 1968 B1 diagnosis but died before he could be interviewed. By 1975, his B1 diagnosis had been removed from statistical calculations yet he was still counted as a non-spectrum relative. Adding to the confusion, he was counted as a screened control record-based diagnosis in Table 4b,[145] but not in Table 3.[146] The omission of this relative from Table 3 made it appear as though B1 diagnoses were significantly concentrated among index biological relatives (5 to 0), when in fact the difference was a non-significant 5 to 1 if he is included.[147] This first-degree biological relative was a 'screened' as well as a B1 matched control. We shall soon discover the probable reason this person was not counted, but it cannot be accepted that he was not — especially since he was counted in some statistical calculations but not in others. Fortunately, this issue has been resolved because in 1988 Kety began counting C9's biological father as a Copenhagen B1 control diagnosis.[148]

Control adoptee C9's biological father provides an example of the way a diagnosis could be counted or not counted in different statistical calculations. His record-based B1 diagnosis was counted in the 1968 study, but the fact that he was the *only* first-degree biological relative diagnosed with schizophrenia was deemed unimportant when compared with the significant spectrum clustering among index biological relatives. The 1975 study failed to count this control biological father's chronic schizophrenia diagnosis, which allowed the index/control B1 comparison among all biological relatives to reach statistical significance. By 1988 this control father had been rehabilitated as a Copenhagen B1 relative, but by then he could be mixed in with the Provincial diagnoses, and the comparison remained statistically significant.[149]

Of the 364 alive and accessible biological and adoptive relatives, 12 refused the interview but 'adequate information' was said to have been available to make a diagnosis. According to the investigators, 'even though the individual persistently refused to give an interview, Dr. Jacobsen nevertheless obtained considerable information in the process.'[150] A description of how 'considerable information' may have been obtained is provided in a 1974 paper by Paikin et al. This paper, which was co-authored by Kety, Rosenthal, and Wender among others, described the problems that the investigators had with subjects who did not want to be interviewed. The following description illustrates a scenario where the researchers believed that they had obtained enough information from a home visit to be able to determine if a subject was in or out of the schizophrenia spectrum:

> In those cases where the psychiatrist was not invited into the house, the face-to-face contact varied between a few minutes to twenty minutes. In general, it was found that if the subject was seen for less than five minutes, the amount of information gained was not sufficient

for a judgment to be made as to whether the subject was psychiatrically inadequate or not . . . It must be stated that much of the information obtained was somewhat superficial; it was necessary to give some weight to a general impression of the subject with particular emphasis being placed on disturbances in emotional contact, language use and thought processes. The diagnostic presentation has, therefore, been restricted to three broad categories: outside the schizophrenia spectrum, suspected schizophrenia spectrum and schizophrenia.[151]

It appears that the Danish-American investigators may have diagnosed people with schizophrenia on the basis of a *five-minute doorstep conversation.* Although Paikin and associates described the interview process of a different Danish-American schizophrenia adoption study, it is possible that several relatives in the Adoptees' Family study were diagnosed in a similar fashion.

Real or fabricated interviews?

Next, we turn to the so-called 'pseudo-interviews,' whose existence was first brought to public attention in 1984. According to Lewontin and colleagues, who had been in correspondence with one of the psychiatrists conducting the interviews,

> In several cases, when relatives were dead or unavailable, the psychiatrist 'prepared a so-called pseudo interview from the existing hospital records.' That is, the psychiatrist filled out the interview form in the way in which he guessed the relative would have answered.[152]

In other words, Danish members of the team *made up* (fabricated) interviews for several relatives. Lewontin and colleagues described the case of index adoptee S11's biological mother, who had received a diagnosis of inadequate personality (category C) in the 1968 study. By 1975 her 'interview' diagnosis had been changed to uncertain borderline (D3). But in fact, this woman 'had committed suicide long before the psychiatrist had attempted to locate her, and so—from the original hospital records—she was "pseudo-interviewed."'[153] It turned out that she had been hospitalized on two different occasions for manic-depression, which was outside the spectrum and considered genetically unrelated to schizophrenia by Kety, Wender, and Rosenthal.[154]

The use of pseudo-interviews, which was left unmentioned in the published Danish-American papers, seems to explain the case of control adoptee C9's biological father. As we recall, he had been diagnosed B1 in the 1968 report but had died and was given no diagnosis in the 1975 follow-up, while still being counted as a case. Apparently, he had been 'pseudo-interviewed' and on the basis of this 'information' was judged non-schizophrenic.

Another description of the pseudo-interviews is contained in a 1984 reanalysis by Kendler and Gruenberg:

Based on an extensive review of hospital records, detailed pseudointerviews were constructed for all of the index adoptees. These pseudointerviews contained more detailed information on the index adoptees than had been available to Kety and co-workers when they made their initial diagnoses. However, although they contained a detailed account of the psychiatric illness, the hospital records did not contain all the information normally present in sections of the real interviews dealing with such factors as personal history or living environment. This difference in information content as well as other differences in format *made it impossible to be 'blind' to whether an adoptee interview was a real interview with a control adoptee or a pseudointerview with an index adoptee* [emphasis added].[155]

It appears that Kety and colleagues' 'blind' diagnoses may not have been as blind as they implied. Kendler and Gruenberg could tell the difference between a 'real' and a 'pseudo' interview, and it is possible that Kety, Rosenthal, and Wender could also tell the difference. Kety and colleagues knew from the 1968 records that there were more spectrum-diagnosed index biological relatives (13 index vs. 3 control), and the recognition of a pseudo-interview meant that the subject was more likely to be an index biological relative. Thus, another bias may have been introduced into the diagnostic process.

As Kendler and Gruenberg noted, *all* index adoptees' diagnoses were re-evaluated by pseudo-interview. There was only one change in index adoptee diagnoses from 1968 to 1975: index adoptee S3 was diagnosed B3 in 1968, but, apparently by pseudo-interview, was changed to B1 in the 1975 report. Adoptee S3 had a biological mother and a half-sibling receiving a 1975 B1 diagnosis. Based on Kendler and Gruenberg's account, it appears that S3's diagnosis was changed without him or her actually having been interviewed. Had this 'pseudo-interviewed' adoptee's 1968 diagnosis been left unchanged, the 1975 study would not have produced a single index first-degree chronic schizophrenia biological relative.

To summarize, the interview process was plagued by many difficulties. Only 64% of biological and adoptive relatives were actually interviewed, suggesting that the subtitle of the 1975 report, 'A Preliminary Report Based on Psychiatric Interviews,' was misleading. The use of 'pseudo-interviews' in place of an a priori decision on whether deceased or unavailable 1968 cases would be counted is a serious methodological error in this study. The procedure should have been discussed and justified in the methods section of all Adoptees' Family reports where it was used.

Reanalysis of the 1975 interview-based study
The index/control biological relative diagnosis difference is calculated here using the same criteria used for the 1968 study, that is, the B1 rate among the first-degree biological relatives of B1 index adoptees versus the B1 rate among the first-degree biological relatives of controls. Looking at affected individuals, there is one person diagnosed with chronic B1

schizophrenia out of 36 index first-degree biological relatives, versus one B1 diagnosis among the 70 first-degree biological relatives of controls. Among families, there is one affected index and one affected control family. It is therefore concluded here that the 1975 study provides no evidence supporting the genetic basis of schizophrenia.

The Danish-American Adoptees' Family study: 1994

Most critics of the Danish-American Adoptees' Family studies have concentrated on the Copenhagen half of the study, and with good reason—the final Provincial study paper was not published until 1994, over 30 years after the beginning of the study. As we begin a reanalysis of the Provincial study, the reader should bear in mind that most of the methodological flaws and potential environmental confounds discussed in relation to the 1968 and 1975 studies are applicable to the Provincial study as well.

The preliminary results of the Provincial study were first published in 1978.[156] The interview process began in 1980, but the first paper reporting interview diagnoses did not appear for another eight years,[157] and the first paper listing diagnoses for each family member did not appear until four years after that.[158] In 1994 the final Provincial study paper was published by Kety and colleagues, which shall be taken as a starting point.

The Provincial adoptee group was selected from a total of 8,944 Danish individuals given up for adoption between 1924 and 1947. These adoptions took place in the Danish provinces outside of the greater Copenhagen area. The Provincial study therefore looked at adoption records in 'the rest of Denmark.' Like the Copenhagen study, the names of Provincial adoptees were checked against the National Psychiatric Register in order to determine how many had been admitted to a psychiatric facility. When the records indicated that an adoptee's symptoms were compatible with a diagnosis of schizophrenia, a summary of the institutional records was translated into English, edited to remove family-related information, and sent to Kety, Rosenthal, and Wender in the United States, who performed blind consensus diagnoses. This process produced 41 index cases, consisting of chronic, latent, and acute diagnoses.

A matched control group was established on the basis of age, sex, socioeconomic status, and time of transfer from the biological family. By the end of the study the control group consisted of 24 adoptees and their 121 biological and 55 adoptive relatives. Because acute schizophrenia was dropped from the spectrum in the late 1970s,[159] the final index group consisted of 29 B1 (now called 'chronic schizophrenia') and 4 B3 (now called 'latent schizophrenia') adoptees. Biological and adoptive parents and their offspring were identified through the use of population registers. The chronic schizophrenia index adoptees had 171 biological and 71 adoptive relatives. The first stage of the relative work consisted of checking these names against the Psychiatric Register, after which blind diagnoses were made in much the same way as they had been made for the index adoptees.

The task of interviewing relatives began in 1980. Although nearly 90% of the available relatives were interviewed, this represented only 63% of all identified relatives.[160] As in the previous studies the researchers used the global diagnostic method, in spite of the publication of the DSM-III and its 'operationalized' diagnostic system.[161] Only two spectrum diagnoses were made on relatives in this study: chronic schizophrenia and latent schizophrenia.

The results were as follows: for chronic schizophrenia, index biological relatives had eight diagnoses (4.7%), and there were no such diagnoses among control biological relatives. For latent schizophrenia, index biological relatives had 14 diagnoses (8.2%), and control biological relatives had three (2.5%), which is also a statistically significant difference. The combined diagnoses (chronic + latent) totaled: index, 22/171 (12.9%); control, 3/121 (2.5%), leading Kety and associates to conclude:

> This study and its confirmation of previous results in the Copenhagen Study speak for a syndrome that can be reliably recognized in which genetic factors play a significant etiologic role. These findings provide important and necessary support for the assumption often made in family studies; observed familial clustering in schizophrenia is an expression of shared genetic factors.[162]

Latent schizophrenia in the provincial study

It has been argued briefly here and in detail elsewhere that 'latent schizophrenia' should not have been counted as schizophrenia in the Danish-American adoption studies.[163] However, in spite of the evidence in favor of this argument the diagnosis was retained as part of the Provincial study's dependent variable. There were 14 latent schizophrenia diagnoses among the index biological relatives; however, only four of these were first-degree relatives. The remaining ten diagnoses were given to half-siblings. Among controls, one first-degree and two second-degree biological relatives received a latent schizophrenia diagnosis. This meant that the latent schizophrenia rate among the first-degree biological relatives of chronic schizophrenia adoptees versus control first-degree biological relatives was not statistically significant.[164] But this is only the beginning of our story.

As we recall, in 1978 the researchers selected matching control subjects for the Provincial study. In a 1983 paper by Kety it was reported that this control group consisted of 42 adoptees.[165] However, by the 1994 final report there were only 24 control adoptees. It turns out that several controls were eliminated through a screening process quite different from that used in the Copenhagen study. We have seen that the 1975 study designated 23 control adoptees as 'screened controls' on the basis of having been interviewed and being found free from the 'suggestion of schizophrenic disorder.'[166] Although separate statistical calculations were performed for index versus controls and index versus screened controls in the 1975 study, the non-screened group's relatives were displayed in a table and they were not eliminated from the study.

The 1994 group constituted screened controls of a quite different type. At an undisclosed point, Kety and colleagues decided that all 13 controls who had been interviewed by a psychiatrist and diagnosed with a 'serious or confounding mental illness' *were dropped from the study.*[167] These 'confounding illnesses' were primarily non-spectrum affective disorders.[168] It is possibly permissible to eliminate a control adoptee on the basis of receiving a diagnosis of schizophrenia, but the removal of non-spectrum affective disordered adoptees was a serious error in the study. From a genetic perspective, the fact that a control adoptee was diagnosed with a non-spectrum psychiatric diagnosis would not 'confound' the results of the study.

An interviewed 1975 control adoptee diagnosed with an affective disorder was not only retained in the study, but qualified as a screened control. The following is a description of interviewed control adoptees in the 1975 report:

> Nine of these we called normal, one was diagnosed neurotic, *one as affective disorder*, and 12 were called personality disorders other than schizoid or inadequate personality. These 23 controls we designate as 'screened' to indicate that there was no suggestion of schizophrenic disorder among them [emphasis added].[169]

Apparently, a diagnosis certifying a control adoptee as 'screened' in 1975 was enough to remove him or her from the Provincial study.[170] In 1992 Kety and Ingraham wrote of their control group:

> In the Provincial sample, we have found latent schizophrenia in 5 of 37 (13.5%) biological relatives of probands screened out of the study due to a diagnosis of major affective disorder, a rate comparable to that in the biological relatives of the schizophrenic adoptees. It is possible that more precise diagnostic criteria could characterize a specifically schizophrenia-related syndrome.[171]

In other words, these controls were excluded because their biological relatives had spectrum diagnoses at rates comparable to index adoptees. The genetic basis of schizophrenia was therefore *assumed* in the process of investigating it. Although diagnosed with a major affective disorder, according to Kety and Ingraham they *may* have been misdiagnosed 'schizophrenics' and therefore had to be dropped from the study.

Had these control adoptees been retained, the latent schizophrenia index-control difference for *all* biological relatives (first- and second-degree) would have been non-significant.[172] This finding is confirmed by Kendler and Diehl, who viewed the final Provincial data while in preparation. They reported that *before* the reduction of the control group, 'latent and uncertain schizophrenia was not found to be significantly more common in the biologic relatives of the schizophrenia adoptees than in those of the control adoptees (6.5% vs. 5.5%, respectively).'[173]

Like the 1975 Kety et al. report, the Provincial sample was subject to a blind reevaluation by Kendler and associates, who looked at rates of DSM-III schizophrenia, schizotypal personality disorder, and paranoid personality disorder among interviewed index and control adoptees and their interviewed first- and second-degree biological relatives. Of the 37 control adoptees, 25 were given no psychiatric diagnosis, six were diagnosed with major depression, five with anxiety disorder, and one with schizotypal personality disorder.[174] Thus, only one control adoptee (schizotypal personality disorder) could be suspected of having a spectrum disorder, and none was diagnosed with chronic schizophrenia.

Five additional control adoptees were dropped from the Provincial study on the grounds that they refused (or were unable) to be interviewed. This was also a questionable move—there were 11 non-interviewed control adoptees in the 1975 study, but they were not removed from the control group. If these adoptees were selected on the basis of matching criteria and were checked against a national register for psychiatric hospitalization, this should have been sufficient to retain them in the control group. At no point are we told what diagnoses were found among their biological relatives, and at no point did Kety et al. indicate that they were unaware of these diagnoses at the time these five adoptees were excluded.

The Provincial study investigators began their work in 1975, published their first paper in 1978, began interviewing relatives in 1980 ... but first reported that control adoptees had been dropped *in 1992*. In other words, between the ascertainment of a control group in 1978 and the announced reduction of this group, *14 years had elapsed*. One could argue that the unjustified tampering with the Provincial control group is reason enough to invalidate the entire study, but let us move on.

There is also the question of how latent schizophrenia was defined in the Provincial study. In 1983 Lidz and Blatt argued that there is little difference between latent schizophrenia, uncertain latent schizophrenia, and (non-spectrum) schizoid personality, and this position received confirmation in the 1994 study. In 1980, the DSM-III was published. The third edition moved away from the older global diagnostic method and towards an 'operationalized' system. A new DSM category was born: schizotypal personality disorder (SPD), which was created from the differentiating symptoms of Kety and colleagues' Copenhagen B3, D3, and C relatives.[175] To receive a diagnosis of SPD using DSM-III criteria, a person must not meet the criteria for schizophrenia and must demonstrate at least four of the following eight symptoms: (1) 'magical thinking,' (2) 'ideas of reference,' (3) 'social isolation, e.g., no close friends or confidants,' (4) 'recurrent illusions,' (5) 'odd speech (without loosening of associations or incoherence), e.g., speech that is digressive, vague, overelaborate, circumstantial, metaphorical,' (6) 'inadequate rapport in face-to face interaction due to constricted or inappropriate affect, e.g., aloof, cold,' (7) 'suspiciousness or paranoid ideation,' (8) 'undue social anxiety or hypersensitivity to real or imagined criticism.'[176]

Kety frequently referred to SPD as comparable to the Danish-American latent schizophrenia diagnosis. For example, 'The components of the spectrum have assumed different names in DSM-III: schizotypal personality disorder takes the place of our latent or uncertain schizophrenia, from which its characteristics were derived.'[177] In Kety and Ingraham's 1992 update, the terms 'schizotypal personality disorder' and 'latent schizophrenia' were used interchangeably.

Schizotypal personality disorder carries the dubious distinction of being, in the words of Gunderson and Siever, 'the first diagnostic category introduced into standard diagnostic usage that is built on a genetic rationale and explicitly gives familial relationship primacy as a validating criterion.'[178] For the architects of the SPD classification, schizotypal personality disorder was 'merely a subdivision of what has for years been referred to as Schizoid Personality Disorder.'[179] By 1983, however, Kety had definitively removed schizoid and inadequate personality from the spectrum: 'There was . . . no justification for believing that schizoid and inadequate personality, as we had diagnosed them in the interview study, were related to schizophrenia, and were therefore excluded from the subsequent analyses.'[180] It therefore becomes even more evident in the Provincial study that SPD/latent schizophrenia is barely distinguishable from non-spectrum schizoid personality disorder. According to the DSM-III, the only symptom differentiating SPD from schizoid personality was the former's 'eccentricities of communication or behavior.'[181] Even Kety and associates admitted that 'it is doubtful that we could demonstrate a significant differentiation between' latent schizophrenia, uncertain schizophrenia, and schizoid/inadequate personality.[182]

In their 1994 paper, Kety and associates listed 'criteria regarded as important in our diagnosis of schizophrenia and their presence or absence in the adoptees for whom we made that diagnosis.'[183] A table was provided for chronic schizophrenia adoptee diagnoses, and another table listed symptoms of relatives diagnosed with latent schizophrenia. The table for latent schizophrenia[184] demonstrated just how few symptoms were necessary to make a spectrum diagnosis, and how often latent schizophrenia was essentially indistinguishable from the schizoid diagnosis. According to my calculations, only five of 17 relatives met the criteria for the supposedly comparable DSM-III schizotypal personality disorder, and two relatives were diagnosed on the basis of *one* major symptom. One of these, a member of family #420, was diagnosed on the sole criterion of 'inappropriate or constricted affect'; the other, from family #936, on the basis of 'odd or digressive speech.' The diagnosis of a member of family #871 is even more illustrative of the lack of any meaningful difference between latent schizophrenia and schizoid personality. This individual was diagnosed with latent schizophrenia on the basis of (1) 'no close friends/seclusive/withdrawn,' and (2) 'suspicious/paranoid ideation.' However, in the same report *schizoid* personality was described as 'most frequently reflect[ing] severe introversion, suspicious or referential symptoms in otherwise psychiatrically unremarkable individuals.'[185] In other words, the difference between spectrum latent schizophrenia and non-spectrum schizoid personality is known only to the practitioners of the 'global diagnostic system.'

How much further could Kety and associates have strayed from the teachings of Eugen Bleuler, the inventor of the schizophrenia concept? In contrast to the Danish-American investigators, Bleuler believed that schizophrenia should be diagnosed conservatively: 'Only a few isolated psychotic symptoms can be utilized in recognizing the disease, and these, too, have a very high diagnostic threshold value.'[186] On the other extreme, the Danish-American investigators were prepared to count as schizophrenia persons demonstrating the single symptom of 'inappropriate or constricted affect,' and by Rosenthal's definition, simply 'cold, distant and inadequate' people.[187]

In Kendler and colleagues' 1994 reanalysis, the SPD rate among all Provincial index biological relatives versus all control biological relatives was not statistically significant.[188] Looking at first-degree relatives only, the difference remained non-significant.[189]

The evidence suggests that 'latent schizophrenia' did not deserve to be counted as schizophrenia in 1994 any more than it did in 1968 or 1975. Kety claimed that its association with chronic schizophrenia was reaffirmed in this study. In fact, the opposite is true. The Provincial study reconfirmed its *non-significant* clustering among index first-degree biological relatives,[190] and also supported Lidz and Blatt's 1983 observation that 'latent schizophrenia' describes non-psychotic people considered odd and reclusive, who are not easily distinguished from people labeled (non-spectrum) schizoid personality.

Chronic (B1) schizophrenia in the Provincial study
Although the B3 latent schizophrenia diagnosis is unworthy of being counted as schizophrenia, the fact remains that Kety and associates diagnosed eight Provincial index biological relatives with chronic schizophrenia, versus zero control biological relatives. As always, these figures require closer examination. According to Kety and colleagues' Figure 1, two of the eight index diagnoses were given to maternal half-siblings, who are second-degree relatives.[191] The remaining six diagnoses consist of two biological fathers, one biological mother, and three biological full-siblings. Regarding the latter, the point was made by Kety et al. that, whereas the Copenhagen study suffered from a lack of biological full-siblings 'that prevented a meaningful estimate of disease prevalence in first-degree relatives' (although Kety et al. studied 131 biological parents), in the Provincial sample,

there were a substantial number of full siblings (24 biological index siblings compared with three in the Copenhagen Sample), probably the result of differences in lifestyle from that of the large city. Schizophrenia was found in three (12.5%) of the 24 full siblings, nearly six times the rate in the half-siblings (two [2.2%] of 90) (one-tailed P = .062).[192]

However, the chronic schizophrenia rate of 3/24 (12.5%) among biological full-siblings is misleading because *all three full-siblings were from the same family*.[193] And what a family this was! Index biological family #069 consisted of a mother, father, their three full-daughters, two full-sons, and three paternal half-siblings. As always, no information was provided on how or where these children were reared. In the likely case that they were reared in the same home, a plausible explanation would be that environmental conditions contributed to the level of psychological disturbance in this family.

In addition to family #069's daughter and two sons diagnosed with chronic schizophrenia, one of the two remaining full-daughters received a diagnosis of non-spectrum 'bipolar illness.' Of the three paternal half-siblings in family #069, two were diagnosed as non-spectrum schizoid. Because the Provincial sample was drawn from the small towns and rural communities of Denmark, one can envision this family living together in the Danish countryside, possibly on a farm. The parents (neither of whom received a psychiatric diagnosis) were perhaps hard-working people tending to their land, but from a psychosocial perspective they raised some pretty screwed-up kids. What kind of a household could this have been in which, of the eight children, three were diagnosed with spectrum chronic schizophrenia, two with non-spectrum schizoid, and one with non-spectrum bipolar illness? And what were the circumstances surrounding the abandonment of a child (the index adoptee) by parents who raised eight biological children together? As is always the case in a Danish-American study, we can only speculate due to a total lack of information about any of these people's *environments*.

Of the remaining 20 index biological full-siblings, who were born to six different sets of biological parents, *none* received a diagnosis of chronic schizophrenia.[194] One could legitimately conclude that family environment, rather than genetic background, best explains the clustering of chronic schizophrenia among the siblings of one family. This underscores the observation by Benjamin, Mosher, and Kety et al. themselves (in 1975) that the counting of affected individuals, as opposed to affected families, is misleading because a common environment violates the assumption of the independence of individual observations.

The three remaining first-degree biological relative chronic schizophrenia diagnoses were made on two fathers and one mother. The chronic schizophrenia adoptee from family #139 demonstrates the lack of symptoms necessary to receive this diagnosis using the global diagnostic method. According to a table listing the course and symptoms of the chronic schizophrenia adoptees, this individual manifested the following five symptoms: (1) 'insidious onset,' (2) 'schizoid features observed premorbidly,' (3) 'chronic course observed,' (4) 'withdrawal from social interaction,' and (5) 'flat affect.'[195] Other symptoms which this adoptee *did not* manifest include, 'autistic behavior,' 'poverty of thought/speech,' 'loose associations,' 'suspicious/ideas of reference,' 'delusions,' 'auditory hallucinations,' and 'other hallucinations.' This adoptee, therefore, presented no uniquely psychotic symptoms, or even one symptom that would have differentiated him from a person receiving a latent schizophrenia or schizoid diagnosis. Nor would he have been diagnosed with chronic schizophrenia by the criteria of DSM-I through IV. This individual's chronic schizophrenia diagnosis is therefore questionable; in fact, it calls into question the validity of every chronic schizophrenia diagnosis in the entire Adoptees' Family series.

Of the original 42 index adoptees (29 of whom were diagnosed with chronic schizophrenia by Kety et al.), Kendler, Gruenberg, and Kinney diagnosed only 19 with chronic schizophrenia.[196] Although it is unclear how many of these overlapped with Kety and colleagues' 29 chronic index adoptees, at least ten of the latter were diagnosed with conditions other than chronic schizophrenia. Among the 28 index first-degree biological relatives of chronic schizophrenia adoptees, Kendler et al. found two cases of chronic schizophrenia, and among the 60 first-degree biological relatives of controls, they found one case.[197]

The 'National sample'

In 1994 Kety and associates tabulated the results of what they called the 'National sample,' meaning the combined diagnoses of the Copenhagen and Provincial studies. The investigators reported highly significant differences for chronic and latent schizophrenia rates among all index biological relatives versus controls.[198] However, as Kety and colleagues acknowledged, the Provincial study constituted a 'replication of the Copenhagen study in the rest of Denmark.'[199] By its very nature, an attempt to replicate must stand on its own merit; otherwise, two statistically non-significant results could be combined into one large significant result, or a non-significant replication could be added to the significant original study and thus lead to a completely different conclusion than it would have, had it stood alone. The 'independence of samples,' wrote Gottfredson and Hirschi, 'is essential to the interpretation of replication research.'[200] Therefore, the results of the National sample are not reanalyzed here.

Reanalysis of the 1994 study

The Provincial study was plagued by serious methodological problems. Almost one-half (43%) of the control adoptees were eliminated from the study on questionable grounds, and there were several latent schizophrenia diagnoses among their biological relatives. How many relatives were diagnosed on the basis of a five-minute doorstep interview? How many by pseudo-interview? How many index adoptive parents had been admitted to a mental hospital? How many deceased hospital-diagnosed relatives were excluded from statistical calculations? In short, as Lewontin and associates noted in 1984 when discussing the Provincial study in its preliminary phase, 'there is no reason to suppose that the more recent work is free of the invalidating flaws we have outlined above.'[201] Most importantly, the results are confounded by the selective placement of adoptees, meaning that the critical theoretical assumption of all adoption studies was violated.

The six cases of first-degree index biological relative chronic schizophrenia (diagnosed by Kety et al.) produce the first statistically significant comparison in the Adoptees' Family series. These diagnoses are found among a group of 82 first-degree relatives. Among controls, no chronic schizophrenia diagnoses were made in the group of 59 first-degree biological relatives.[202] If adoptee #139 and his two first-degree relatives are removed, however, the comparison does not reach statistical significance.[203] The first-degree relative chronic schizophrenia rate among affected *families* is not significant.[204] On the basis of these calculations, and of the previously discussed flaws in this study, it is concluded that, like the 1968 and 1975 investigations, the results of the Provincial study do not support a genetic basis for schizophrenia.

Rosenthal and Associates' Danish-American Adoptees study

Rosenthal and colleagues' 1968 Adoptees study began with a group of 5,483 Danish adoptees placed between 1924 and 1947, and from that group identified about 10,000 of their biological parents. The names of these parents were checked against the Psychiatric Register of the Institute of Human Genetics. Case reports were assembled in Denmark and were sent to Rosenthal, Wender, and Kety in the United States. When a consensus B1, B2, or B3 diagnosis was arrived at, that person became an index case. The investigators then retrieved the information on the child this index parent had given up for adoption. This information was sent to the examiners in Copenhagen. Of the 69 index adoptees identified in the 1968 report, 30 could not be interviewed for reasons such as refusal, death, and emigration. This left a total of 39 index adoptees, who were interviewed by two Danish psychiatrists involved in the study (J. Welner and F. Schulsinger). A control group consisting of the adopted-away biological offspring of 47 people with no history of admission to a psychiatric hospital was established. These adoptees were matched with the index group on the basis of sex, age, age at transfer to the adopting family, and socioeconomic status.

According to Rosenthal and colleagues, the study was based on four assumptions: (1) 'That heredity was an important contributor to schizophrenia,' (2) 'That this inherited factor manifested itself in the behavior or personality of persons who were not frankly schizophrenic,'

(3) 'That these manifestations could be detected by tests or in one or two interviews,' and (4) 'That we would know which questions to ask or which tests to detect these manifestations.'[205] Although the 1968 study reported a statistically non-significant index/control spectrum diagnosis difference, Rosenthal concluded, 'The data provide strong evidence indeed that heredity is a salient factor in the etiology of schizophrenic disorders.'[206] By 1971, Rosenthal and colleagues were reporting statistically significant results on the basis of a larger sample of 76 index and 67 control adoptees. Of these, 24 index and 12 control adoptees received a 'schizophrenia spectrum' diagnosis, a difference that just reached the .05 significance level.[207]

Critique of the study
The inclusion of manic-depressive subjects in the study
Of the original 69 index biological parents, 11 were diagnosed by one or all of the judges with manic-depressive disorder.[208] This is surprising when we consider that Rosenthal believed that schizophrenia and manic-depressive psychosis are 'genetically distinct and different disorders.'[209] In 1976 Kety, Rosenthal, and associates wrote that 'Manic-depressive illness was never thought to be in the schizophrenia spectrum by us.'[210] (This is a curious claim because the Rosenthal et al. 1971 final report listed 'manic-depressive psychosis' as one of the diagnoses 'that we are tentatively including in the "schizophrenia spectrum."'[211]) Why then did this 'genetically distinct and different disorder' qualify a parent as an index case? Rosenthal offered two explanations:

> These [manic-depressive] cases were included for two reasons. *The first was one of expediency.* There were periods when we simply did not have enough schizophrenic parents processed and the staff in Copenhagen had no subjects to examine. The second and more important reason derived from this question: What if we should find differences between our Index and Control groups . . . If we had a comparison pathology group, we might be able to learn something . . . about the possible genetic relationship between schizophrenia and manic-depressive psychosis [emphasis added].[212]

Rosenthal and colleagues must be criticized for including an admittedly non-related diagnosis in a study of the genetics of schizophrenia *for reasons of expediency*. Imagine a study on the genetics of heart disease that added cancer patients because the staff 'had no subjects to examine.' Of course, as noted by Lidz, Blatt, and Cook, it may have been permissible to study the offspring of people diagnosed with manic-depression in order to contrast the two conditions, but it was erroneous to include them as part of the index parent group.[213] According to Pam, the inclusion of manic-depressive subjects rendered the study 'invalid on its face.'[214]

Although Rosenthal maintained that learning about the relationship between schizophrenia and manic-depression was the primary reason for including the latter disorder, the evidence suggests that 'expediency' was a more important factor. Had Rosenthal been able to obtain enough spectrum index parents to complete his sample, there would have been no need to look outside of the spectrum for additional cases. At this point, observed Lidz and colleagues, 'the study could no longer properly be termed a study of adopted-away offspring of schizophrenic parents.'[215] It is also difficult to reject the idea that manic-depressive psychosis was placed 'tentatively' in the spectrum because (as we will see) the index/ control spectrum diagnosis difference would not have reached statistical significance without it.

Selective placement

The problem of selective placement discussed in relation to the Kety and colleagues studies applies to all studies using adoptees placed in mid-twentieth century Denmark, and is particularly problematic in Rosenthal's Adoptees study.

Because 84% of adoptees were born before their biological parent's first admission to a psychiatric hospital, Rosenthal and colleagues argued in 1968 that knowledge of the biological parent's condition did not play a major role in the adoption process. However, it is not enough to state that only 16% of adoptees were placed when their parent's status was known; the diagnostic status of these 16% must also be revealed. Although the 84% figure might appear to have minimized the possibility of selective placement on the basis of a parent's mental health status (in direct contrast to Heston's 1966 study, where *all* of the adopted children were born to institutionalized women diagnosed with schizophrenia), we have seen that the Danish adoption agencies typically checked the psychiatric records of the *biological families* of potential adoptees, not just the parents. Therefore, Rosenthal and associates needed to demonstrate that the existence of psychiatric disorders among index biological family members did not create conditions leading to the placement of index adoptees into less qualified adoptive homes. That such placements likely occurred is based on the fact that *from either a genetic or environmental perspective*, the biological family of a person diagnosed with schizophrenia — or a person later diagnosed with schizophrenia — would be expected to have more psychiatrically diagnosed members than a control biological family. This means that in societies (such as Denmark, with its national psychiatric register) where adoption agencies use the psychiatric status of an adoptee's biological family as a placement criterion, the Adoptees method is confounded by selective placement and therefore constitutes an invalid research model. In 1994 Kety and associates wrote,

> The possibility of knowledge on the part of adoptive parents, adoptees, and mental health personnel regarding the presence of mental illness in the biological parents and the effect such knowledge could have on the occurrence, perception, and diagnosis of mental illness in the adoptee ... could be of some significance in studies where the adoptee

sample represented children born of mentally ill mothers.[216]

Thus, Kety and colleagues acknowledged that the results of schizophrenia adoption studies using the Adoptees method could be influenced by the adoptive parents' knowledge of the adoptee's biological background.

Diagnostic procedure

Welner and Schulsinger interviewed and tested subjects in Copenhagen and provided a 'thumbnail diagnostic formulatin' for each By the end of the interview process, Rosenthal and colleagues would claim that 'none of the examiners knew if the subject before him was an index or contol case,'[217] although four years later it would develop that in 'perhaps a few cases' they did know the group status of the subject.[218] All diagnoses were based on Welner's and Schulsinger's thumbnail formulations. As noted by Lewontin et al. (who had been in personal correspondence with several of the collaborators), these formulations did ot specify whether an individual warranted inclusion in the spectrum and the decision of whether these diagnoses were in or out of the spectrum 'was made in a manner and by parties unknown.'[219] Thumbnail diagnoses were preliminary and were made in lieu of consensus . Some examples of thumbnail diagnoses from the 1971 study include, 'possible paranoid borderline,' 'paranoid character,' 'almost pseudoneurotic borderline,' 'moderately schizoid,' 'pronounced preschizophrenic diathesis,' and 'conceivably paranoid borderline.'[220]

Assumptions

It is necessary to address the ambiguity surrounding the goals of this study. The investigators claimed that the study's purpose was to test the 'assumed diathesis' of schizophrenia, and as we recall from the study's four assumptions, the first was that heredity is 'an important contributor to schizophrenia.' This position was reaffirmed in a 1983 paper by Kety:

> It is important . . . to explain what sometimes appear to be differences in the prevalence of one or another diagnosis between our 'adoptee' studies (Rosenthal et al., 1968, 1971) and the present studies on families (Kety et al., 1968, 1975). In the former studies *the emphasis was on the components in the diathesis for schizophrenia,* and we adopted a low threshold for their notation. Diagnosis was a secondary consideration . . . In the family studies, on the other hand, because of the hypothesis being tested, the emphasis was placed on diagnosis [emphasis added].[221]

Thus according to Kety, Rosenthal's Adoptees study looked at the components of an assumed predisposition for schizophrenia, which explains the inclusion of diagnoses in Rosenthal's spectrum that the Kety-led studies did not use. This is a perfectly acceptable explanation, provided that Rosenthal's Danish Adoptees study made no claim about providing evidence in favor of the genetic basis of schizophrenia. But the conclusion of the 1971 report (of which Kety was a co-author) stated that 'the evidence supports the theory that heredity plays a significant role in the etology of schizophrenia spectrum disorders.'[222] This is not a valid conclusion of an investigation whose purpose was to study the components of an *assumed* genetic component of schizophrenia. The study thus revolved around the investigators' circular argument that they assumed that which they concluded, and concluded that which they assumed.

Kety also claimed in 1983 that 'we did not make independent and consensus diagnoses in the "adoptee" studies but merely cited the investigator's diagnostic impressions.'[223] This was true in the 1968 and 1971 reports but, as we will see, consensus diagnoses were reported in a 1978 paper of which Rosenthal and Wender were co-authors. By 1983 Kety was clearly attempting to downplay the importance of Rosenthal's1971 study. Neertheless, the study has been cited in countless textbooks as providing important evidence in favor of the genetic basis of schizophrenia.

Generalizabilty

It is possible that the status of being a mid-twentieth century Danish adoptee, in itself, led to psychological disturbance (although not to a diagnosis of schizophrenia, as we have seen). Evidence supporting this position can be found in the Danish adoption study literature, and two examples are cited here. Rosenthal and colleagues reported that 18 out of 148 (12%) index and control adoptees had been admitted to a Danish psychiatric facility at some point in their lives.[224] Although some would argue that genetic factors explain this high rate for the index adoptees, over 10% of controls (7/67) had been admitted to a psychiatric facility, and 25% of controls received a consensus spectrum diagnosis.[225] This high rate could be explained on the basis of psychologically harmful aspects in the adoption process, which psychodynamically oriented commentators would identify with terms such as abandonment, and attachment disturbance. As a second example, Mednick and colleagues' Danish criminality adoption study data show that Danish male adoptees born between 1924 and 947, who lacked a criminally registered biological parent, had about twice the risk of becoming an adult registered criminal as non-adoptees also lacking a criminally registered biological parent (see Chapter 8).[226] In addition, another Danish study by Raine and colleagues found a correlation between early maternal rejection and later violent behavior as an adult.[227] It is therefore questionable whether the results of any Danish psychiatric adoption study can be generalized to the non-adoptee population.

The failure to find statistically significant index-control differences

Although a statistically significant index/control spectrum diagnosis difference was claimed in the 1971 study, the removal of non-spectrum manic-depressive index parents and their biological children produces different results. By excluding the 'M' and 'B2/M' manic-depressive index parents and their six 'spectrum' children, Lidz and colleagues found that the study yielded non-significant results.[228] And it is clear that further reanalysis is in order. The 'B/D/M/A' category cannot be included because three of the four judges diagnosed these parents as either 'not in the schizophrenia spectrum,' 'manic-depressive,' or 'doubtful schizophrenia.'[229] 'Paranoids' have never been part of the Danish-American spectrum, and must also be excluded as adoptee spectrum cases.

Furthermore, we have seen that the Danish-American researchers ultimately removed category C (schizoid and inadequate personality) from the spectrum,[230] although it was retained in the Kety et al. 1975 spectrum in spite of being omitted from certain comparisons. This put Rosenthal in the interesting position of dropping the schizoid diagnosis from these comparisons at the exact same time (and in the same book) as he was defending its inclusion in the spectrum when applied to his Adoptees study:

> In the Kety study, which we call the Family study, the findings strongly favor the view that diagnostic categories such as uncertain schizophrenia and borderline schizophrenia are genetically related to classical process schizophrenia, but schizoid personality is not. Thus, we now talk about a hard spectrum and a soft spectrum . . . Undaunted, I want to show you why I think the soft spectrum, which includes a number of syndromes that we call schizoid, is indeed genetically related to process schizophrenia.[231]

In their jointly authored publications of the late 1970s, Kety and Rosenthal wrote that the evidence suggested that the schizoid diagnosis was unrelated to chronic schizophrenia, but that the evidence was not conclusive.[232] This allowed Kety to drop the diagnosis from certain comparisons in his study, and Rosenthal to retain it in his, and allowed both to claim statistical significance for the comparisons they considered most important. Contrary to Rosenthal's reasoning, however, a 'spectrum' diagnosis cannot be appropriate for one study and not the other (both of which, it should be stressed, were co-authored by Rosenthal). Schizoid personality is either genetically related to schizophrenia, or it is not.

Additionally, the B2 parents (and the schizophreniform adoptees) must also be eliminated as spectrum diagnoses because by 1978 the Danish-American investigators decided that this diagnosis did not belong in the schizophrenia spectrum.[233] The remaining diagnoses are chronic schizophrenia and borderline schizophrenia; all other categories have been eliminated by the Danish-American investigators' own criteria. This means that the results of the Rosenthal et al. 1971 investigation, based on the Danish-American investigators' post-1978 definition of the schizophrenia spectrum, were not statistically significant (12.5% vs. 7.5%).[234] Therefore, apart from the methodological problems associated with this study, the index/control 'thumbnail diagnosis' comparison was not statistically significant.

Later reanalyses
Haier, Rosenthal, and Wender, 1978
In subsequent years there appeared two articles purporting to sustain Rosenthal and associates' previous findings. The first was authored by Haier, Rosenthal, and Wender, and appeared in 1978.[235] This study reported an analysis of MMPI scores for 128 adoptees from the Rosenthal et al. 1971 study (64 index, 64 control). All index adoptees' biological parents had been hospitalized for 'schizophrenia,' so apparently the manic-depressive index parents and their adopted-away children were not included in this study. There is no indication of whether parents diagnosed as B2 were included.

Haier, Rosenthal, and Wender made several comparisons of index and control MMPI profiles, all of which failed to differentiate the two groups. They produced a table of MMPI clinical scale raw scores in which neither the index nor the control group had a mean T score above 61 on Scale 8 (Schizophrenia). There were no significant differences on two MMPI-derived scales looking at psychosis — 'Eichman Schizophrenia Signs' and 'Peterson Signs.' Finally, the researchers devised a formula that produced statistically significant results: they looked at the difference between index and control adoptees who had received spectrum diagnoses *and* had an elevated T score (≥ 70) on *any clinical scale*. Using this formula they were able to show that 66% of the index adoptees met these criteria, whereas only 25% of controls did. The difference is statistically significant and permitted Haier et al. to conclude that their study produced evidence which 'continued to support the genetic hypothesis,' in spite of the admission that 'on the basis of the MMPI data, there are no overall personality differences between the index and control groups.'[236]

Haier and associates' methods provide an example of the unfortunate practice of 'data dredging' (or 'data mining'). There is no indication that any of these comparisons were conceived *before* the data were collected and analyzed, and Haier, Rosenthal, and Wender clearly described a process in which they found no differences between index and control adoptees. Finally, they devised a comparison that provided results supporting the genetic position, and their conclusions were based on *this* comparison. 'The authors,' wrote Boyle, 'were by now arbitrarily searching for anything which would discriminate index and control groups.'[237]

The Haier et al. paper featured the first published report of consensus spectrum diagnoses among index and control adoptees. As we have seen, only thumbnail diagnoses were reported in the Rosenthal et al. 1968 and 1971 papers. Although the results from these studies were preliminary, countless reviewers have focused on them and ignored (or were unaware of) the consensus diagnoses reported by Haier and associates. According to the results, there was no significant difference (33% vs. 25%) between index and control groups on the basis of an investigator-defined spectrum of chronic schizophrenia, borderline schizophrenia, and 'schizophrenic personality' (which included schizoid and paranoid personalities).[238] The importance of this *negative result* was not discussed by Haier et al., who mentioned the consensus diagnoses only in the context of whether 'the MMPI can . . . sharpen the comparison between index and control groups.'[239]

Eight years prior to the publication of the Haier et al. reanalysis, Rosenthal wrote, 'Through psychological tests and intensive psychiatric interview, we hope to be able to discriminate the two groups [index and control] with respect to a number of psychological and behavioral characteristics.'[240] As the results of the 1978 study clearly show, Rosenthal and associates were *unable* to discriminate the two groups on the basis of tests and interviews.

Lowing, Mirsky, and Pereira, 1983
A final reanalysis, this time using DSM-III diagnostic criteria, was published by Lowing, Mirsky, and Pereira in 1983. Here, the index and control groups were reduced to 39 subjects each by eliminating all adoptees whose parents did not have a diagnosis of 'schizophrenia,' or whose biological father was not identified. The diagnoses of these adoptees' index parents were: B1 = 29, B2 = 3, B3 = 5, and 'mixed schizophrenia symptoms' = 2.[241] Two raters blindly reviewed the data on these adoptees and made diagnoses using a new version of the schizophrenia spectrum even broader than the 1968 original. This new spectrum included chronic schizophrenia, schizotypal personality disorder, paranoid personality disorder, borderline personality disorder, and a new hybrid—'mixed spectrum disorder,' which was defined as overlapping 'at least two of the five spectrum disorders but did not meet the full criteria for any one of the five.'[242] Adoptees whose diagnoses fell into this broadly defined spectrum were called 'disordered subjects.' No justification for widening the spectrum was provided by Lowing and associates, nor was it explained why three parents diagnosed with B2 'acute schizophrenia' (a diagnosis excluded from the spectrum by Kety et al. in 1978) were retained in the study.

Lowing and associates found 15 disordered subjects among the index adoptees and five among controls, a statistically significant difference. However, this comparison was based on a spectrum containing several categories that were excluded from Kety et al.'s original 1968 spectrum. When Lowing et al. narrowed their comparison to chronic schizophrenia, schizotypal personality disorder (the DSM-III equivalent of borderline schizophrenia), and schizoid personality, they found the difference significant at the .04 level. But this finding was dependent on the schizoid diagnosis, which means that it cannot stand. If the comparison is limited to chronic schizophrenia and schizotypal personality disorder, the results are not statistically significant.[243] In fact, Lowing and colleagues diagnosed only one case of chronic schizophrenia in the entire sample. In 1988

Ingraham and Kety wrote,

> Using DSM-III criteria, Lowing et al. found three times as many schizophrenia spectrum disorders in the index group as in the control group. These authors specified DSM-III schizophrenia, schizotypal personality disorder, and schizoid personality disorder as the schizophrenia spectrum.[244]

But because he had dropped the schizoid diagnosis in 1983, Kety himself *did not* define the spectrum this broadly. Thus, Kety should have concluded that Lowing and colleagues found no statistically significant differences between Rosenthal's index and control groups.

According to Lowing and associates, 'On the basis of the findings reported here of the significantly greater risk for schizophrenia spectrum disorders among the index subjects, we conclude that the hypothesis has been strongly supported.'[245] But the fact remains that they identified only one case of DSM-III schizophrenia among the adoptees, and, even with the inclusion of DSM-III schizotypal personality disorder, no significant results were found. Their report, in fact, confirmed the *negative* results of the previous studies.

Additional problems with Rosenthal's study

It is a testament to the problems with Rosenthal and associates' study that there are several other problems which, in the interest of brevity, I chose not to examine. These include the failure to find index schizophrenia rates significantly higher than the general population expectation as a prerequisite for being able to generalize the results to the non-adoptee population; the failure to provide life-history information for the people under study; the problems that may be associated with adoptees who spent months or years with their biological mothers before being placed; and the fact that homosexuality was a diagnostic criterion for the B1, B2, and B3 diagnoses,[246] and that being diagnosed as a 'pervert' qualified as an index spectrum diagnosis.[247] Rosenthal's Adoptees study was methodologically unsound to the extreme, and the investigators' bias is seen at every step and in every conclusion. This bias also blinded the investigators to the confounding influence of selective placement in their study.

The 'Finnish Adoptive Study of Schizophrenia'[248]

Pekka Tienari has made the study of schizophrenia his life's work — beginning with his well-known twin study, and culminating in a longitudinal study of adoptees who had been given away by their schizophrenia-diagnosed biological mothers. Tienari is unique among schizophrenia adoption researchers because in addition to genetics, he places significant importance on the psychological environments of the adoptive families whom he has studied. The Finnish study is, by a large margin, the most comprehensive and methodologically sound schizophrenia adoption study to date. Therefore, the investigators' finding that genetic factors influence schizophrenia must be carefully examined.

Tienari has criticized the Danish-American studies for having looked at their subjects' family environments 'in a very limited manner,'[249] and as having been 'misinterpreted as a final proof of genetics, though they only propose suggestions and trends.'[250] On the other hand, his study is the product of an era in which the genetic basis of schizophrenia is rarely questioned. Tienari and associates believe that the genetic component of schizophrenia is real but overemphasized, and that familial environmental factors have been given insufficient attention. 'The major goal of the Finnish Adoptive Family Study,' wrote Tienari and colleagues, 'is to reassess genetic contributions to schizophrenia and to add measures of the adoptive family rearing environment.'[251]

The general approach of the investigators has been:
1. To present evidence that the adopted-away children of mothers diagnosed with schizophrenia manifest schizophrenia spectrum disorders at a significantly higher rate than their matched controls, thereby establishing the genetic component, *then*,
2. To note that virtually all adoptees diagnosed with schizophrenia have been raised in chaotic or disturbed adoptive families. Thus, 'genetically predisposed' children may be more vulnerable when raised in such families, whereas healthy rearing families may serve to protect them from schizophrenia. Both genetic background *and* family environment are seen as predictor variables of schizophrenia.

Method
The Finnish Adoptive Family Study began in the late 1960s. There is no national adoption register in Finland, and the arduous task of collecting the names of hospitalized women diagnosed with schizophrenia netted a total of 9,832 such patients between 1969 and 1972. Because this sample was deemed insufficient to provide enough children adopted away at an early age, a second sample was collected, which included the entire psychiatric population of Finland, plus consecutive admissions from 1960–1979. These two samples yielded information for a total of 19,447 women diagnosed with schizophrenia or 'paranoid psychosis.' Through the use of local population registers it was determined which of these women had given up a child for adoption. A list of 291 adopted-away offspring of 264 biological mothers diagnosed with either schizophrenia or paranoid psychosis was obtained in this way. All offspring had been placed in non-relative homes and had gone through formal adoption procedures. From this group, 127 adoptees were excluded from the study for various reasons, leaving a total of 164 index adoptees.

A matched control group was established on the basis of sex, age of placement, age of adoptive parents (ten year maximum difference), and social status. Adoptees were eligible to be controls if there was no record that their biological mother had been treated for psychosis. About one-half of the biological fathers were identified, and adoptees were excluded if these fathers were found to have been treated for psychosis. The matching of index and control adoptees was performed independently by persons not involved in the study. By 2000, 197 control adoptees were matched against the index group.

Joint interviews were performed with the entire family, after which separate interviews were undertaken with the adoptive couple only. All interviews were tape-recorded for the purpose of future blind evaluation. Both parents and adoptees were given a battery of psychological tests, which included the Rorschach and the WAIS. In addition, all index adoptees completed an MMPI. Where possible, the biological parents were also interviewed and tested. The interviewers and testers were blind to the status of the families with whom they worked.

Tienari and associates outlined five different types of adoptive families, whose status was determined through interviews and testing: (1) 'healthy,' (2) 'mildly disturbed,' (3) 'neurotic,' (4) 'rigid, syntonic,' and (5) 'severely disturbed.'[252] The descriptions of each type of family could be criticized on several grounds (for example, there is no mention of violence, child abuse, alcoholism, etc.), but they are a huge step forward when compared with the Danish-American practice of ignoring family environments (other than to record psychiatric diagnoses). An analysis of the adoptive family environment is important because, as a longitudinal study, the researchers are able to observe family interaction before a child is diagnosed with schizophrenia. This is one of the unique aspects of the study. As Tienari observed,

> The Finnish program will be the first attempt in schizophrenia research to combine the direct adoptive family strategy and the risk research strategy to study adoptive families and adoptees prospectively, beginning prior to the onset of illness in the offspring.[253]

Results

The following results are based on a 2000 publication by Tienari and colleagues.[254] Of the 164 index adoptees, whose biological mothers were diagnosed with DSM-III-R schizophrenia or paranoid psychosis, 11 (6.7%) received a DSM-III-R schizophrenia diagnosis, versus four (2.0%) such diagnoses among the 197 control adoptees. The difference is statistically significant.[255] Tienari and colleagues created a 'narrow spectrum' of supposedly related disorders, which included 'schizoaffective disorder,' 'schizophreniform disorder,' and 'schizotypal personality disorder.' They also made diagnoses for a 'broad spectrum' which included the above diagnoses plus 'paranoid personality disorder,' 'schizoid personality disorder,' 'delusional disorder,' 'bipolar psychosis,' and 'depressive psychosis.' The index 'narrow' and 'broad' spectrum rates were highly significant when compared with controls ($p < 0.001$ for both narrow vs. controls, and broad vs. controls).[256]

The investigators concluded, 'The genetic liability to "typical" DSM-III-R schizophrenia is decisively confirmed. Additionally, the liability also extends to a broad spectrum of other psychotic and non-psychotic disorders.'[257] However, these results are dependent on the inclusion of disorders other than schizophrenia. According to unpublished results graciously provided by Tienari, DSM-III-R schizophrenia was diagnosed in seven out of 137 adoptees (5.1%) whose biological mothers also were diagnosed with DSM-III-R schizophrenia, versus three of 192 (1.6%) control DSM-III-R schizophrenia diagnoses. [258] Therefore, looking only at schizophrenia, the index-control difference is not statistically significant.[259] In another paper coming out of the study, Wahlberg and associates compared scores of a subsample of index and control adoptees on the 'Thought Disorder Index' (TDI), which according to the investigators is 'the most widely used test measure of thought disorder.'[260] The results showed no differences between index and control adoptee TDI scores, and that both groups' scores were about the same as normal subjects. However, on the basis of other comparisons Wahlberg and colleagues concluded that genetic factors influence some aspects of thought disorder.

Tienari and associates have concluded that both genes *and* adoptive family rearing environment are 'predictor variables' for schizophrenia. According to Tienari, index adoptees who grew up in seriously disturbed adoptive families are diagnosed with schizophrenia more often than control adoptees reared in these types of adoptive families. Thus,

> the results were consistent with the hypothesis that healthy rearing families have possibly protected the vulnerable children, whereas in disturbed rearing families the vulnerable children have been more sensitive to dysfunctional rearing.[261]

The correlation between the psychological health of the adoptive family and the rate of index schizophrenia cases was so great that, at least through 1987, Tienari could report that 'all adoptees who had been diagnosed either as schizophrenic or paranoid had been reared in seriously disturbed adoptive families.'[262] In a later report, Tienari et al. wrote that 'the individual psychopathology of the adoptive parents, assessed for all subjects in the study, can be used as an environmental predictor variable.'[263]

Selective placement
Tienari has acknowledged that a 'major assumption' of adoption studies 'is that genetic background and the rearing environment are not correlated, in terms of factors relevant to the development of the specific psychopathology in question.'[264] Here we will review evidence that, like the Oregon and Danish studies, the perceived genetic background of many adoptees *was* correlated with their rearing environment in ways that could have made them more vulnerable to receiving a schizophrenia diagnosis.

Like Denmark and Oregon, Finland had a long history of eugenics-inspired legislation aimed at curbing the reproduction of people labeled mentally retarded and mentally ill.[265] Eugenic ideas took root in Finland during the 1920s, and a government commission was created in 1926 to look into the desirability of promoting the sterilization of people seen as 'mentally retarded,' 'mentally ill,' or epileptic—although at that stage, few were calling for mandatory sterilization. In 1935 the Finnish parliament passed the Sterilization Act, which allowed the *compulsory* sterilization and castration of 'idiots,' 'imbeciles,' and the 'insane,' which included people diagnosed with schizophrenia and manic-depression.[266] The decision to sterilize was made by the Finnish National Board of Health.[267] The law permitted compulsory sterilization if there was reason to believe that a person's condition could be genetically transmitted to his or her children.

The widespread support for this law is evidenced by the fact that only 14 out of 200 members of parliament voted against it. The year 1950 saw the passage of the Castration Act, which permitted the compulsory castration of criminals, the mentally retarded, and the 'permanently mentally ill.' It was not until the Abortion Act of 1970 that compulsory sterilization was legally abolished in Finland.

The purpose of this discussion is to highlight the conditions under which most of the children in the Finnish Adoptive Family Study were placed. The adoptees in Tienari's study were born between 1927 and 1979—and therefore most were placed during a period when eugenic ideas were widespread in Finland and sterilization for eugenic purposes was permitted by law. A child born to a person with a family history of schizophrenia was seen as a person who should 'have never been born,' and as someone who represented a threat to the 'purity' of the Finnish gene pool. Those who did make their way into the adoption process were seen as carriers of the 'hereditary taint of mental illness.' It is into this type of atmosphere that Finnish children with a known family history of mental disorder were born.

In the Finnish Adoptive Family Study, up to one-third of index adoptees were placed *after* their mother was diagnosed as psychotic.[268] Because two-thirds were placed before their mothers' psychosis, Tienari and colleagues argued that in 'most cases . . . it is not likely that information about psychosis could have influenced the placement,'[269] as it undoubtedly did in Heston's study. Although Tienari and colleagues considered two-thirds of their index adoptees to be 'most cases,' this still leaves many index adoptees who were adopted at a time when their mother's psychiatric status was known by the adoption agency, and probably also by prospective adoptive parents.

The confounding effect of the adoptive parents' knowledge of a child's biological heritage is evidenced in one of the few case histories presented by the research team. In 1987 Tienari and colleagues described the family environment of an adoptee named 'P,' who was raised in a 'severely disturbed' family. The investigators did not state whether P received a psychiatric diagnosis, but the case history suggests that she was not diagnosed as psychotic. Her status, in fact, is of secondary importance to the apparent impact that the knowledge of her biological family's psychiatric history had on the adoptive family:

> The adoptive father of P, a girl of 19, was a primary school teacher; the mother was a kindergarten teacher. They had been married during the war in 1943 after hardly any contact except by correspondence. The father had threatened to commit suicide unless the mother married him. The parents had been emotionally distant from each other throughout their marriage, devoting themselves to P and four older, biologic children. Before P's adoption, the youngest biologic child had started school, and the marriage seemed about to break down. The adoption of P was a conscious effort to save the marriage. Ever since her infancy, P had been the bond between her adoptive parents, understanding both of them and solving their conflicts, while both parents had leaned upon and parentified her. *At the same time, however, both parents attributed to P's biologic background some of the potential madness they had explicitly feared in themselves. They spoke of their strenuous efforts to protect her from becoming crazy, and their relief that they had been able to preserve her sanity for so long* [emphasis added].[270]

There are several points worth making about P's adoptive family. First, these parents do not appear to have been typical candidates for the adoption process, and their turbulent history may have prevented them from being able to adopt an 'untainted' child. It is also possible that this couple intentionally sought out a potentially psychotic child in order to help them resolve their own 'explicit fears' of going mad. Second, the fact that they made 'strenuous efforts to protect her from becoming crazy' implies that P's biological family history of schizophrenia was not only a factor in the parent-child relationship, but that it had become a central preoccupation of the family. From a family systems' perspective, we might say that P's history was a major factor in her becoming 'triangled' into the couple's relationship. A control adoptee, free from the perceived predisposition to madness, would not face this type of potentially 'crazy-making' family environment.

A third point is that P's parents *expected* her to 'become crazy.' For them it was a question of when, not if. Given the prevailing attitudes in Finland, one might ask the same question posed by Kringlen in 1987 in relation to Heston's study: What type of parents would want to adopt such a child? As we have seen, up to one-third of Tienari's index adoptees were born to mothers already diagnosed with a psychotic disorder. It is not stated how many of the remaining two-thirds had a known history of mental disorder in their biological families, which could also have influenced the adoption process. We have already seen that the use of the Adoptees method is problematic in societies (such as Denmark) where genetic ideas were strong and psychiatric records were available. If adoptees' biological families were similarly checked in Finland, this would constitute a serious problem for Tienari's study as well. As we saw in relation to Rosenthal's study, on either environmental or genetic grounds, a psychotic or prepsychotic person would be expected to have more psychiatrically diagnosed biological relatives than a 'normal' person.

At an earlier point, Tienari recognized that the adoptive parents' knowledge of the biological mother's psychosis could influence the results of his study. In 1975 Tienari wrote that separation from the biological mother

> ought to have been complete, in the sense that neither the child nor his adoptive parents should have had any contact with the child's biological mother, nor should they even have been aware of her psychosis. The group meeting these criteria was still too small.[271]

This passage demonstrates both Tienari's understanding of the impact of the adoptive parents' 'awareness,' and the likely reason he was compelled to violate his standard. Like Kety and colleagues, Tienari found it necessary to change his criteria when confronted with the danger of small sample sizes. Critics could sympathize with researchers who have invested enormous amounts of time and effort in their search for the causes of schizophrenia. The same critics would be abdicating their responsibility should they also fail to point out the potentially confounding and invalidating nature of such changes in criteria.

Genes and environment
Tienari and colleagues failed to see the larger picture of the data they have recorded. Instead, they reached conclusions about genetic influences without first assessing whether environmental factors may have influenced the results upon which their conclusions were reached. As an example of how this could occur, I present a hypothetical study of non-adoptive populations where non-randomized placement could lead to unwarranted conclusions about genetic factors. Suppose that we want to study two populations of elementary school children, Group A and Group B. *Group A* consists of the children of affluent, upper-middle-class families. These children attend private schools with excellent instruction and small class sizes. Their highly educated parents often help them with their homework and attempt to stimulate their interest in learning. These children are well nourished and receive excellent health care. *Group B* consists of the children of poor, working-class, or unemployed parents. These children attend overcrowded and poorly financed public schools. Their parents are either unavailable, too tired, or unable to help them with their schoolwork. These children may also be involved in activities which are helping the family to survive economically. They are undernourished and are the recipients of second-rate or non-existent health care.

Groups A and B will be compared for abilities in reading and mathematics. After being matched for age, sex, and grade level, they are given math and reading achievement tests. After the tests are scored and sorted by group, it is determined that the children from Group A scored significantly higher than Group B children. Using the method of Tienari and colleagues, the significant differences between the two groups would be considered 'clear-cut' evidence that the higher Group A scores are influenced by genetic factors. *Then* each group's family environment would be studied, and it would be argued that the child's socioeconomic background 'also' predicted achievement test scores. Would such an analysis permit us to conclude that differences in these children's math and reading scores are influenced by both genes and environment?

Clearly, the answer is no. If we find that both genes and environment predict scores, then we have failed to consider the existence of a third factor influencing the results of the study. By looking at the larger picture of the circumstances of these children's lives, score differences could be completely accounted for by the unfavorable learning environment experienced by Group B children. In this example it is clear that Group A and B children were not randomly placed into their respective learning environments, but rather their environments were a direct result of the socioeconomic status of their parents. But adoption studies require the random placement of adoptees as a prerequisite to assessing genetic influences on the trait or condition under study. If the adopted-away children of schizophrenia-diagnosed mothers were more likely to be placed into inferior adoptive homes, this would suggest the same type of environmental placement bias as we saw in the hypothetical example of Groups A and B. Selective placement would now be the confounding variable — just as socioeconomic status was in the study of the two groups of children. The failure to gauge the interaction of all aspects of the Finnish adoption process has led Tienari to the unwarranted conclusion that genetic factors are a cause of schizophrenia.

As another a way of illustrating the problems raised by Tienari's conclusions, I would point to *favism* (G6PD deficiency), which is Gottesman's preferred example of a disease requiring both genetic and environmental factors. Gottesman and Shields described the gene-environment interaction in this condition as follows:

> Favism, a hemolytic anemia that follows the eating of fava or broadbeans, provides a textbook example of a genotype X environment interaction. Only those persons with the particular X-linked G6PD enzyme variant develop favism, and then, only after eating the bean. Both the gene and the bean are necessary for the disease to appear, neither alone is sufficient, and the disease is *both* a genetic and an environmental one.[272]

Gottesman has correctly pointed out that a favism adoption study looking at genetically vulnerable children adopted by residents of a 'non-bean' community could erroneously conclude that favism lacks a genetic component.[273] It could be said that both genetic predisposition and the consumption of fava beans are 'predictor variables' of favism—with one or none, there is no favism; with both, there is.

If the principles of this 'textbook example' are applied to Tienari's data, a problem becomes apparent. Tienari has concluded that genetic background and dysfunctional rearing environments are predictors of schizophrenia, in the same way as 'genes and beans' are predictors of favism. A problem arises when we look at index adoptees raised in severely disturbed homes, because in Tienari's reports they are diagnosed with schizophrenia or other 'spectrum psychotic' disorders only 15–18% of the time.[274] One would think that adoptees exposed to two major theorized predictor variables of schizophrenia spectrum disorders would have a much higher rate than that. It can only follow that most of the prediction variability of schizophrenia and the spectrum disorders remains unknown. What accounts for the fact that over 80% of the 'at risk' adopted-away biological offspring of people diagnosed with a spectrum disorder, who were also raised in severely disturbed adoptive homes, did not receive a schizophrenia spectrum diagnosis? Clearly, the possibility exists that a large and unknown factor or set of factors, related to both the biological background and the adoptive family environment of the child, is wholly responsible for an adoptee receiving a schizophrenia diagnosis.

As Tienari has acknowledged, it is difficult to determine the direction of causality of adoptive parent and offspring dysfunction. About this problem, Kendler wrote,

> Most of the adoptive families [in Tienari's study] were evaluated when their adoptees were older than 15. It may have been the disturbed adoptees who affected their adoptive families, not the other way around. Only prospective studies examining families when the adoptees are quite young can hope to disentangle the direction of causality of disturbances in the adoptees and the adoptive family.[275]

This is an interesting argument when we realize that Kendler defends the validity of the twin method in schizophrenia research because he believes that family environments do not contribute to the causes of schizophrenia (see Chapter 3). Although when speaking of *adoptees*, Kendler acknowledges that family members can contribute to each other's 'disturbances,' he refuses to acknowledge that parental behavior can lead to disturbances in *twins*. He takes counterposed positions on this question depending on whether he is talking about twins or adoptees and, as always, the position taken supports the genetic argument.

Age of transfer (abandonment)
Like Rosenthal's Danish Adoptees series, Tienari's study is flawed because some adoptees were not placed until they were four years old.[276] While Tienari and associates have presented evidence that adoptee mental health

ratings are not associated with the age at adoption,[277] the trauma of parental abandonment and placement in temporary homes or institutions could have been greater for those late-separated adoptees. It appears that a fair number of adoptees spent considerable time with their biological parents or in foster homes.[278] Tienari has acknowledged that children adopted away at age four 'were doubtful as to their suitability for a study discriminating between genetics and environment,'[279] and 12 years earlier he had written that an adoptee's separation from the biological mother should have occurred 'in no case later than the age of 3.'[280] Although the number of late-placed adoptees apparently is not large, in adoption studies every case is important because it can potentially make the difference between statistically significant and non-significant results.

Attachment theorists point to the negative psychological impact of the rupture of the parent-child bond—even in the case of less-than-nurturing parents. One could argue that a child separated from his or her mother at or near birth would not be greatly damaged in the process, but a child ripped from its caregivers at age one, two, three, or four is truly an 'abandoned child.' The matching of controls on the basis of age-of-placement cannot eliminate this problem.

Conclusion

Tienari is the creator of the most well-planned and executed schizophrenia adoption study. However, diagnostic differences which might exist between index and control adoptees are confounded by environmental factors such as selective placement and late placement. In addition, the finding of significantly higher rates of schizophrenia among index adoptees is, like the Danish-American studies, dependent on the questionable use of a 'schizophrenia spectrum of disorders.' The authors of the Finnish Adoptive Family Study of Schizophrenia have concluded that rearing environment is not only important, but that severely disturbing family environments can predict which children will be diagnosed with schizophrenia. However, the evidence points to the rejection of the additional finding of a genetic component for schizophrenia.

Schizophrenia adoption studies: summary and conclusions

At the beginning of this chapter I mentioned two broad categories of problems with the schizophrenia adoption studies: selective placement, and unsound methodology and bias. In these concluding comments I will briefly summarize the argument around these issues.

If we look at all schizophrenia adoption studies in the context of the social and political environments from which they obtained their subjects, we can conclude the following: *the great majority of adoptees under investigation by the schizophrenia adoption studies were given up for adoption at a time when the compulsory sterilization of 'schizophrenics' for eugenic purposes was permitted by law in the country or state in which their adoptions took place (Denmark, Finland, Oregon).* Leaving aside all other problems, the evidence suggesting that the selective placement of adoptees occurred in these studies is reason enough to reject any conclusions about genetic factors, until convincing evidence can be produced that placements were not influenced by index adoptees' 'tainted' biological backgrounds.

Unfortunately, selective placement is rarely discussed by adoption researchers other than to briefly dismiss its impact on their results. This is also true of the way these studies are discussed in textbooks. Typically, the original investigators' conclusions are accepted, and few textbook authors submit these studies to a critical examination or assess the influences of selective placement, which happens to be the most important factor potentially confounding adoption studies.[281] In fact, most psychiatry and psychology textbooks present misleading and often inaccurate accounts of the 'genetics of schizophrenia' topic. In Thomas Kuhn's classic work, *The Structure of Scientific Revolutions*, he observed that textbooks and the popularizations and philosophical works modeled on them 'record the stable *outcome* of past [scientific] revolutions and thus display the bases of the current normal-scientific tradition.' According to Kuhn,

> To fulfill their function they need not provide authentic information about the way in which those bases were first recognized and then embraced by the profession. In the case of textbooks, at least, there are even good reasons why, in these matters, they should be systematically misleading.[282]

There is a parallel between adoption researchers' denial that selection factors influenced their results and twin researchers' handling of the equal environment assumption. Most twin researchers now admit that identical twins share a much more similar environment than fraternals. Yet they claim that, after all, it doesn't really matter. The claims of some adoption researchers are similar. Although Mednick recognized that the agencies had a 'policy of selective placement,' the Danish-American schizophrenia adoption researchers acted as if it didn't matter. One of the purposes of this chapter has been to show that, in fact, it *did* matter.

If genetic inferences from these studies are invalidated by selective placement, the Oregon and Danish-American studies are equally invalidated by unsound methodology and bias. Investigators such as Kety, Rosenthal, Wender, and Schulsinger were intent on confirming their strong genetic views. *As seen clearly in their published works*, they changed definitions, comparisons, and ways of counting to ensure that they would find what they were looking for. It is not a matter of fraud, but rather of how conclusions drawn from ostensible scientific experiments are primarily a statement of the investigators' *beliefs*. Genetic research has a long history of these types of conclusions, going all the way back to Galton.

If the early schizophrenia adoption studies helped to 'break a paradigm,' then the paradigm they helped gain acceptance must be immediately re-evaluated.

Tienari's study cannot and should not be dismissed as easily as its predecessors. Tienari and colleagues were the first to make diagnoses on the basis of clearly defined criteria, and they made a conscious effort to assess adoptees' rearing environments. They also put in place important safeguards that were not used in the previous studies. However, the four main errors of the Finnish investigators have been, (1) the failure to understand the influence of selective placement on their results; (2) the inclusion of late-adopted children in the study; (3) the unwarranted acceptance of a broad definition of schizophrenia, which included disorders not traditionally recognized as schizophrenia in Finland itself;[283] and (4) the failure to correctly interpret the meaning of their data. Although Tienari and colleagues' study is sound in many respects, their conclusions can be countered with alternative explanations unrelated to genetics.

In conclusion, the results from family, twin *and* adoption studies do not support the position that genes influence the appearance of a set of behaviors given the name 'schizophrenia.' It is much more likely that 'schizophrenic' behavior is the way that some people respond to having experienced 'seriously disturbed families' and seriously disturbing social and political environments. 'All symptoms of schizophrenia,' wrote psychologist Bertram Karon, 'may be understood as manifestations of chronic terror and defenses against terror.'[284] According to Szasz,

> Every 'mental' symptom is a veiled outcry of anguish. Against what? Against oppression, or what the patient experiences as oppression. The oppressed speak in a million tongues . . . What of the psychiatrist or of others who wish to help such a person? Should they amplify the dissent and help the oppressed shout it aloud? Or should they strangle the cry and reoppress the fugitive slave? This is the psychiatric therapist's moral dilemma.[285]

Calling this response a 'disease' or a 'genetic disorder' delays the discovery of environmental factors and perpetuates the belief that people diagnosed with schizophrenia are a problem for society. In reality, society is a problem for them.

Notes

[1] Andreasen, 1998, p. 1659.
[2] E. Bleuler, 1950, p. 294.
[3] Rosenthal, 1972, p. 68.
[4] Kety, 1974, p. 961.
[5] Rowe & Jacobson, 1999, pp. 14–15.

[6] L. F. Richardson, 1913, p. 394.

[7] Wender et al., 1974, p. 122.

[8] Kety et al., 1976, p. 416.

[9] Boyle, 1990; Lewontin et al., 1984; Lidz, 1976.

[10] Van Dyke et al., 1975, p. 227.

[11] Wender & Klein, 1981, p. 175.

[12] Pam, 1995, p. 31.

[13] Gottesman & Shields, 1976a, p. 364.

[14] Plomin, DeFries, et al., 1990, p. 354.

[15] Heston, 1966, p. 821.

[16] Ibid., p. 823.

[17] Ibid., p. 819.

[18] Rosenthal, 1974a, p. 168.

[19] Ibid., p. 168.

[20] Heston, 1966, p. 821.

[21] Ibid., 1966, p. 821.

[22] Cassou et al., 1980.

[23] Heston & Denney, 1968, p. 368.

[24] Gottesman, 1991.

[25] If one fewer experimental adoptee had been diagnosed with schizophrenia the result would have been: experimental, 4/47 vs. control, 0/50, p = .051; if one control adoptee had been diagnosed with schizophrenia the result would have been: experimental, 5/47 vs. control, 1/50, p = .09, Fisher's Exact Test, one-tailed.

[26] Rosenthal, 1975b, p. 20. In a review article published in *Science* four years after the original investigation, Heston obscured the fact that he had participated in the diagnostic process while not blind to the status of his adoptees. In a discussion of the adoption studies performed in the 1960s, Heston (1970, p. 250) wrote, 'in these newer studies diagnoses either were made by raters who did not know the genetic background of the subjects or were taken unchanged from medical records.' Heston then cited his 1966 investigation as one of these 'newer studies.' However, Heston *was* aware of the genetic background of his subjects, and diagnoses in his study were not simply 'taken unchanged from medical records.' Had this been the case, there would have been no reason to conduct interviews or to have reached diagnoses on the basis of a conference of psychiatrists.

[27] Heston, 1966, p. 821.

[28] Heston & Denney, 1968, p. 368.

[29] Heston, 1966, p. 822.

[30] Heston & Denney, 1968, p. 369.

[31] See Heston, 1966, p. 823 for a description of this case.

[32] APA, 1965, p. 27.

[33] Largent, in press.

[34] Olson, 1920, p. 1487.

[35] Ibid., p. 1487.

[36] Ibid., p. 3176.

[37] Largent, in press.

[38] Anonymous, 1925, p. 71.

[39] Largent, in press.

[40] Quoted in Largent, in press.

[41] Kringlen, 1987, pp. 132–3.

[42] Reilly, 1991.

[43] Heston, 1966, p. 822.

[44] Eight cases were reported; Heston & Denny, 1968, p. 370.

[45] Heston & Denney, 1968, p. 366.

[46] Ibid.

[47] Heston et al., 1966, p. 1110.

[48] See Plomin & DeFries, 1985.

[49] Heston & Denney, 1968, p. 374. In striking contrast to Heston and Denney's original description, biological psychiatrist Nancy Andreasen (2001, p. 199) wrote that in Heston's study, 'adoptees were reared in families that were considered to be "normal" or "healthy."'

[50] Heston et al., 1966.

[51] This discussion of the Kety et al. schizophrenia adoption studies is based on a revised version of a previously published article (Joseph, 2001a).

[52] The index group consisted of 34 subjects but only 33 cases because two of the index adoptees were B3 identical twins.

[53] According to Häfner and der Heiden (1986, p. 29), the Danish Psychiatric Register was established in 1937 and registered 'all mental disorders; psychiatric inpatient services and partial hospitalization.' The world's first psychiatric register was established in Norway in 1936. Note that these registers were founded at a time when eugenic ideas were at their high point in these countries (see Broberg & Roll-Hansen, 1996), which suggests that they may have been intended for use in eugenic programs. It is therefore ironic that psychiatric registers, which enabled the Danish-Adoption studies to be performed, may have been used in determining adoption placements.

[54] Kety et al., 1968, pp. 351–2.

[55] Kety, 1974.

[56] Kety, 1987, p. 424.

[57] Kety et al., 1968, p. 353.

[58] Ibid., p. 361.

[59] Kety, 1978. Also see Munk-Jørgensen, 1985.

[60] Joseph, 2000a.

[61] Strömgren, 1993.

[62] Rosenthal, 1970b; Slater & Cowie, 1971. Few people over 45 receive their first schizophrenia diagnosis.

[63] Quoted in Slater & Cowie, 1971, p. 13. In Denmark, only chronic 'process' cases were counted as schizophrenia (Kety, 1978).

[64] Sarbin & Mancuso, 1980, p. 138.

[65] Rosenthal, 1972, p. 68.

[66] Rosenthal, 1971b, p. 194.

[67] Ibid., p. 194.

[68] This scenario was suggested by Lewontin and colleagues in 1984, and Pam in 1995. In a rare acknowledgment by a supporter of the Danish-American studies of the real purpose of the spectrum, in 1986 Häfner and der Heiden wrote in relation to the Rosenthal et al. study, 'As the findings were not significant because of the small sample size, the investigators extended the case definition to the so-called "schizophrenia spectrum disorder," a rather imprecise diagnostic group' (p. 37). An unbiased investigator does not assume that no differences were found simply because the sample was small. Rather, he or she is obligated to conclude that no genetic influences on schizophrenia were found.

[69] Rosenthal, 1972, p. 68.

[70] Rosenthal, 1975b, p. 19.

[71] APA, 1968, p. ix.

[72] See Joseph, 2000a.

[73] Ibid.

[74] Kety, 1985b.
[75] E. Bleuler, 1950, p. 294.
[76] See, for example, Kety, Rosenthal, & Wender, 1978.
[77] Gould, 1981, p. 242.
[78] Kety, 1987, p. 424.
[79] Rosenthal, 1970b, p. 57.
[80] Lidz & Blatt, 1983, p. 426.
[81] Ibid., p. 427.
[82] Kety, 1983a, p. 721.
[83] Rosenthal, 1967, p. 25. I am unaware of any other published description of the design and method of the 1968 study prior to its publication. Apart from Kety's 1959 paper discussing the need to study adopted children, none are cited in any Danish-American Adoptees' Family publication.
[84] Kety et al., 1968, p. 346.
[85] Ibid., 1968, p. 355 (p = .076, Fisher's Exact Test, one-tailed).
[86] Psychiatric geneticists Faraone and Tsuang (1995, p. 92) described the Adoptees' Family method, which they call the 'adoptee-as-proband design,' as follows:

> As its name suggests, the adoptee-as-proband design starts with ill and well adoptees and examines rates of illness in both biologic and adoptive relatives. If the biologic relatives of ill adoptees [IB] have higher rates of illness than the adoptive relatives of ill adoptees [IA], then a genetic hypothesis is supported. In contrast, if the adoptive relatives show higher rates of illness, then an environmental hypothesis gains support.

The likelihood that Kety and colleagues (1968) abandoned this comparison in favor of another which produced significant results did not hinder Faraone and Tsuang, two paragraphs later, from citing this study as evidence in favor of the genetic basis of schizophrenia.

[87] Kety, 1966, pp. 230–1.
[88] Ibid., p. 231.
[89] Kety, 1970, p. 240.
[90] Tienari, Sorri, et al., 1987, p. 482.
[91] The mythological status of the Danish-American adoption studies continues unabated in psychiatry, as seen in the following examples. (1) The authors of a 1995 American Psychiatric Association *Annual Review of Psychiatry* article claimed that the 1968 Kety et al. study 'found that the prevalence of schizophrenia was significantly higher in the biological parents,' and that the 'Danish adoption studies also found that the biological relatives of schizophrenic persons had elevated rates of "borderline schizophrenia" . . . ' (Byerley & Coon, 1995, p. 366). In fact, the Kety et al. 1968 study diagnosed *zero* biological index parents with chronic schizophrenia, and only one was diagnosed with borderline schizophrenia (see Kety et al. 1968, p. 354, Table 4a). (2) In a 2001 tribute to Kety (who died in 2000), his colleague Philip S. Holzman wrote that one of the two 'incontrovertible' conclusions of the Danish American adoption studies was that 'children born to a schizophrenic mother and reared in an adoptive family become schizophrenic at the same rate as those reared by the biological mother, who is schizophrenic' (Holzman, 2001). However 'incontrovertible' this conclusion may have appeared to Holzman, the fact remains that no such comparisons were made by the Danish-American investigators. The Kety et al. Adoptees' Family

studies *began* with adoptees diagnosed with a spectrum disorder, and *then* recorded diagnoses for their biological and adoptive relatives. As it turned out, only 2 of these adoptees' 66 biological mothers (3%; 1975 N = 1, 1994 N = 1) were diagnosed with chronic schizophrenia. Even in reference to Rosenthal and colleagues' Adoptees study, which *did* study 'children born to a schizophrenic mother and reared in an adoptive family,' Holzman's statement remains incorrect because Rosenthal did not study the children of these adoptees' biological parents. (3) According to Neale and Oltmanns (1980, p. 197), 'Kety (1974) suggested that a comparison of the prevalence of schizophrenia in the paternal half-siblings' would eliminate the possibility that *in utero* experiences led to greater index biological relative spectrum rates. However, because Kety made this suggestion *after* he had collected the data (see Kety, 1974, p. 961; Neale & Oltmanns, 1980, p. 197) there is no evidence that the importance of the paternal half-sibling comparison was determined *before* the collection of data. (4) According to Lyons and colleagues (1991, p. 131), 'Kety et al. . . . predicted that "if schizophrenia were to some extent genetically transmitted, there should be a higher prevalence of disorders in the schizophrenia spectrum among the biological relatives." . . . Kety did, indeed, find a concentration of schizophrenia spectrum diagnoses in the biological relatives of schizophrenic adoptees.' Kety and associates were thus credited with 'predicting' and 'finding' such a concentration in the same 1968 publication. (5) The Tenth Edition of *Modern Clinical Psychiatry* listed the authors of the 1968 Adoptees' Family study — one of the most famous investigations in the history of psychiatry and performed by Kety, Rosenthal, Wender, and Schulsinger — as 'Kety, Rosenfeld, Winther, and Scholfinger' (Kolb & Brodie, 1982, p. 351). One wonders how many psychiatry textbook authors actually read the original adoption study source material they cite (see Joseph, 2000b; Leo & Joseph, 2002; Paul, 1998).

[92] Gottesman & Shields, 1976a, p. 370.
[93] Kringlen, 1976, p. 430.
[94] Kety et al., 1976, p. 416.
[95] Boyle, 1990, p. 144.
[96] Benjamin, 1976, p. 1131.
[97] Kety, 1976, p. 1135.
[98] Kety, 1983a.
[99] Rosenthal, 1974a, p. 168.
[100] Kety was once asked the following question after giving a talk on the genetics of schizophrenia and the results of his studies: 'In your data on the paternal half siblings, how can we be sure that those fathers might not have sought out a schizophrenogenic mate repeatedly?' Kety replied to this question as follows:

> That is a very good question. Frankly, we cannot rule it out; but this hypothesis requires the assumption that the fathers have some uncanny ability, greater than that of any psychiatrist I know, to pick out schizophrenogenic mothers. Furthermore, if the father had the propensity for picking schizophrenogenic mothers, this would still not be as effective as *being* the schizophrenogenic mother. On the basis of the hypothesis you propose, we would expect the biological maternal half siblings to have more schizophrenia than the paternal half siblings. That is not what we find . . . so this hypothesis finds no support (Kety, 1978, p. 67).

Kety's response contained one clear error: the rate of schizophrenia among the *biological* maternal half-siblings is irrelevant to the questioner's hypothesis because they were not raised by the same mother who raised the adoptee. Kety should have responded that children in the *adoptive homes* were the group of interest, because this is where the theorized schizophrenogenic environment was located. There were only four full- and two half-siblings raised in the adoptive homes of B1 index adoptees, so this question could not be looked at in the 1975 study. However, the evidence from schizophrenia family studies indicates that the siblings of people diagnosed with schizophrenia are more likely to be diagnosed with schizophrenia than randomly selected members of the population. Theoretically, this increased risk could be due to genes, environment, a combination of both, or methodological error. From a psychosocial perspective, family studies demonstrate that schizophrenogenic environments do exist, and nothing in Kety's data suggests that such environments did not also exist in the homes (or institutions?) of the paternal half-sibs.

[101] Gottesman & Shields, 1982, pp. 144–5.
[102] Rosenthal, 1974b, p. 589.
[103] Benjamin, 1976, p. 1130.
[104] Kety, 1976.
[105] Kety et al., 1975, p. 163.
[106] As noted by Lidz and Blatt (1983), spectrum category C was omitted from this comparison. If category C is included, the results were: 23/33 index families affected vs. 16/34 affected control families. Using a Fisher's Exact Test the probability for this difference is not significant (p = .05108). In another paper, Kety et al. (1976, p. 418) acknowledged that the comparison was not significant, writing 'NS' below the totals (see my discussion of the 1975 study).
[107] Mosher, 1975, p. 3.
[108] Hutchings & Mednick, 1975, p. 115.
[109] Teasdale, 1979, p. 108.
[110] Kety et al., 1994, p. 452.
[111] Joseph, 2000d.
[112] Kety et al., 1994, p. 452.
[113] Ibid., p. 453. No mention of the adoptees' biological parents' diagnostic status at the time of adoption was made in the Kety et al. 1968 or 1975 papers.
[114] Mother's Aid Organization for Copenhagen, Copenhagen County and Frederiksborg County, Annual Report for 1946–1947. Quoted in Mednick & Hutchings, 1977, p. 163.
[115] S. Hansen, 1925, p. 81.
[116] Quoted in B. Hansen, 1996, p. 38.
[117] Ibid., p. 39.
[118] Ibid., p. 41.
[119] Ibid., p. 45.
[120] M. Bleuler, 1978, p. 473.
[121] Mednick et al., 1987, p. 78.
[122] Mednick & Hutchings, 1977, p. 161.
[123] Mednick, 1996, p. 134.
[124] Lewontin et al., 1984, p. 223.
[125] Mednick, 1996, p. 134.
[126] Kety et al., 1975, p. 150.
[127] Ibid., p. 155.
[128] Ibid., p. 154.
[129] Ibid., p. 156.

[130] Ibid., p. 156.

[131] Joseph, 2000b.

[132] Kety et al., 1975, p. 156.

[133] Ibid., p. 156.

[134] Ibid., p. 156.

[135] When the comparison is limited to the B or 'definite' cases, Kety and colleagues reported that the rate of affected index families was significantly higher than the control rate (index 14/33 vs. control 3/34, p = .002; Kety et al., 1975, p. 163). However, this comparison reflects an error on the part of Kety et al. because there were only eight index biological families with a member receiving a 1968 or 1975 B diagnosis, rendering the comparison statistically non-significant (8/33 vs. 3/34, p = .08, Fisher's Exact Test, one-tailed). See Boyle (1990) for more details.

[136] Kety et al., 1975, p. 154.

[137] Ibid., p. 155.

[138] Kety et al., 1976, p. 418, Table 2a.

[139] Lidz & Blatt, 1983, p. 430.

[140] Joseph, 2000b.

[141] Rosenthal, 1975a, p. 201.

[142] See Kety et al., 1975, p. 151.

[143] Ibid., p. 153.

[144] Lidz & Blatt, 1983.

[145] Kety et al., 1975, pp. 160–1.

[146] Ibid., p. 154.

[147] Index 5/173 vs. control 1/174, p = .10, Fisher's Exact Test, one-tailed.

[148] Ingraham & Kety, 1988, p. 122; Kety et al., 1994, p. 452.

[149] Ingraham & Kety, 1988, p. 122.

[150] Kety et al., 1975, p. 150.

[151] Paikin et al., 1974, pp. 308–10.

[152] Lewontin et al., 1984, p. 225.

[153] Ibid., p. 224.

[154] Kety et al., 1976.

[155] Kendler & Gruenberg, 1984, p. 556.

[156] Kety, Rosenthal, & Wender, 1978; Kety, Rosenthal, Wender, Schulsinger, & Jacobsen, 1978.

[157] Ingraham & Kety, 1988.

[158] Kety & Ingraham, 1992.

[159] Kety, 1985a; Kety, Rosenthal, & Wender, 1978.

[160] Kety et al., 1994, p. 445.

[161] APA, 1980.

[162] Kety et al., 1994, p. 442.

[163] Joseph, 2000a.

[164] 4/82 index vs. 1/59 control, p = .30, Fisher's Exact Test, one-tailed.

[165] Kety, 1983b.

[166] Kety et al., 1975, p. 155.

[167] Kety et al., 1994, p. 446.

[168] Kety & Ingraham, 1992.

[169] Kety et al., 1975, p. 155.

[170] This did not prevent the 1975 Copenhagen screened controls (with an affective disordered adoptee) from being combined with Provincial screened controls (affective disordered adoptees removed) to constitute the 47 National sample 'Control Adoptees With No History of Major Mental Illness' (Kety et al., 1994, p. 452).

[171] Kety & Ingraham, 1992, p. 250.

[172] 14/171 index vs. 8/158 control, p = .18, Fisher's Exact Test, one-tailed.

[173] Kendler & Diehl, 1993, p. 265.

[174] Kendler, Gruenberg, & Kinney, 1994, p. 458.

[175] Spitzer & Endicott, 1979.

[176] APA, 1980, p. 313.

[177] Kety, 1983a, p. 724.

[178] Gunderson & Siever, 1985, p. 532.

[179] Spitzer & Endicott, 1979, p. 98.

[180] Kety, 1983a, p. 723.

[181] APA, 1980, p. 310.

[182] Kety, Rosenthal, & Wender, 1978, p. 220.

[183] Kety et al., 1994, p. 445.

[184] Ibid., p. 447.

[185] Ibid., p. 445.

[186] E. Bleuler, 1950, p. 294.

[187] Rosenthal, 1979, p. 23.

[188] Index 10/140 vs. control 5/162, p = .09, Fisher's Exact Test, one-tailed.

[189] Index 7/51 vs. control 3/60, p = .10, Fisher's Exact Test, one-tailed.

[190] See Joseph, 2000a.

[191] Kety et al., 1994, p. 448.

[192] Ibid., pp. 447-448.

[193] Ibid., p. 448, Figure 1.

[194] Although Kety et al. counted 24 index biological full-sibs, the pedigree chart on page 448 of their 1994 paper indicates that there were 25.

[195] Ibid., p. 446.

[196] Kendler, Gruenberg, & Kinney, 1994, p. 458.

[197] Ibid., p. 460; p = .24, Fisher's Exact Test, one-tailed.

[198] Kety et al., 1994, p. 452.

[199] Ibid., p. 442.

[200] Gottfredson & Hirschi, 1990, p. 55.

[201] Lewontin et al., 1984, p. 225.

[202] 6/82 vs. 0/59, p = .036, Fisher's Exact Test, one-tailed.

[203] 5/80 vs. 0/59, p = .06, Fisher's Exact Test, one-tailed.

[204] Index 4/29 vs. control 0/24, p = .08, Fisher's Exact Test, one-tailed.

[205] Rosenthal et al., 1968, p. 380.

[206] Rosenthal, 1970b, p. 129.

[207] Rosenthal et al., 1971, p. 310.

[208] Rosenthal et al., 1968, p. 382.

[209] Rosenthal, 1971a, p. 124.

[210] Kety et al., 1976, p. 417.

[211] Rosenthal et al., 1971, p. 309.

[212] Rosenthal et al., 1968, p. 382.

[213] Lidz et al., 1981.

[214] Pam, 1995, p. 31.

[215] Lidz et al., 1981, p. 1064.

[216] Kety et al., 1994, p. 453.

[217] Rosenthal et al., 1971, p. 308.

[218] Rosenthal, 1975b, p. 21.

[219] Lewontin et al., 1984, p. 226.

[220] Rosenthal et al., 1971, pp. 309–10.

[221] Kety, 1983a, p. 724.

[222] Rosenthal et al., 1971, p. 310.

[223] Kety, 1983a, p. 724.

[224] Rosenthal et al., 1971, p. 311.

[225] Haier et al., 1978.

[226] Mednick et al., 1984.

[227] Raine et al., 1997.

[228] p =.075; see Lidz et al., 1981, p. 1066.

[229] Rosenthal et al., 1971, p. 309.

[230] Kety, 1983a; Kety et al., 1994.

[231] Rosenthal, 1975a, p. 201.

[232] For example, see Kety, Rosenthal, Wender, Schulsinger, & Jacobsen, 1978.

[233] Kety, Rosenthal, & Wender, 1978.

[234] Index = 6/48 (12.5%), control = 5/67 (7.5%), p = .35, Fisher's Exact Test, one-tailed.

[235] Haier et al., 1978.

[236] Ibid., p. 175.

[237] Boyle, 1990, p. 153.

[238] Haier et al., 1978, p. 174, Table 3.

[239] Ibid., p. 174.

[240] Rosenthal, 1970a, p. 255.

[241] Lowing et al., 1983, p. 1168.

[242] Ibid., p. 1169.

[243] Index 7/39 vs. control 3/39, p = .16, Fisher's Exact Test, one-tailed.

[244] Ingraham & Kety, 1988, p. 125.

[245] Lowing et al., 1983, p. 1170.

[246] Kety et al., 1968, p. 352. Also see Joseph, 2000a.

[247] Rosenthal et al., 1968, p. 387. This index adoptee's thumbnail diagnosis read 'border-line schizophrenia or pervert.'

[248] This section is based on an updated and revised version of a previous publication (Joseph, 1999a).

[249] Tienari et al., 1985, p. 21.

[250] Tienari et al., 1994, p. 25.

[251] Tienari, Sorri, et al., 1987, p. 477.

[252] Tienari, Lahti, et al., 1987, pp. 40–1.

[253] Tienari, 1991, p. 465.

[254] Tienari et al., 2000.

[255] P = .027, as reported by Tienari et al., 2000.

[256] Tienari et al., 2000, p. 440.

[257] Ibid., p. 433.

[258] P. Tienari, personal communication of 8 June, 2001.

[259] 7/137 vs. 3/192, p = .065, Fisher's Exact Test, one-tailed.

[260] Wahlberg et al., 2000, p. 128. Also see Wahlberg et al., 1997, 2001.

[261] Tienari et al., 1994, p. 23.

[262] Tienari, Sorri, et al., 1987, p. 482.

[263] Tienari et al., 1994, p. 23.

[264] Tienari, 1992, p. 53.

[265] The discussion of eugenics in Finland is based largely on the historical research of Hietala, 1996.

[266] Hietala, 1996, p. 232.

[267] Hemminki et al., 1997, p. 1877.

[268] Tienari, Lahti, et al., 1987, p. 44.

[269] Ibid., p. 45.

[270] ibid., p. 43.

[271] Tienari, 1975, pp. 34–5.

[272] Gottesman & Shields, 1976b, p. 447.

[273] Gottesman, 1978.

[274] For example,Tienari et al. (1991, p. 49, Table 2) reported that only eight (17.8%) of the 45 index adoptees raised in 'severely disturbed' adoptive homes were diagnosed as 'spectrum psychotic.'

[275] Kendler, 1986, pp. 36–7.

[276] Tienari, Sorri, et al., 1987, p. 479.

[277] Tienari et al., 1985.

[278] Breggin, 1991; Jacobs, 1994; Lehtonen, 1994.

[279] Tienari, Sorri, et al., 1987, p. 478.

[280] Tienari, 1975, p. 34.

[281] See Joseph, 2000b for a discussion of textbooks' handling of the schizophrenia adoption studies.

[282] Kuhn, 1996, p. 137.

[283] Tienari et al., 2000, p. 441.

[284] Karon, 1999, p. 3. See also Karon & VandenBos, 1981.

[285] Szasz, 1968, p. 52

CHAPTER EIGHT

Is Criminal Behavior in the Genes? A critical review of twin and adoption studies of criminal and antisocial behavior

> Crime is a medical problem in at least half of its extent; and I look forward to the day when the physician will be looked on as the criminologist of the country, and when our police courts and our police administration will be put largely under medical control . . . We have the humiliating spectacle of the defective criminal insane of our community . . . because we have never taken the steps to stop this at its fountain head by preventing the reproduction of defectives.
>
> *American physician J. N. Hurty in 1909.*[1]

> Expunge the criminal, but leave the environment untouched: this has always been the message of the biological or genetic determinist point of view.
>
> *Biologist and historian Garland E. Allen in 2001.*[2]

In today's 'everything is genetic' atmosphere, even traits usually thought of as being the product of environmental influences are subject to the claims of hereditarians. It should therefore come as no surprise that, increasingly, criminal behavior is seen as the result of defective genes.[3] This chapter focuses on twin and adoption studies of 'criminal,' 'antisocial,' and 'psychopathic' behavior as an example of the way that genetic research is used in the study of human behavioral differences.[4] Topics beyond the scope of this chapter include the 'XYY' controversy of the 1960s and 1970s, and molecular genetic research. Suffice it to say that, apart from claims made about Brunner and colleagues' 1993 study of one Dutch family, no specific genes have been found to cause criminal or antisocial behavior.[5]

It is necessary to look at the genetics of criminality and antisocial behavior in light of the increasing influence of behavior genetic thinking on this subject. When members of the general public read about a supposed genetic link to crime, the conclusion is often reached that 'crime is genetic.' This leads to the belief that criminality is the result of genetic makeup—and not of racism, poverty, and other oppressive social conditions—thereby influencing public attitudes toward ethnic groups having a relatively high conviction/incarceration rate. As a prelude to a discussion of twin and adoption studies, it is worthwhile to look at the historical background of the 'genetics of criminality' question. This is the subject of the following section.

Some early views of the genetics of criminality

Like other types of socially disapproved behavior, the inherited basis of criminality was widely accepted long before adoption, twin, or even family studies had been performed. In the nineteenth century, Italian physician Cesare Lombroso argued that criminals could be identified by their physical features and that they represented, to use Gould's description in *The Mismeasure of Man*, 'evolutionary throwbacks in our midst.'[6] According to Lombroso,

> Many of the characteristics presented by savage races are very often found among born criminals. Such, for example, are: the slight development of the pilar system; low cranial capacity; retreating forehead; highly developed frontal sinuses . . . the thickness of the bones of the skull; enormous development of the maxillaries and the zygomata . . . greater pigmentation of the skin; tufted and crispy hair; and large ears. To these we may add the lemurine appendix; anomalies of the ear; dental diastemata; great agility; relative insensitivity to pain; great visual acuteness; ability to recover quickly from wounds; blunted affection; precocity as to sensual pleasures . . . laziness; absence of remorse; impulsiveness; physiopsychic excitability; and especially improvidence, which sometimes appears as courage and again as recklessness changing to cowardice.[7]

For Charles Davenport, criminals represented 'our . . . ape-like ancestors' from 'animalistic strains,' who should be rooted out of the American breeding stock:

> The acts of taking and keeping loose articles, of tearing away obstructions to get at something desired, of picking valuables out of holes and pockets, of assaulting a neighbor who has something desirable or who has caused pain or who is in the way, of deserting a family and other relatives, of promiscuous sexual relations – these are crimes for a twentieth-century citizen but they are the normal acts of our remote, ape-like ancestors . . . Imbecility and 'criminalistic' tendency can be traced back to the darkness of remote generations in a way that forces us to conclude that these traits have come to us directly from our animal ancestry and have never been got rid of . . . If we are to build up in America a society worthy of the species *man* then we must take such steps as will prevent the increase or even the perpetuation of animalistic strains.[8]

Today, the publicly stated views of Lombroso and Davenport are an embarrassment to genetic researchers, but the view of criminals as 'evolutionary throwbacks' remains in some quarters. For example, in his book *The Antisocial Personalities*, twin researcher David Lykken, allegedly in jest, described different criminal types as a 'zoo of caricatures,'[9] distinguished by their various 'species and subspecies.'[10]

The role of genetic factors in crime was widely accepted during the late nineteenth and early twentiethth centuries.[11] In 1914 an American physician wrote, 'That a criminal father should beget a child pre-destined to criminality is a foregone conclusion. The father exerts a hereditary influence equal to all the previous ancestors in the paternal line.'[12] A physician at Sing Sing prison in New York held this opinion:

> My own observations, which have been practically unlimited along lines of information connected with the male offender, have led me to believe, in the last few years, that criminal character depends in the first instance on heredity . . . So the born criminal is the product, mind and body, of the forces of heredity. Not only his body, but his mind is deeply impressed with the character of the parentage. And few indeed are the criminals who come to our prison at Sing Sing with minds that were at birth *tabula rasa*, whose mental powers at birth were not already thickly sown with seeds of crime.[13]

There were frequent calls for the sterilization of criminals on eugenic grounds during this period. Vasectomy was proposed in the late nineteenth century by A. J. Ochsner as a 'humane' alternative to castration.[14] Ochsner justified the procedure on the grounds that it was 'demonstrated beyond a doubt that a very large proportion of all criminals, degenerates and perverts have come from parents similarly afflicted.'[15] He claimed that the large-scale sterilization of criminals 'would do away with hereditary criminals from the father's side,' and recommended the same treatment for 'chronic inebriates, imbeciles, perverts and paupers.'[16]

In 1907 Indiana became the first of many U.S. states to pass a law permitting compulsory eugenic sterilization. The law sanctioned sterilization 'to prevent procreation of confirmed criminals, idiots, imbeciles, and rapists' residing in a state institution who had been judged as 'unimprovable' by a panel of physicians.[17] A prison physician, Harry Sharp, had performed vasectomies on inmates in Indiana since 1899. According to Sharp, 'There is no disputing the fact that mental as well as physical defects are transmitted to the offspring . . . The decidedly defective individual is very easily recognized, as the mental abnormality is usually accompanied with prominent physical defects, described by Lombroso.'[18] In a 1909 discussion of Sharp's article in the pages of the *Journal of the American Medical Association*, we find the following comment by Dr. J. N. Hurty, who had recently visited the estate of a wealthy family:

I was standing near the man who was in charge of the beautiful collies at the kennels; one of them (a female) came up to me, and she looked so pleasant that she seemed to me to have a laugh on her face. I patted her on the head, and she was duly grateful for the attention. I asked him, 'Do you have any vicious dogs here?' He said, 'Do you suppose that we would breed from viscious animals? *If a vicious animal appears here we kill it*; we have nothing to do with them at all; and the result is that we have no biting animals, but only those amenable to instruction.' *Why cannot we apply this to the human family?* [emphasis added].[19]

While the call to kill (and thereby, as imagined, eliminate) 'vicious' strains in the 'human family' was an extreme view even for that time, a glance at the bibliography in Fink's 1938 review of the period's literature on the causes of crime demonstrates the common concern over the procreation of 'criminals' and 'defectives.' The consensus of the time was that society could greatly reduce criminality and antisocial behavior by means of controlled breeding.

The period 1877–1919 saw the publication of several histories of American 'degenerate families,'[20] the most well-known being Richard Dugdale's *The Jukes*, and Henry Goddard's *The Kallikak Family*. For many years these 'studies' were cited as proof that criminality and 'feeble-mindedness' were hereditary conditions. For Lombroso, Dugdale's book provided 'the most striking proof of the heredity of crime.'[21] According to Davenport, Dugdale showed that criminal family pedigrees could be the result of 'a single focal point' of bad heredity, which in this case was 'traced back to Max [Juke] living in a lonely mountain valley.'[22] Davenport created pedigrees of his own and concluded, 'The foregoing cases are samples of scores that have been collected and serve as fair representations of the kind of blood that goes into the making of thousands of criminals in this country.'[23]

Today, 'studies' like the Jukes and the Kallikaks are largely discredited because, among other reasons, it is now acknowledged that poverty, criminality, and illiteracy can run in families for social and environmental reasons. That being said, it should be noted that both Dugdale and Goddard recognized the importance of environmental conditions. According to Dugdale:

Where the organization is structurally modified, as in idiocy and insanity, or organically weak as in many diseases, the heredity is the preponderating factor in determining the career; but it is, even then, capable of marked modification for better or for worse by the character of the environment . . . Where the conduct depends on the knowledge of moral obligation (excluding idiocy and insanity), the environment has more influence than the heredity . . . For instance, where hereditary kleptomania exists, if the environment should be such as to become an exciting cause, the individual will be an incorrigible thief; but if, on the contrary, he be protected from temptation, that individual may lead an honest life, with some chances in favor of the entailment stopping there.[24]

And Goddard wrote that 'there is every reason to conclude that criminals are made and not born,' while adding that the 'best material out of which to make criminals, and perhaps the material from which they are most frequently made, is feeble-mindedness.'[25]

Few modern proponents of the hereditary basis of crime believe that there are specific 'crime genes.' Rather, they argue that people inherit predispositions for personalities that make them more likely to commit crime. As behavior geneticists Goldsmith and Gottesman have written,

> Notions such as 'genes for crime' are nonsense, but the following notion is reasonable: there may be *partially* genetically influenced *predispositions* for basic behavioral tendencies, such as impulsivity, that in certain experiential contexts make the *probability* of committing certain crimes higher than for individuals who possess lesser degrees of such behavioral tendencies.[26]

Still, despite disclaimers of this type, the general conclusion drawn from the results of genetic research is that criminal behavior is caused by heredity.

Defining criminal and antisocial behaviour

Investigators studying the causes of crime are confronted with the problem of how crime and criminality are defined. Not surprisingly, political and moral considerations are bound up with the definition of criminal and antisocial behavior. As eugenicist Paul Popenoe pointed out in an article written during the Great Depression, 'A few years ago the man who had a bag of gold in his safe was a thrifty and praiseworthy citizen; today he is a criminal.'[27] Criminality, as critics Hubbard and Wald observed, is a social construct that depends on context. 'Killing,' they wrote, 'can be heroism or murder,' and 'taking someone's property can be confiscation or theft.'[28] Even Lombroso would have likely applauded the deeds of his 'criminal men,' had their 'savagery' only been directed at the opponents of the Italian army's colonial campaigns.

Another difficulty in determining who should be studied is that being a registered or convicted criminal is dependent on one's being apprehended and charged. According to Rutter, 'Many surveys have shown that, at some time, almost all boys commit acts that fall outside the law and which could have led to prosecution if they had been caught.'[29] Surveys in Scandinavian countries such as Norway, Sweden, and Denmark, where some of the most frequently cited twin and adoption studies were performed, show that most young men, when answering anonymously, admitted to committing one or more criminal offenses at some point.[30] If these surveys are representative of the population, one could conclude that most Scandinavian males are genetically predisposed to commit criminal acts! It might be argued that genetic studies of registered or convicted criminality look at possible genetic factors in *the apprehension for*, rather than the commission of, criminal acts. According to sociologist Troy Duster, 'if one looks at the record in 250 years of U.S. history, no white man ever committed the crime of rape on a black woman in twelve southern states.' Another critic wrote that something that is illegal one day can become legal the next, and that the reporting of certain types of crime varies widely. He also noted that wealthy people usually have better legal representation than the poor. His conclusion: 'In using any sort of recorded crime figures, researchers are taking on something that is very messy indeed, and that is one reason why confusion will follow.'[31]

This chapter is limited to an analysis of the two research methods most frequently cited in favor of a genetic basis for criminal and antisocial behavior: twin and adoption studies. We have seen that family studies can at best demonstrate 'scientifically' what society already knows — that the relatives of criminals are more likely to be criminals than are the relatives of non-criminals. According to adoption researchers Brennan, Mednick, and Gabrielli, 'In terms of genetics, very little can be learned from . . . family data alone,' because 'The parents have a major influence on the child's environment as well as on his/her genetic makeup; family studies cannot disentangle these hereditary and environmental influences.'[32]

Criminal twins: Blood brothers or partners in crime?

The twin method and the study of criminality
The first criminal twin study was published in 1929 by the German investigator Johannes Lange, only a few years after the twin method had been developed. Lange identified 13 identical and 17 fraternal pairs where one twin had been imprisoned. Because Lange (who was often referred to as 'Kraepelin's favorite student'[33]) found that ten identicals but only two fraternals had a co-twin who was also imprisoned, he concluded that genetic factors 'play a predominant part.'[34]

In 1934 Rosanoff, Handy, and Rosanoff published their study of 97 twin pairs, finding male concordance rates of 22/33 identical (67%) and 3/23 same-sex fraternal (13%). The investigators concluded that criminal behavior is caused by 'pre-germinal [genetic] rather than germinal' factors.[35]

Two criminal twin studies were published in Nazi Germany in 1936. Friedrich Stumpfl, who produced the most influential criminal biology research during the Nazi era,[36] studied 18 identical and 19 fraternal pairs, finding concordance rates of 65% and 37% respectively.[37] While opposing the blanket sterilization of criminals, Stumpfl called for extending the sterilization law to include all 'incorrigible recidivists.'[38] Kranz found concordance rates of 66% identical (21/32) and 54% same-sex fraternal

(23/43). This comparison is not statistically significant, which did not prevent Kranz from calling for racial hygienic measures against criminals (see Chapter 2).

A few smaller studies appeared between 1937 and 1976. Then, two major Scandinavian criminal twin studies were published in the 1970s by Christiansen, and Dalgard and Kringlen. The newer Scandinavian studies were based on national registers and studied an entire population of twins unselected for criminality.

Dalgard and Kringlen studied 49 male identical and 89 same-sex male fraternal pairs, finding no significant concordance rate difference using either a 'broad' or 'strict' definition of crime. The identical twin pairwise rate for broadly defined crime was 22%, and 26% for strictly defined crime. On the basis of these results and of the more similar environments experienced by identical twins, the authors concluded, 'These findings support the view that *hereditary factors are of no significant importance in the etiology of common crime*' (emphasis in the original.)[39]

Christiansen studied 85 identical and 147 same-sex fraternal pairs (and an additional 196 pairs of opposite-sex fraternals), reporting pairwise concordance rates of 33% identical (28/85) and 12% same-sex fraternal (17/147).[40] Christiansen's sample included all twins born between 1881 and 1910 in the Danish islands east of the Little Belt in which both twins were alive past the age of 15. These twins were then checked against Danish police records. The final results of Christiansen's study were reported by Cloninger and Gottesman in 1987, who found a pairwise identical rate of 48% (56/116) and a same-sex fraternal rate of 28% (56/202).[41]

Lyons and associates studied pairs obtained from the Vietnam Era Twin Registry (VETS). Diagnostic interviews were administered by telephone to 3,226 pairs of male twins.[42] The researchers concluded that several antisocial traits were 'significantly heritable.' However, the method of determining antisocial and criminal behavior by self-report is certainly questionable. For example, most of the respondents in a companion study denied any 'early criminal behavior' such as 'swip[ing] things from stores or from other children or steal[ing] from your parents or from someone else.'[43] As a critic asked, 'Who hasn't swiped something before the age of 15?'[44] In fact, two-thirds of the respondents denied *all* antisocial behaviors. The very title of Lyons's 1996 investigation, 'A Twin Study of Self-Reported Criminal Behavior,' illustrates a serious problem with studies of this type. Unless answering anonymously, people tend to refrain from telling other people about their past and present criminal behavior – including telephone interviewers who know their identity.

There have been other twin studies of antisocial behavior performed in the last 20 years, which will not be reviewed here.[45] As discussed later in this chapter (and in Chapter 3), a determination of whether the underlying assumptions of the twin method are valid is the crucial question at hand.

Twins reared apart

While there are several individual case studies of *reared-apart* criminal twins in the literature, there has been no systematic study of reared-apart criminal twins. There has been one published twins-reared-apart study of antisocial behavior, by Grove and his MISTRA colleagues in 1990.[46] Although Grove et al. claimed to have found 'significant heritability' for childhood and adult antisocial behavior, we saw in Chapter 4 that reared-apart twin studies are plagued by methodological problems and the failure to control for the common environmental influences shared by reared-apart twins.

An example of the problems with case histories of allegedly reared-apart twins is evidenced in the report of Schwesinger in 1952, who discussed 18-year-old Mexican-American identical twins Esther and Elvira. Although the girls were separated at nine months, Elvira soon returned to her mother's home, while Esther was sent to live with another relative. For eight years following their separation the twins lived 'in neighboring houses, perhaps a couple of hundred feet apart.'[47] Their mother reported that the girls played together, had the same type of clothing and toys, and 'were aware of their twin relationship.'[48] The girls became geographically separated only at the age of nine, but even then they had contact with each other including yearly visits in the summertime. As an adolescent, Elvira was beaten by her 'brutal' stepfather, who threw her out of the house. She subsequently ran the streets and was sent to a correctional institution because of her delinquency. Around this time, Esther returned to her mother's home determined to be reunited with her twin: 'Her one desire was to be with her twin, just as the latter's was to be with her.' Schwesinger noted that 'the bond between the girls was a close one.'[49] Soon, Esther became delinquent and was committed to a correctional school different from the one to which Elvira had been sent. She was able to escape from this school, while Elvira was on release from her school 'very much upset, and determined to find Esther.' The girls were picked up and sent to the same school where they 'proved devoted to each other.'[50] At the age of 18, Esther, who unlike Elvira, 'had not basked in warm mother love throughout her formative years,'[51] was found dead of a morphine overdose in the hallway of a big city hotel.

Clearly, these 'reared-apart' twins were raised in different branches of the same family and had considerable contact and a strong twinship bond. Although genetically-oriented reviewers such as Adrian Raine considered the handful of reported reared-apart concordant pairs as 'clear evidence for the role of genetic factors,' Susan Farber correctly viewed the case of Esther and Elvira as typical of the 'dubious separation' found in many cases of reared-apart twins.[52] She concluded, 'There is no substantive evidence in the twin reared-apart data to support the claim that genetic determination is significant in this area [crime], and there is much to support the idea that environment is potent.'[53]

Does greater identical twin concordance point to a genetic pre-disposition to crime?

The validity of the equal environment assumption has been questioned by criminal twin researchers themselves. According to Dalgard and Kringlen, the twin method requires that the 'environmental conditions are in general similar for MZ and DZ pairs,' but they concluded, 'This assumption is obviously not true.'[54] Later, they wrote that the EEA is 'an assumption which today cannot be accepted.'[55]

Dalgard and Kringlen found that 86% of their identical pairs had felt an extreme or strong interdependence, while only 36% of fraternals felt this way.[56] When the researchers grouped twins on this basis, there was virtually no concordance rate difference among pairs with an 'extreme or strong' level of interdependence (identicals: 6/26, or 23.2%, vs. fraternals: 3/14, or 21.4%).

According to Christiansen,

> The fundamental assumption underlying conclusions about heredity and environment that have usually been drawn from criminological twin studies is that the relevant environment of the two twins is (and has been) equally similar or equally different, regardless of zygosity. Stated in another way: intrapair environmental variations must be the same for MZ co-twins as they are for DZ co-twins. *It is open to serious doubts whether this condition is fulfilled with respect to social behavior* [emphasis added].[57]

Christiansen wrote that 'it is generally accepted that the experienced environment of MZ twins is more similar than that of DZ twins,'[58] citing several studies as evidence. Thus, the authors of two large population-based Scandinavian criminal twin studies had serious doubts about the validity of their research method.

The findings of behavior geneticist Gregory Carey confirmed the doubts of Dalgard, Kringlen, and Christiansen. After analyzing the Danish criminal twin study data, Carey concluded in 1992,

> The assumptions of the traditional twin method may be violated for phenotypes related to externalizing antisocial behavior . . . If MZ twins influence each other more than do DZ twins — a hypothesis that cannot be rejected in this analysis — genetic effects for criminal liability may actually be small.[59]

Carey, a proponent of the twin method in general, believed that the equal environment assumption was likely invalid for criminal and antisocial phenotypes.

Lyons also recognized that identical twins experience more similar environments, but argued that the twin method is valid on the basis of the trait-relevant EEA:

> The twin method does not assume that the environment is the same for both members of the twin pair . . . What the twin method does assume is that the environment is not more similar for MZ twins than it is for DZ twins in a way that would inflate the observed resemblance for MZs relative to DZs . . . the evidence is quite supportive of the 'equal environments' assumption.[60]

The following case illustrates the likely effect of the twinship bond on identical twin concordance rates. In 1951 British investigator Lorna Wheelan chronicled the lives of a pair of reared-together British identical twins whose behavior differed greatly. The 'patient' and his twin brother (whom I will call 'Twin A' and 'Twin B') had been raised in an abusive and alcoholic family environment. Twin A, who was later diagnosed as an 'aggressive psychopath' and was convicted of larceny, was the acknowledged 'leader of the twins' from an early age.[61] The twins were investigated at age 27, and their divergent personalities are described as follows:

> The patient [Twin A] has no friends, and he soon tires of acquaintances whom he makes easily; he is described as being cold-hearted, selfish and unpredictable; he never heeds advice; borrows money, is dishonest, and [is] a shiftless worker; he shows no affection or consideration for anyone — even his four children, for whom he has never accepted responsibility. His brother [Twin B] is steady, stable, modest, less quick tempered and more ambitious. In contrast to his brother's agnosticism, he recently joined the R. C. Church.[62]

We see that Twin B, upstanding citizen that he was, was nonetheless convicted as an accessory to a jewelry theft. His brother Twin A, who was living in a different part of the country, had asked him to pawn stolen jewelry for him. Twin B agreed and was subsequently arrested and convicted for the deed. Thus, vastly different identical twin brothers were concordant for criminality on the basis of the criteria used in most family, twin, and adoption studies. We should recall that Twin A was the leader of the pair, and it was probably difficult for Twin B to turn down his request in spite of his likely distaste at being involved in illegal activities. Like most of the twins recorded in the various studies, concordance for criminal behavior in this case was the probable result of association, not common blood.

Contrary to genetic theory, there is a marked difference between same-sex and opposite-sex fraternal twin concordance rates in the various studies. In Rosanoff and associates' 1934 study the adult criminality concordance rate for same-sex fraternals was 18% (5/28), but only 3% (1/32) for the opposite-sex fraternal pairs. Kranz reported same-sex fraternal rates of 23/43 (54%) and opposite-sex fraternal rates as 7/50 (14%), and Stumpfl reported rates of 7/19 (37%) and 2/28 (7%). Large differences were found in the final report of Christiansen's study. The rates were: same-sex fraternal 56/202 (28%) and opposite-sex fraternal 14/228 (6%).[63] The Eley et al. 1999 study also reported higher same-sex versus opposite-sex correlations. Dalgard and Kringlen believed that the higher rate among same-sex fraternal twins 'emphasizes the significance of environmental factors.'[64] This statement is true but requires clarification: the difference emphasizes the role of environmental factors *affecting concordance rates*, which suggests that the identical-fraternal concordance rate difference is also affected (or explained entirely) by environmental factors.

In his 1983 study of teenage twins, behavior geneticist David Rowe attempted to test the validity of the equal environment assumption, and concluded that the assumption is valid.[65] According to Rowe, the equal environment assumption is supported because delinquency among his twins was not predicted by their level of association, and because identicals did not commit delinquent acts together more frequently than fraternals. Not surprisingly, however, these findings have little meaning. Problems include: (1) The study depended on mailed responses from twins, who were asked to self-report delinquent acts. It is unlikely that these teenage respondents were willing to honestly report antisocial (and sometimes criminal) acts. (2) Only 50% of the twins returned questionnaires, which likely biased the sample in the direction of better behaving pairs. (3) A statistically significant association between shared activities and DZ males' delinquent behavior *was* found. However, as Rowe described it, he inspected the scatterplot, removed an outlier, and declared the association non-significant.[66] (4) Rowe's test involved only a few environmental variables, while many others went unchecked. The shared activities questionnaire contained a nine-item Likert scale. Rowe calculated mean scores from the Likert scale responses, which potentially obscured important associations between answers on the ends of the scale and concordance.[67]

Conclusion
The genetic basis of criminal or antisocial behavior, or any other type of behavior, is not supported by a greater concordance rate or correlation among identical versus fraternal twins (see Chapter 3). Like schizophrenia, several researchers have studied adoptees in an attempt to disentangle possible genetic and environmental influences on crime. It is to these studies that we now turn.

Adoption studies of criminality, psychopathy, and antisocial behavior

According to adoption researchers Sarnoff Mednick and Elizabeth Kandel, family and twin studies suffer from possible environmental contamination: 'To address this problem, adoption studies have been utilized. These are natural experiments in which the effects of genetic and rearing influences may be separated to a relatively high degree.'[68] Like schizophrenia, Scandinavian adoption studies are frequently cited in support of a genetic predisposition for criminal behavior. Not surprisingly, however, these studies suffer from many of the invalidating flaws discussed in Chapter 7.

There have been five adoption studies of criminality, antisocial personality, or psychopathy: two minor North American studies (Crowe, Cadoret) and one minor Danish study (Schulsinger),[69] plus two major Scandinavian investigations (Bohman and colleagues in Sweden, Mednick and colleagues in Denmark). It is important to note that *in no adoption study did the researchers claim to have found evidence in support of a genetic basis for violent crime.* The investigators claimed only to have found a genetic component for 'petty' or 'property' offenses, or for a vaguely defined notion of antisocial or psychopathic behavior.

Here, I subject the Bohman and Mednick studies to a critical reanalysis in an attempt to determine whether they provide evidence in support of genetic influences on crime. The studies of Crowe, Cadoret, and Schulsinger contain numerous invalidating flaws which have been documented in detail elsewhere.[70] Regarding the Crowe study, a pair of critics wrote that it 'is so far from the minimum standards of scientific adequacy that it deserves only minimal comment.'[71]

The Bohman and associates Swedish study

In 1978 Swedish investigator Michael Bohman reported on the rates of criminality and alcoholism in a large group of Swedish male and female adoptees given up by their biological parent(s) and placed into non-relative adoptive homes during the first three years of life. The names of these adoptees and their biological and adoptive parents were checked against the records of the Swedish Criminal Register and of the Excise Board, which maintained records on people fined for intemperance.

Bohman et al. sought to determine whether alcoholism and criminality rates were elevated among adoptees who had a criminal or alcoholic biological parent. Bohman found that the adopted-away sons of biological parents with a criminal record were themselves criminal at rates comparable to the expected population rate. He also performed a control study of 50 male and 50 female adoptees whose biological fathers 'were among those with the most serious criminal records . . . Most of these men had been sentenced to long terms in prison.'[72] Only 8% (4/50) of the male adopted-away biological children of these men, and 8% of controls (4/50), had criminal records. Bohman concluded, 'The results suggest that there is a genetic determinant for alcoholism but not for criminality (defined as repeated offenses with long prison sentences).' Thus Bohman found no evidence of a genetic predisposition for criminality, while stating that the results 'must be regarded as preliminary.'[73]

The year 1982 saw the publication of a reanalysis of the study by Bohman and his associates. In this paper Bohman et al. reported the criminality rate of 862 Swedish men from the same adoptee cohort as the

1978 study. They reanalyzed the data and argued that, although there was no association between criminality in the biological parents and their adopted-away sons, genetic influences were detected if the type of offense and the use of alcohol were controlled for. The investigators claimed that 'nonalcoholic petty criminals had an excess of biologic parents with histories of petty crime but not alcohol abuse.'[74] They concluded in favor of genetic influence for petty offenses — a claim that cannot go unchallenged — but *not* for violent crime.

Critique

As noted, in 1982 Bohman et al. studied 862 male adoptees. However, in 1978 Bohman had studied 1,125 male adoptees. Because the 1982 paper was essentially a reanalysis of the 1978 data, one must ask why the male adoptee group was reduced from 1,125 to 862, a drop of 23.4%. The investigators offered an explanation:

> The subjects included all 862 men born out of wedlock in Sweden from 1930 through 1949 who were adopted at an early age by nonrelatives. Other subjects included by Bohman in preliminary analyses (Bohman, 1978) were excluded because of incomplete data, late placement, or intrafamilial adoption. The age at adoption was less than 3 years in all cases and 8 months on the average. The adoptees ranged in age from 23 to 43 years at the time of last information.[75]

The dramatic reduction of the male adoptee group is puzzling because, as Bohman reported in his 1978 paper, *all* 1,125 male adoptees had been placed with non-relatives before the age of three, and were checked against the Criminal Register. Thus, there is no indication that the 1982 adoptee group was subject to more stringent criteria than the 1978 group, and no more specifics were given by the authors. Nor was it stated that the researchers were unaware of the Criminal Register status of the 263 adoptees (and their relatives) who were dropped from the study.

In Table 1 of the Bohman and associates' 1982 paper we see that the criminality rate of the adopted-away biological sons of parents with a record of criminality (with or without alcohol abuse) is 13%. This figure is comparable to the Swedish male population risk of 11%.[76] Contrary to genetic expectations, the fact that an adoptee had a criminally-registered biological parent did not lead to an elevated crime rate when compared to the population rate. The authors presented several pages of complex statistical procedures which they claimed suggested a genetic pre-disposition for petty offenses. In addition to the problems already discussed, there are two factors receiving insufficient attention from the investigators: (1) the rate of index petty criminality was not stated as being significantly higher than the population expectation, and (2) the selective placement of adoptees.

In any adoption study, researchers must demonstrate that the experimental group rate of the trait in question is significantly higher than the general population expectation, or, more accurately, than the non-adoptee population expectation.[77] This is true even if a control group is utilized. For example, Bohman stated that the population risk for registered male criminality in Sweden was 11%. Suppose that Bohman, in his 1978 study, had found that 11% of his male adoptees with seriously offending fathers had criminal records, whereas only 2% of the control adoptees had a criminal record—a statistically significant difference. Contrary to the beliefs of most adoption researchers, no conclusions about genetic factors can be reached from these figures because the experimental group adoptees had the same rate as one would expect in a randomly selected group of Swedish males. The significant experimental/control difference could indicate that controls experienced environments less conducive to producing criminality, but the difference would not provide evidence for the operation of genetic factors. In the Bohman at al. 1982 reanalysis there is no indication that any group of adoptees had a rate of petty criminality higher than the population prevalence, which could have been determined through the records of the Criminal Register. Groups of adoptees are compared with each other, but the crucial population base rate comparison was not examined.

Bohman and colleagues' 1982 study was essentially a post hoc reanalysis of Bohman's 1978 data, from which the authors drew a different set of conclusions more in line with genetic thinking on the subject. Ironically, the opposite approach was utilized by the Danish-American schizophrenia researchers. In their studies from the 1960s and 1970s, there were no statistically significant elevations among index biological relatives versus controls for any one diagnostic category (which is especially true if the comparisons are limited to first-degree relatives). This is evidenced by the words of Kety himself, in a 1988 paper co-authored with Ingraham:

> At the prototypical end of the spectrum, chronic schizophrenia is found exclusively in the biological relatives of chronic schizophrenia patients where it occurs at a low prevalence (approximately 3%), whereas the prevalence in the biological relatives in the normal controls is negligible. The same is true for uncertain chronic schizophrenia. Latent or borderline schizophrenia was found at a 4–5% prevalence in the biological index relatives and 1% to 1.5% in the biological relatives of controls. This is also true where the symptoms are less distinct and the diagnosis is designated uncertain. *Since neither in chronic nor in latent schizophrenia the results for the definite or uncertain diagnoses are statistically different, it appears justified to combine them* [emphasis added].[78]

Thus, Kety needed to combine his spectrum diagnoses into one total in order to find significant results—and his investigation is frequently cited as having produced the most compelling data in favor of a genetic basis for schizophrenia. Contrast this to Bohman and associates' 1982 reanalysis. In this case the definition of crime *was narrowed* in order to find a significant genetic effect. And in neither the Kety and colleagues schizophrenia investigation nor the Bohman et al. criminality study were the criteria for inclusion or exclusion published *before* the appearance of papers stating how the dependent variable would be defined. Thus, in both cases the investigators performed post hoc reanalyses of the data — using opposite approaches — in order to find a statistically significant genetic effect for the trait under study. And in both studies, the researchers could just as easily have concluded that they found no genetic influences on schizophrenia or criminality.

Regarding selective placement, there is reason to believe that the biological relatives of registered criminals were not placed into the same types of environments as adoptees lacking such a family history. Adoptees in the Swedish study were born between 1930 and 1949 — a period which coincided with the widespread belief in the inherited nature of most mental abnormalities, including antisocial behavior.[79] The world's first 'racial biology' institute, the *Uppsala Institute for Race Biology*, was established in Sweden in 1922. The opening of this Institute is said to have inspired the creation of similar institutions in Germany.[80] Sweden passed its first eugenic sterilization law in 1934, and another law was passed in 1941 permitting eugenic sterilization for those demonstrating 'an antisocial way of life.'[81] The existence of such laws suggests that children with a criminal family background, easily checked through the use of registers, would not have been attractive candidates for adoption because of their perceived 'hereditary taint.' According to Bohman et al., Swedish children 'thought to be at high risk for heritable disorders were unlikely to be considered eligible for adoption.'[82] But what about children thought to be at moderate genetic risk? They might well have been included in the adoption process but would have been considered less desirable adoptees. According to Bohman and colleagues, Swedish 'adoptee[s] and the adoptive parents were never informed [by the adoption authorities] about the identity or behavior of the biological parents,'[83] although in an earlier paper Bohman indicated that a child's biological background *was* a factor influencing the adoption process. In a discussion of a subgroup of children from his 1971 study, Bohman wrote,

> This group may be regarded as a negative selection from the primary series, in that many of the children were considered at birth, or while at the infants' home, to be difficult to place on account of retarded development, *poor heredity*, or somatic complications. *It often took longer to place these children in their ultimate home environment than it did the other children in this study;* more than half of them spent over nine months at an institution before being placed [emphasis added].[84]

It appears that a potential adoptee's 'poor heredity' was a factor in the adoption process. This could have been manifested in two ways: (1) prospective adoptive parents were informed of the biological background of the child, or (2) the authorities were reluctant to place children with 'poor heredity' into the homes of 'good' adoptive families. In any case, Bohman clearly indicated that these children were placed later and spent more time in an institution than other children.

In Bohman's 1978 control study, the biological male children of fathers and mothers registered for criminality *but not for alcohol abuse* were recorded alcohol abusers at rates significantly higher than controls.[85] Among these adopted-away biological offspring of registered criminal fathers, nine of 50 (18%) were registered alcohol abusers, versus two of 50 controls (4%).[86] Among the offspring of registered mothers, the rate of alcohol abuse is nine of 48 (19%), versus two of 48 controls (4%).[87] Combining both sets of offspring versus controls, the significance level reaches .001. Of course, one could argue that these figures suggest a genetic relationship between criminality and alcohol abuse, but a far more plausible explanation would be that the children of criminally registered parents were placed into more psychologically damaging environments than were control adoptees, leading to significantly more alcoholism.

According to Bohman and Sigvardsson, 'children were selected and placed . . . according to the social and occupational status of their biological parents,'[88] and in 1978 Bohman wrote that adoptees with antisocial or alcoholic biological parents were placed on average two to three months later than controls. He found that 'later placement is associated with selective factors that contributed independently to poorer social adjustment later in life and hence to an increased risk of appearing in the registers.'[89] If true, then placement policies might account for any possible significant elevation in so-called petty offenses.

To summarize, the Swedish adoption study of Bohman and colleagues, as acknowledged by the investigators themselves, found no evidence for genetic influences on violent crime. The finding of a genetic influence for petty offenses was based on a post hoc reanalysis of the data, and the study suffers from several serious problems and environmental confounds.

Mednick and associates' Danish study

The 1984 Mednick and associates Danish adoption study is perhaps the most well-known and most frequently cited study in support of a genetic basis for criminal behavior.[90] The investigators utilized a register of 14,427 Danish adoptees placed between 1924 and 1947. This register had been compiled by Kety and associates for their schizophrenia studies. Mednick and colleagues, who used the Adoptees' Family method, identified 13,194 adoptees (6,129 male, 7,065 female). Like most genetic studies of criminality and antisocial behavior, Mednick and associates concentrated on the results among male subjects, which will be the focus of attention here.

As discussed by Hutchings and Mednick in 1975, criminality was determined by checking the names of adoptees against the records of the Danish Police Record Office and a separate criminal record, called the *Personalia Blad,* which kept information on all persons who had been convicted of offenses. Criminal behavior was defined as those subjects with a record of court conviction. The investigators checked the conviction records of the biological and adoptive relatives of convicted adoptees. According to Mednick and colleagues:

> The size of the population permits segregation of subgroups of adoptees with combinations of convicted and non-convicted biological and adoptive parents in a design analogous to the cross-fostering model used in behavior genetics. If neither the biological nor the adoptive parents are convicted, 13.5 percent of the sons are convicted. If the adoptive parents are convicted, and the biological parents are not, this figure rises only to 14.7 percent. However, if the adoptive parents are not convicted and the biological parents are, 20.0 percent of the sons are convicted. If the adoptive parents as well are convicted, 24.5 percent of the sons are convicted. These data favor an assumption of a partial genetic etiology.[91]

Mednick and colleagues also performed a 'sibling analysis,' which compared rates among split-up half- and full-sibling pairs placed into different adoptive homes. Genetic theory predicts that full-sibling pairs would be more concordant because they are more similar genetically. Of the 126 male-male *half-sibling* pairs, 31 had at least one convicted member and four pairs were concordant (concordance rate = 4/31, 12.9%). Of the 40 male-male *full-sibling* pairs, 15 had at least one convicted member and three of these pairs were concordant (concordance rate = 3/15, 20%). Although the full- versus half-sibling difference is not statistically significant, Mednick and colleagues concluded, 'The numbers are small but indicate that as the degree of genetic relationship increases, the level of concordance increases.'[92]

The authors wrote that although the evidence pointed to a genetic predisposition for property offenses, 'This was not true with respect to violent crimes.'[93] This finding was elaborated upon in a subsequent paper:

> A significant relationship exists between parents' convictions and property offending. A significant relationship does not exist for violent offending . . . Genetic factors predispose to property offending but not to violent offending. *If a biological predisposition to violence does exist, then it must be a result of other postconception factors* [emphasis added].[94]

Like Bohman and associates in Sweden, the investigators believed they had found evidence of a genetic component for petty or property offenses, but not for violent crime.

Danish adoptees at greater risk for registered criminality

Like the other criminality adoption studies, there are major problems with the Danish investigation. To begin, Danish adoptees in general were significantly more susceptible to criminal conviction than non-adoptees.

According to Mednick and associates, 84.1% of the identified male adoptees had no record of being convicted, meaning that the adoptee conviction rate was 15.9%.[95] The researchers stated that the prevalence rate for male conviction in Denmark was about 9%,[96] while Hurwitz and Christiansen put the figure at 8%.[97] We have seen that Mednick and associates found a conviction rate of 13.5% for male adoptees whose biological and adoptive parents had no record of conviction,[98] which means that about 827 such adoptees were registered criminals (13.5% of 6,129). This figure can be compared to the expected rate among 6,129 members of the Danish male general population, yielding a figure of 490 (8% of 6,129). Thus, Danish adoptees placed during the era under study were far more likely to be convicted when compared to the general population prevalence, and thereby constituted a distinct population with regard to criminal conviction.

An 8% Danish population prevalence is actually a high figure to compare to the 13.5% rate among adoptees lacking a convicted biological and adoptive parent, for two reasons. The first is that the 8% population prevalence rate is slightly inflated by the inclusion of adoptees, who have a higher rate. The second reason is the likelihood that a fair percentage of the non-adoptee convicted criminals had a parent with a conviction record, yet Mednick and colleagues found a 13.5% prevalence among adoptees whose adoptive (and biological) parents *were not convicted*. Although the Danish figures are not available, it is probable that the conviction prevalence rate among non-adoptees whose parents had no conviction record was well below 8%.[99] Thus, Danish adoptees placed between 1924 and 1947 having non-convicted biological and adoptive parents were at *roughly twice the risk* of being convicted, versus non-adoptees with non-convicted parents.[100] Therefore, with regards to criminality, *findings among this Danish adoption cohort cannot be generalized to the non-adoptee population, Danish or otherwise.* Conclusions can be applied only to the population of Danish adoptees.

A controversy in the schizophrenia twin study literature is relevant to this discussion. We recall that in his devastating critique of these studies, Jackson theorized that the unique aspects of the identical twinship itself might lead to higher rates of schizophrenia. If this theory were true, then identical twin schizophrenia concordance rates would have been inflated and identical-fraternal differences would not be generalizable to the single-born population. In response, Rosenthal and others argued that identical twins were no more likely to receive a schizophrenia diagnosis than members of the single-born population, and that the schizophrenia twin studies were therefore valid.[101] As we have seen, Danish *adoptees* were about twice as likely to be registered criminals as non-adoptees, meaning that the argument used by schizophrenia twin researchers in defense of their studies restricts Mednick's generalizations to the Danish adoptee population.

Selective placement

The next important problem with Mednick's study is the likelihood that, as we saw in the Swedish study, the biological children of convicted criminals were placed into inferior rearing environments when compared to other adoptees.[102] In Chapter 7 we saw that the biological background of children put up for adoption in mid-twentieth century Denmark was an important factor in the placement process. A few more points should be added to this discussion as they relate to criminality, which was also viewed as being the result of bad heredity.

In 1935 a law was passed in Denmark allowing the compulsory sterilization of mentally 'abnormal' people on eugenic grounds. In the same year, a panel of leading Danish medical authorities reviewed the results of the 1929 law, writing,

> [The psychopaths] are often — to a larger extent than for example, the mentally retarded — asocial or antisocial (criminal); and their erotic activity and inventiveness, considered together with their fertility — often extramarital — is considerable . . . With respect to hereditary tainted progeny the psychopaths are comparable to the more well-defined mental diseases, even though the pattern of inheritance is still unknown.[103]

As this document suggests, the children of criminals and 'psychopaths' were regarded as 'hereditary tainted progeny.' Historian Bent Hansen noted that 'the leading [Danish] medical experts in 1935 were ready to go very far in their pursuit of eugenic goals and social control of the marginal groups of society.' He pointed out that 'there were no adverse reactions' to the 1935 document, and that 'the medical world seemed to agree with the conclusions.'[104]

Between 1940 and 1945 Denmark was under German occupation, which represents about a fifth of the period when the adoptees in Mednick's study were placed. The war years saw a dramatic rise in the Danish conviction rate,[105] and these convictions are counted in the study. The period was likely marked by an even greater emphasis on eugenics, since Denmark was under the occupation of the armies of the German government of 'applied racial science.' Mednick noted on several occasions that the twin studies of Stumpfl and Kranz were 'tainted by their origins in Nazi Germany,'[106] which he considered 'a politically unfortunate period.'[107] However, Mednick typically failed to discuss the importance of the fact that about 20% of the period in which his Danish adoptees were placed, and criminals were registered, occurred during a similarly 'unfortunate period' in Danish history.[108]

Thus, the period when adoptees were placed in the Mednick and colleagues study (1924–1947) coincided with a period of Danish history when undesirable traits were seen largely as the product of bad genes; so much so that the authorities passed laws in an attempt to prevent the alleged carriers of these genes from reproducing. According to Mednick and Hutchings, a 'potential problem with the adoption method is the possibility that the adoptive family is informed by the adoption agency of deviance in the biological family,' and (as seen in Chapter 7) they noted that a social worker who had read the older adoption journals 'formed the impression that serious deviance in the biological parents was routinely reported to the prospective adoptive parents unless they refused the information.'[109]

Mednick and colleagues stated that 37% of the biological parents had their first conviction before their child was adopted, whereas 63% were convicted after the adoption. They produced figures showing that there was no difference between the conviction rates of adoptees placed before and after their biological parents' conviction, and concluded, 'The fact that the adoptive parents were informed of the biological parents' criminality did not alter the likelihood that the adoptive son would be convicted of a crime.'[110] However, this comparison looks only at the parents and leaves out knowledge of the criminal status of other family members which, as we saw in Chapter 7, was of interest to adoption authorities.

Mednick and associates failed to understand the significance of the differing environments of desirable and undesirable children. For them, prospective adoptive parents' knowledge of the adoptee's biological family history of criminality might cause them to expect the adoptee to commit criminal acts when they grow up, which could affect the way the adoptee is perceived and reared. But what is more important is that a child's background, under the conditions of Danish society during the era in which he or she was placed, meant that the child was not a desirable adoptee in the eyes of the most qualified adoptive parents who, as we have seen, were routinely provided information on the criminal records of the biological family. Into what kinds of adoptive families were the adopted-away biological children of families with convicted criminals eventually placed? It is likely that these adoptees were placed into the homes of people who, for various reasons, were not qualified to adopt the most desirable children. These homes may have been more chaotic or potentially exploitative than the others. In a 1975 control study by Hutchings and Mednick, the biological and adoptive fathers of criminal adoptees (N = 143) were compared to the biological and adoptive fathers of adoptive (non-criminal) controls (N = 143). Among the criminal adoptees, 33 (23%) had a criminally registered *adoptive* father, whereas only 14 (9.8%) of the control adoptive fathers were registered.[111] The comparison is statistically significant,[112] and Hutchings and Mednick added that the difference 'was also reflected in the various indices of criminality such as number of recorded cases and total length of sentence.'[113]

The rate among these adoptive fathers further illustrates the different family environments experienced by the criminal adoptees. In fact, a major aspect of the Adoptees' Family model compares the prevalence of the trait in question among index and control *adoptive* relatives. According to Rosenthal's 1970 description of how the model is used with schizophrenia, 'A higher incidence among the adoptive relatives of index cases than of controls supports the view that rearing by, of, or with schizophrenics contributes to the development of the disorder.'[114] Commenting on the results of the Danish-American Adoptees' Family study, Rosenthal noted that.

> among adoptive relatives, there was no appreciable difference between the two groups . . . Genetic theory would predict such a finding. Most environmentalist theories would predict a higher incidence of such disorders among the adoptive parents of the index cases as compared with the controls, but the findings do not support this prediction.[115]

It appears that the conviction rate difference between index and control adoptive fathers in Mednick's study is, as Rosenthal outlined, is consistent with the environmental prediction.

Placement factors probably influenced the previously discussed Mednick et al. 'sibling analysis,' which compares concordance rates among adopted-away half- and full-sibling pairs. Because the full-siblings had a more similar biological family background, it is likely that they were placed into more similar environments than the half-sibling pairs. As noted, the concordance rate difference between the pairs was not statistically significant. Contrary to basic statistical principles, however, Mednick and colleagues claimed that a non-significant difference (based on an admittedly 'small' sample) showed that a relationship existed between genetic similarity and concordance.[116]

In a footnote to their 1984 *Science* article, Mednick et al. wrote,

> Among males, there was a statistically significant association between adoptee criminality and the amount of time spent in the orphanage waiting for adoption. This effect, which was not true for females, may be due to institutionalization. Or it may be a function of selection bias ('less desirable' boys adopted later and also being convicted).[117]

Like Bohman et al., Mednick and colleagues found an association between time spent in an institution and subsequent conviction, and realized that the association might be due to selection bias or the effects of institutionalization. The psychological damage inflicted on children spending significant periods of time under the typically appalling and nurtureless conditions of an orphanage (particularly under the conditions of the Great Depression, foreign occupation, and war) was greater for convicted than non-convicted adoptees, and was not controlled for in this study.

Conclusion
Despite having found no evidence supporting a genetic basis for violent crime, the Mednick group typically emphasizes the theme of alleged

genetic influences on criminality in general. The investigators could have given their papers titles such as, 'No Genetic Basis for Violent Crime,' or 'Environmental Causes of Violence Must be Identified,' but instead used potentially misleading titles such as 'Genetic Influences in Criminal Convictions: Evidence from an Adoption Cohort,' 'Genetic Correlates of Criminal Behavior,' and 'Predisposition to Violence.'[118] Although the investigators believed they had found evidence for property (but not violent) crime, the high registered criminative rate among adoptees in general means that the results cannot be generalized to the non-adoptee population. In addition, the evidence suggests that Mednick's study was confounded by selection factors in the Danish adoption process, and it is therefore unlikely that children with a criminal family background were placed into the same types of environments as children lacking such a history.[119]

Discussion

We have seen that the idea of criminality as a hereditary condition is an old one, whose 'scientific' basis dates back at least to Lombroso. In the United States, the sterilization of criminals on eugenic grounds was permitted in several states. Long before the publication of Lange's German twin study, the idea of 'crime as destiny' was widespread in the North American population, including especially its most educated layer. Studying the genetics of criminality and antisocial behavior fell into disfavor in the years after World War II, with the revelations of National Socialist genocide in the name of 'racial hygiene.'

Since the late 1960s, however, genetic theories for most human traits have made a comeback. So too have ideas about genetic factors in criminal behavior, although statements in support of the genetic position are more cautious than for other behaviors, in part because of the commonly held view that social conditions such as racism, unemployment, and poverty contribute greatly to criminal behavior. Most genetic researchers also recognize that these and other environmental factors play a role. Why then is it necessary to study the genetics of criminality? According to Brennan, Mednick, and Gabrielli,

> We must try to identify the specific biological mechanisms through which heritable predispositions toward criminal behavior are expressed. By identifying these mechanisms we can learn how to successfully treat and prevent criminal behavior.[120]

The authors failed to discuss the types of treatment and preventive measures they had in mind, or to distinguish these from the 'treatments' of previous generations. Perhaps Brennan and colleagues had in mind relatively benign measures such as early intervention programs,[121] but they rarely discuss the possibility of improving the socioeconomic environment as a way of eliminating crime. Why change the person and not the environment? This is not a scientific but rather a political question.

The call by Brennan, Mednick, and associates for treatment is even more puzzling when we realize that one of the main findings of their adoption study was that *there is no genetic basis for violent crime*, which is the type of offense of greatest concern to the general public. Thus, they might have concluded that social conditions must be changed in order to reduce or eliminate violent crime. Furthermore, how can there be a genetic predisposition for property crime but not for violence? How can people be predisposed to steal but not to shoot; to write bad checks but not to rape? The explanation becomes clear when the invalidating flaws of the research are exposed, which is the purpose of this chapter.

But suppose that future researchers present compelling evidence in support of a genetic predisposition for criminal behavior. What significance would this finding have? One could argue that it would not be an important finding, because the proper environmental conditions would eliminate most criminal behavior (even among the genetically predisposed), just as the expression of a genetic disorder such as PKU is prevented with the proper diet. While the practical significance of finding a genetic predisposition for any psychiatric condition or behavioral trait is open to debate (see Chapter 6), the finding of a genetic predisposition for criminal and antisocial behavior would accomplish little more than diverting society's attention from eliminating the social conditions leading to these behaviors. But diverting attention from environments causing crime is needed by privileged economic classes and politicians seeking to absolve themselves of responsibility for the dreadful social and material conditions of a large part of the population. Their objective, as well as the objective of those whose research they underwrite, is to locate the causes of social problems within the bodies, minds and genes of the oppressed.

Conclusion

The title of a 1989 review article by Walters and White asks whether the body of evidence on the genetics of criminality records 'bad genes or bad research.' The answer is that we have a body of bad research—not only because of the way it was carried out, but because the methods used are hopelessly confounded by environmental factors. Quite erroneously, authors such as Wilson and Herrnstein promote the twin method as a 'natural approximation to a controlled experiment for estimating the genetic involvement in a trait,'[122] and adoption studies of criminal and antisocial behavior have ben promoted as 'natural experiments in which the effects of genetic and rearing influences may be separated to a relatively high degree.'[123] On paper, an adoption study might appear to be just such a natural experiment. In the real world of selective placement, socioeconomic differences, and attachment disturbance, thesestudies typcally are confounded by environmental factors—to a lesser degree than family and twin studies perhaps — but confounded nonetheless. At bottom, the criminal adoption studies attempted (and failed) to confirm Galton's belief (see Chapter 2) that the adopted-away children of 'wild, untamable savages' raised in 'civilized' environments tend to return to a life of 'contented barbarism, without a vestige of their gentle nature.'

Goddard's study of the 'Kallikak' family traced two lines of the descendants of Martin Kallikak. The 'good' line began with the progeny of Martin and his 'good Quaker wife,' whereas the 'defective strains' were produced by Martin's liaison with a 'feebleminded tavern girl.' For Goddard, a comparison of these two lines was pure science: 'We have, as it were, a natural experiment with a normal branch with which to compare our defective side.'[124] Few people of Goddard's day disagreed with the notion that family pedigrees (compiled by non-blided investigators) were a true 'natural experiment' of the influences of heredity and environment. One day we will look upon twin studies and most adoption studies in much the same way as we now view the folly of Goddard's logic.

In summary, the evidence from twin and adoption studies does not support the existence of a genetic predisposition for any type of 'criminal,' 'psychopathic,' or 'antisocial' behavior, however it may be defined at any given time or in any given society. Finally, given (1) the potential social impact of criminal genetic research, (2) the well-known social factors leading to crime, and (3) the political aspects of deciding who is and is not labeled a criminal, it is questionable whether this type of research should even be performed.

Notes

[1] Discussion in Sharp, 1909, pp. 1900–1901.

[2] Allen, 2001, p. 210.

[3] This chapter is based largely on a previous publication in *The Journal of Mind and Behavior* (Joseph, 2001c).

[4] The terms 'antisocial' and 'psychopath' are political and moral judgments that have been transformed into 'mental illness'. Were it not for the fact that it would make for more difficult reading, I would place these terms in quotation marks every time they are used.

[5] The study purporting to have found a gene for criminal behavior is by Brunner et al. (1993). For a critique of biological explanations of crime, see Katz & Chambliss (1991). For a review of the 'XYY fiasco,' see Hubbard & Wald (1993).

[6] Gould, 1981, p. 124.

[7] Lombroso, 1968, pp. 365–6.

[8] Davenport, 1911, pp. 262–3. See Allen, 2001, for a critical discussion of Davenport's views on crime.

[9] Lykken, 1995, p. 17.

[10] Ibid., p. 24.

[11] See Fink, 1938, for a review.

[12] Hall, 1914, p. 87.

[13] Irvine, 1903, p. 750.

[14] Ochsner, 1899.

[15] Ibid., p. 867.
[16] Ibid., p. 868.
[17] Quoted in Reilly, 1991, pp. 46–7.
[18] Sharp, 1909, pp. 1897–1898.
[19] Ibid., p. 1901.
[20] See Rafter, 1988.
[21] Lombroso, 1968, p. 161.
[22] Davenport, 1911, p. 183.
[23] Ibid., p. 92.
[24] Dugdale, 1910, p. 65. Commenting on Dugdale's book, Fink (1938, p. 179) wrote, 'Perhaps no one book in the field of criminology in America has lent itself to suh partisan interpretation as has Richard Dugdale's *The Jukes*, published in 1877. Unread, misread, or willfully distorted, it has been used by hereditarians and environmentalists alike to assert and supposedly to prove their respective positions.'
[25] Goddard, 1927, p. 54.
[26] Goldsmith & Gottesman, 1996, p. 9.
[27] Popenoe, 1936, p. 388.
[28] Hubbard & Wald, 1993, p. 105.
[29] Rutter, 1996, p. 2.
[30] Hurwitz & Christiansen, 1983, pp. 42–3.
[31] P. Taylor, 1996, p. 137. The quote about rape statistics in the Old South is from Duster, 1990, p. 97.
[32] Brennan et al., 1991, p. 232.
[33] Wetzell, 2000, p. 157.
[34] Lange, 1930, p. 46.
[35] Rosanoff, Handy, & Rosanoff, 1934, p. 929.
[36] Wetzell, 2000.
[37] Cited in Christiansen, 1977b, p. 72.
[38] Quoted in Wetzell, 2000, p. 277.
[39] Dalgard & Kringlen, 1976, p. 231 (emphasis in the original).
[40] $P < .0001$. Christiansen, 1977a.
[41] Cloninger & Gottesman, 1987.
[42] Lyons et al., 1995.
[43] Lyons, 1996, p. 63.
[44] M. Daly, discussion in Lyons, 1996, p. 71.
[45] e.g., Eley et al., 1999; Rowe, 1986.
[46] Grove et al., 1990.
[47] Schwesinger, 1952, p. 40.
[48] Ibid., p. 40.
[49] Ibid., p. 42.
[50] Ibid., p. 42.
[51] Ibid., p. 46.
[52] Raine, 1993, p. 59; Farber, 1981, p. 51.
[53] Ibid., p. 229.
[54] Dalgard & Kringlen, 1976, p. 214.
[55] Ibid., p. 223.
[56] Ibid., p. 224. In a discussion of the Dalgard and Kringlen study, Cloninger and Gottesman (1987, pp. 98–9) wrote, 'As expected, a slightly greater proportion of MZ twins than of DZ twins were psychologically close (84% of 31 vs. 74% of 54).' However, they were looking at the concordance figures of twins' level of 'intra-pair interdependence' in Dalgard and Kringlen's Table 12 (1976, p. 224), which

compared concordance rates among a select group. When information was reported on the *entire* sample of interviewed pairs (Dalgard & Kringlen, 1976, p. 224, Table 11), we find that 42 of 49 identical pairs (86%) had an 'extremely strong' or 'strong' level of closeness, which was true for only 32 of the 89 fraternal pairs (36%). Seven identical pairs (14%) were as close as ordinary siblings, which was true for 57 (64%) of fraternals.

[57] Christiansen, 1977a, p. 93.
[58] Ibid., p. 94.
[59] Carey, 1992, p. 21.
[60] Lyons, 1996, p. 72.
[61] Wheelan, 1951, p. 134.
[62] Ibid., p. 136.
[63] Reported in Cloninger & Gottesman, 1987, p. 99.
[64] Dalgard & Kringlen, 1976, p. 217.
[65] Rowe, 1983.
[66] Ibid., p. 478.
[67] The practice of deriving mean scores from ordinal scales is potentially misleading because differences between responses are not equivalent. According to Bradley and Schaefer (1998, pp. 118–19), 'It is frequently forgotten that arithmetic is not meaningful on these scales . . . a mean is not an appropriate statistic to calculate for the Likert scale or, in fact, for any ordinal data. Mathematically, the use of a mean requires a scale that is unique up to positive linear transformation, that is, at least an interval scale . . . The appropriate statistic to use for this data is a frequency count.'
[68] Mednick & Kandel, 1988, p. 103.
[69] Cadoret, 1978; Crowe, 1972, 1974; Schulsinger, 1977.
[70] Gottfredson & Hirschi, 1990; Joseph, 2001c; Walters & White, 1989.
[71] Gottfredson & Hirschi, 1990, pp. 59–60.
[72] Bohman, 1978, p. 272.
[73] Ibid., p. 276.
[74] Bohman et al., 1982, p. 1233.
[75] Ibid., p. 1234.
[76] Ibid., p. 1235.
[77] Boyle, 1990.
[78] Ingraham & Kety, 1988, pp. 121–3.
[79] Broberg & Tydén, 1996. It is also known that there were close contacts between Swedish and Nazi-era German scientists and eugenicists (Weingart, 1999).
[80] Broberg & Tydén, 1996, p. 87.
[81] Ibid., p. 108.
[82] Bohman et al., 1982, p. 1234.
[83] Ibid., p. 1234.
[84] Bohman, 1971, p. 6.
[85] Bohman, 1978, p. 274, Table 6.
[86] $p = .026$, Fisher's Exact Test, one-tailed.
[87] $p = .025$, Fisher's Exact Test, one-tailed.
[88] Bohman & Sigvardsson, 1980, p. 348.
[89] Bohman, 1978, p. 275.
[90] Mednick et al., 1984.
[91] Ibid., p. 892.
[92] Ibid., p. 893.
[93] Ibid., p. 891.
[94] Mednick et al., 1988, p. 29.

[95] Mednick et al., 1984, Table 1.

[96] Hutchings & Mednick, 1975.

[97] Hurwitz & Christiansen, 1983, p. 41.

[98] Mednick et al., 1984, p. 892.

[99] In another paper by the investigators (Kandel et al., 1988) which looked at a consecutive series of Danish children born between 1936 and 1938 who were *not* adopted away, 39% of the sons of 'severely sanctioned' criminals were themselves recorded criminals by 1972, while 7% of the sons of non-registered fathers had been registered (p. 225).

[100] The higher rate among adoptees is consistent with the results of a 1997 paper from the investigators which found a correlation between early maternal rejection and adult violence (Raine et al., 1997).

[101] For a more detailed discussion of this point, see Joseph, 2001b, pp. 41–3.

[102] Kamin, 1985.

[103] Quoted in B. Hansen, 1996, pp. 42–3.

[104] Ibid., p. 43.

[105] Hurwitz & Christiansen, 1983.

[106] Mednick & Volavka, 1980, p. 92.

[107] Ibid., p. 95.

[108] Not all researchers of the genetics of criminality considered studies published during the National Socialist period to be 'tainted.' For example, Fini Schulsinger wrote, 'The very special eugenic ideas of the Third Reich involved some German psychiatric geneticists in the classical type of family studies on relatives of psychopathic probands . . . Their work was carried out in the same neat way as other, respectably intended, genetic work from the famous Munich school. The results of these studies unanimously indicated that heredity plays a role in the etiology of psychopathy' (Schulsinger, 1977, pp. 112–13).

[109] Mednick & Hutchings, 1977, p. 161.

[110] Mednick et al., 1984, p. 893.

[111] Hutchings & Mednick, 1975, p. 109; also discussed by Kamin, 1985.

[112] $p = .002$, Fisher's Exact Test, one-tailed.

[113] Hutchings & Mednick, 1975, p. 109.

[114] Rosenthal, 1970b, p. 57.

[115] Ibid., p. 127.

[116] Mednick et al., 1984, p. 893.

[117] Ibid., p. 894.

[118] Mednick et al., 1984; Gabrielli & Mednick, 1983; Mednick et al., 1988.

[119] See Kamin, 1985, 1986.

[120] Brennan et al., 1991, p. 243.

[121] As discussed in Gabrielli & Mednick, 1983.

[122] Wilson & Herrnstein, 1985, pp. 90–1.

[123] Mednick & Kandel, 1988, p. 103. More recent attempts to use twin and adoption data in support of genetic influences on criminal and antisocial behavior include Rhee & Waldman, 2002, and Rowe, 2002.

[124] Goddard, 1927, p. 68.

CHAPTER NINE

Genetics and IQ

> In effect, then, Galton's aim, and that of his followers, became simply an attempt to reproduce an existing set of ranks (social class) in another, the test scores, and pretend that the latter is a measure of something else. This is, and remains, the fundamental strategy of the intelligence-testing movement.
>
> *Psychologist Ken Richardson in 2000*[1]

We now arrive at the most studied and debated aspect of the nature-nurture debate: the genetics of intelligence as allegedly measured by IQ tests. In previous chapters we saw how IQ tests were used by eugenicists and behavior geneticists to demonstrate that 'intelligence' is determined largely through inheritance. Historically, IQ studies of twins and adoptees have been a central preoccupation of these fields. The issues of racial and class differences in IQ have been the subject of intense controversy for almost 100 years. Eugenically oriented psychologists' use (and creation) of IQ tests to be able to proclaim a 'scientific' basis for racism has been documented in several books, which include Gould's *The Mismeasure of Man*, Kamin's *The Science and Politics of IQ*, Chase's *The Legacy of Malthus*, Tucker's *The Science and Politics of Racial Research*, and Guthrie's *Even the Rat was White*.[2] Eugenic sterilization laws in the United States, Germany, Scandinavia and elsewhere targeted 'mental defectives' and the 'feebleminded.' A low IQ score for a German child in the late 1930s was sometimes a death sentence.[3]

The claim that standardized IQ tests such as the Stanford-Binet and Wechsler scales measure innate intelligence (or general intelligence, represented as g) has been the subject of heated debate. I will list some of the most frequent objections to this claim:[4]

- The psychometric emphasis on differences in abilities is only one of many approaches to the study of human intelligence.
- General intelligence is merely the product of a mathematical formula. It has no physical reality.
- There is no consensus definition of what 'intelligence' is.
- The tests measure school learning more than innate intelligence.
- IQ tests are biased against non-Caucasians and people of lower

economic classes.
- The tests measure only narrow abilities, and ignore 'real world' intelligence.
- Standardized IQ tests have been normed on non-representative populations.
- Intelligence is not normally distributed (in a bell curve), as psychometrists assume.

A large body of literature has been devoted to the 'heritability of IQ' topic, with estimates ranging from 0% to 90%. In Chapter 5 we saw why heritability estimates are not appropriate for estimating the magnitude of genetic influences on a trait, and this is particularly true for IQ. Thus, even the notion of 'the heritability of IQ' is invalid. Of course, one might legitimately ask whether (or how much) genes influence intelligence. This question, however, cannot be answered with a heritability coefficient.

Given the large amount of literature on the genetics of IQ, my intention is to briefly review a few of the important issues and to show how the IQ test was created, and continues to be used in some quarters, more as a tool of oppression than as measurements of 'intelligence.'

The psychometric approach to human intelligence

As an American Psychological Association task force acknowledged in 1996, 'The psychometric approach [to cognitive ability] is the oldest and best established, but others also have much to contribute.'[5] And it is self-evident that 'oldest and best established' does not equal 'most useful and valid.' One of the best critiques of the psychometric view of intelligence is found in Fischer and colleagues' 1996 *Inequality by Design*. The authors point out that the psychometric approach is only one of several ways that human intelligence can be looked at. Another approach is information processing, which seeks to understand the way that *all* humans learn and process information. Psychometrics, on the other hand, is concerned with *differences* among people. Fischer and colleagues argued that psychometric IQ tests 'are refined to magnify differences' and that the famous bell-shaped curve is an invention of psychometrists who *assume* that intelligence is normally distributed.[6] In fact, the test data collected by Herrnstein and Murray for *The Bell Curve* did not resemble a bell curve at all! It was only after a 'good deal of statistical mashing and stretching,' as Fischer et al. put it, that Herrnstein and Murray were able to create a bell-shaped distribution.[7] The Gaussian or bell-shaped curve, as Julian Simon observed, 'is seldom if ever observed in nature.' Rather, it is the 'result of scientists gradually isolating and factoring out the large elements in any situation.'[8] Even human height distribution, as Simon noted, is 'skewed toward little things like babies.'[9]

Lewontin, Rose, and Kamin described the 'grand illusion of psychometry' in which scales are refined to produce desired results.[10]

The scale chosen 'depends on whether one wants to make differences of scale appear large or small, and these decisions are those that psychometry arbitrarily makes.'[11] A classic example is a line graph. These days, people with retirement plans frequently log onto the Internet to check the day's stock market activity. A graph may give the initial impression that the market has fluctuated widely all day. That is, until the viewer notices that the range of the graph is only 50 points. The same events captured on a graph with a range of 1000 points would elicit a much different initial reaction from the viewer. I have in front of me a psychological assessment textbook by authors who support the psychometric position. Similar to the example I just gave, the authors present two graphs giving 'vastly different' impressions of the same data, under the title 'Consumer (of graphed data) Beware.'[12] Yet, one cannot truly say which scale is more appropriate. Psychometric scales create the impression of large differences between people and groups, whereas other scales (based on the same data) would show that humans have a great deal in common. The consumer of psychometric data, therefore, should be aware of its tendency to create the appearance of large differences.

Frequently cited evidence in support of the genetic basis of IQ

Genetic theories of IQ are based largely on the results of kinship correlations, which include family, twin, and adoption studies. In previous chapters we saw that these methods are flawed instruments for detecting genetic influences on psychological traits, and in Chapter 4 we examined the invalidating flaws of reared-apart twin studies of IQ. On the other hand, factors which definitely invalidate genetic conclusions for psychiatric disorders and behavior are not as relevant when discussing abilities, cognitive or otherwise. For example, the identity confusion of identical twins would be much more likely to affect schizophrenia concordance rates than it would the distance twins could throw a javelin. Although in both cases important environmental similarities exist, the blurring of psychological boundaries would not be a major factor influencing javelin throwing, as it would a schizophrenia diagnosis.

Twin studies

Behavior geneticists generally put the reared-apart identical twin (MZA) IQ correlation (omitting Burt's figures) at about .77. In Chapter 4 we saw that there are many reasons why these figures have little meaning. For reared-together twins, in 1981 Bouchard and McGue pooled the world literature and came up with an identical twin rate of .86, and a fraternal rate of .60.[13] However, these figures are certainly inflated by the inclusion of many poorly performed and biased studies, as well as the problems found in meta-analyses in general.[14] Still, like other traits, both environmentalists and hereditarians would expect to find a higher identical twin correlation (see Chapter 3), which leads us back to the questions of the methods and assumptions of twin research in general.

Adoption studies

IQ adoption studies report two seemingly contradictory findings: although the IQs of adopted-away children tend to correlate more with their biological parents than with their adoptive parents, the typical IQ point rise for children of working class biological parents who are reared in middle- or upper-class adoptive homes is 12–14 points. In general, hereditarian interpreters of adoption studies focus on the IQ correlation of adoptees and their biological parents, while environmentalists emphasize the large gains made by working-class children adopted into middle-class families. There is reason to believe, however, that the correlations are an artifact of sampling, while the IQ score gain is the statistic of value.

In his 1981 debate with hereditarian Hans Eysenck in *The Intelligence Controversy* (published in the UK as *Intelligence: The Battle for the Mind*), Kamin discussed the 'restrictive variance' of adoptive families who are similar in SES and have been screened by adoption agencies. This means that most adoptive parents were above average in IQ and provided above-average environments. Thus, the 'necessary statistical consequence' of the restricted environmental variance of the adoptive families means that 'child IQ correlation in adoptive families cannot be very high.'[15] Kamin used boxing as an analogy. He pointed out that, if weight classifications were not used, high correlations would be observed between boxers' weights and their won-loss records. 'To avoid such a correlation,' wrote Kamin,

> definite weight divisions have been established by boxing authorities. Fights can only take place between boxers of reasonably similar weight, and the correlation between weight and boxing success is consequently very low. We are suggesting that in terms of the environments provided for their children almost all adoptive parents — unlike biological parents — are in the heavyweight division. That would account for the lower parent-child IQ correlation observed in adoptive families. The correlation would presumably be much higher if parents who would provide poor environments wanted to, and were allowed to, adopt more often.[16]

Thus, one could erroneously conclude that the lack of a correlation between boxers' win-loss records and their weights means that increased weight would not improve a boxer's chance of winning, even if weight divisions were abolished. A recent article also pointed to the problems presented by range restriction in adoption studies.[17]

Another aspect of the question of correlation versus mean difference in IQ adoption studies is that, as Schiff and Lewontin demonstrated, 'adopted children, even though they may correlate *individually* with their biological parents more than with their adoptive parents, are, in fact, more similar as *a group* to the adoptive parents than to their biological ones.'[18] Schiff and Lewontin presented a table of hypothetical IQ data to illustrate their point. The data showed that adoptees' IQs were perfectly correlated with those of their biological parents, and there was no correlation between adoptees and their adoptive parents. Yet, adoptees *as a group* had the same mean IQ as the group of adoptive parents, and differed from their biological parents' mean IQ by seven points. In assessing the meaning of these figures, we must keep in mind that correlation does not measure similarity, but only how traits vary together.

Recent IQ adoption investigations include a study carried out in France, and the Colorado Adoption Project (CAP), performed by leading behavior geneticists. The French study of Schiff and colleagues found 32 adoptees who had been separated near birth from their parent (who was an unskilled worker), and were reared in the homes of families in the top 13% of the socio-professional scale. The IQs of these adoptees were compared to the scores of 20 biological half-sibling controls, who were reared in the homes of their lower-class parents. The investigators found that adoptees' IQ scores averaged 14 points higher than the control group, suggesting that being reared in a professional family as opposed to a family of unskilled workers could raise one's IQ almost one full standard deviation.[19]

The Colorado Adoption Project is an ongoing longitudinal adoption study that began in 1974, when the adoptees were infants. As discussed briefly in Chapter 7, this study's design is far superior to the relatively crude schizophrenia adoption studies. We also saw in Chapter 4 that, in sharp contrast to genetic expectations, the investigators found no personality scale correlation between birth mothers and their 245 adopted-away biological offspring. The study created a control group of non-adopted children matched on several factors. For IQ, Plomin and colleagues concluded that genetic factors are important and that 'environmental factors correlated with parents' general cognitive ability have little effect on children's cognitive ability.'[20] However, their conclusions were based on correlations, not mean differences. It is not likely that the CAP biological and adoptive parent IQ scores would be dramatically different, since the socioeconomic status of both adoptive and biological parents were similar. As a commentator noted, 'There is an enormous degree of range restriction' in the CAP adoptive homes.[21]

These problems, along with others I have discussed in earlier chapters such as selective placement and researcher bias, cast doubt on the claim that family, twin, and adoption studies provide important information about possible genetic influence on intelligence as measured by IQ tests. More importantly, there is no standard definition of intelligence (which is what the tests purport to measure), and there are many reasons to reject the claim that 'intelligence' can be measured with IQ tests. Yet, several studies have shown that a superior environment can raise scores almost one standard deviation.[22]

IQ and group differences

The claim that IQ score differences demonstrate the genetic inferiority of particular races and classes is almost as old as the tests themselves. In the 1910s and 1920s the view that races and classes differed in intelligence was axiomatic in American psychology. Jensen's 1969 article in the *Harvard Educational Review* began the modern revival of this position, which was continued by Herrnstein and Murray's incendiary *The Bell Curve*, published in 1994. These authors' arguments have been decisively refuted by numerous commentators, and I merely refer readers to several of these critiques.[23] An argument put forward by a majority of the critics is that, while supposedly all sides acknowledge that American blacks score 12–15 points lower than American whites on standardized IQ tests, the difference is explainable by environmental factors such as poverty and racism. Although this argument is solid even if we accept that IQ tests are unbiased and actually measure intelligence, there is good reason not to accept this position.

Racial and class bias built into IQ tests
Intelligence testing has a long history of being used for reactionary social and political purposes. The testing method developed by Alfred Binet in France was brought to the United States and developed by eugenicists such as Lewis Terman for the purpose of quantifying what they saw as people's innate mental capacities. Indeed, in 1924 Terman would write that his fellow psychologists and their IQ tests were 'the beacon light of the eugenics movement.'[24]

The historical association of IQ testing with eugenics and racism, as described by several authors, does not mean that all of the early pioneers of mental testing were driven by these motivations or that most contemporary mental testing is performed for this purpose. Still, modern proponents of racial and class inferiority continue to use the results of genetic studies of IQ in support of their positions. It therefore becomes necessary to demonstrate that one of the original *purposes* of IQ tests was to prove that blacks and other minorities, southern European immigrants and Jews, and the working class were genetically inferior to upper-class whites of Northern European descent.

Related to this point, it is critically important to understand that IQ tests can produce any result its creator desires. For those who doubt this, I call on the most well-known American defender of IQ testing, Arthur Jensen. 'It is claimed,' wrote Jensen, 'that the psychometrist can make up a test that will yield any kind of score distribution he pleases. This is roughly true, but some types of distributions are much easier to obtain than others.'[25] Although Jensen is a leading proponent of the idea that racial differences in IQ scores are at least partly explained by genetics, he tacitly admits that psychometrists *could* create tests in which blacks and whites scored equally. Interestingly, most *choose* not to.

For the American pioneers of mental testing there was no doubt that blacks and the working class were genetically inferior to upper-class whites. In Terman's 1916 Stanford-Binet manual, he wrote that 'dullness' among blacks and latinos 'seems to be racial, or at least inherent in the family stocks from which they come.'[26] Although Terman suggested that research was needed, he believed that 'when this is done there will be discovered enormously significant racial differences in general intelligence, differences which cannot be wiped out by any scheme of mental culture.'[27] Psychologist Carl Brigham wrote in his influential 1923 book *A Study of American Intelligence*, which discussed the large-scale testing of World War I army recruits, that 'These army data constitute the first really significant contribution to the study of race differences in mental traits. They give us a scientific basis for our conclusions.'[28] This passage is illuminating, since Brigham admitted that he and his colleagues reached conclusions about racial differences before the 'scientific' evidence had come in. Lacking a scientific basis, Brigham, who viewed 'the importation of the negro' as the 'most sinister development in the history of this continent,'[29] was actually describing *prejudices*, not conclusions. Brigham wrote that 'Our own data from the army tests indicate clearly the intellectual superiority of the Nordic race group.'[30] Another pioneer of mental testing, Henry H. Goddard, lectured Princeton students in 1919 as follows:

> Now the fact is, that a workman may have a ten year intelligence while you have a twenty. To demand for him such a home as you enjoy is as absurd as it would be to insist that every laborer should receive a graduate fellowship. How can there be such a thing as social equality with this wide range of mental capacity?[31]

My purpose is to document the beliefs of the people who created the tests. The results of their tests merely confirmed — and were a product of — their pre-existing views on racial and class inequality.

When psychometrists *do* assume equality, they simply create tests to reflect this assumption. Items on standardized IQ tests do not fall from the sky; they are carefully selected (out of a large pool of potential items) to produce results desired by the test creator. In Terman and Merrill's 1937 revision of the Stanford-Binet test, the authors wrote that 'A few tests in the trial batteries which yielded largest sex differences were early eliminated as probably unfair.'[32] Because they assumed that males and females are equal in intelligence, Terman and Merrill created a test whose results reflected this assumption. Apparently, for Terman and Merrill it was not at all 'unfair' if races and classes scored differently, or if people reached conclusions about genetic inferiority on the basis of their scores.

David Wechsler, who developed the most widely used IQ test, followed Terman in eliminating sex differences from his scales, although he had a '"sneaking suspicion" that the female of the species is . . . more intelligent than the male.'[33] Wechsler described the problem of sex differences as follows:

In trying to arrive at an answer as to whether there are sex differences in intelligence much depends upon how one defines intelligence, and on the practical side, on the types of tests one uses in measuring it. The contemporary approach, contrary to the historical point of view, adopts a sort of null hypothesis. Unfortunately this procedure turns out to be a circular affair since the nature of the tests selected can prejudice or determine in advance what the findings will be.[34]

As Wechsler acknowledged, the 'null hypothesis' for male-female differences in intelligence was that the sexes are equal. Like his predecessors, however, another implicit assumption of his tests was that *races and classes* are unequal.[35] This reflects nothing more than the beliefs and prejudices of the test creators and their backers. According to Jensen, 'The practice of eliminating and counterbalancing items to minimize sex differences is based on the assumption that the sexes do not really differ in general intelligence.'[36] What Jensen failed to articulate is the assumption upon which racial and class differences are allowed to remain, which could be stated, '*The failure to remove and counterbalance items in order to eliminate racial and class differences is based on the assumption that races and classes really do differ in general intelligence.*' As Ken Richardson observed,

> While 'preferring' to see sex differences as undesirable artifacts of test composition, other differences between groups or individuals, such as different social classes or, at various times, different 'races,' are seen as ones 'truly' existing in nature. Yet these too could be eliminated or exaggerated by exactly the same process of assumption and manipulation of test composition.[37]

In other words, the belief that races and classes are genetically inferior is built into most standardized IQ tests.[38] It is not simply a matter of whether individual test items are culturally biased; more important is the fact that psychometrists know that individual test items, regardless of how culturally biased they may appear, discriminate between various groups. As IQ critics Mensh and Mensh observed, 'there is no distinction between crassly biased IQ-test items and those that appear to be non-biased.'[39]

The fact that the sexes, but not races and classes, were assumed equal in 'native intelligence' reveals the racial and class bias of the tests. On what grounds, we might ask, did people like Terman and Wechsler anoint women with the same level of native intelligence as men? After all, during the time when they were creating their tests most women were apparently so innately deficient that they rarely worked outside of the home. Almost all of the captains of industry and most professionals, including college professors and psychologists, were men, as were most of history's great leaders, thinkers, poets, military commanders, and inventors. In fact, the innate cognitive impairment of women was so widely recognized that they were not given the right to vote in the United States until 1918!

(Undoubtedly, granting this right was a demonstration of the male cognitive elite's goodwill.) By now the reader has realized that I have slipped into facetiousness – but only to make a point. Using the standards applied for determining the relative worth and intelligence of races and classes, women should have been viewed by Terman, Goddard, Brigham and others as being innately inferior to men. But white upper-class women had two characteristics that set them apart from non-white races and the working class: they were of the same race and class as those for whom the tests were designed to find 'scientifically' superior! Could the test creators actually decide that their own mothers, wives, and daughters were hereditarily inferior? The pioneers of mental testing decided to create cognitive equality between men and women in order to eliminate a distraction from what they really wanted to show – that with regards to measured intelligence, blacks were inferior to whites, Southern Europeans and Jews were inferior to Northern Europeans, and the working class was inferior to the capitalist class.

I recall an undergraduate university lecture by a political science professor who, I later learned, was very interested in the question of racial differences in intelligence. He said that some people argued that blacks score lower than whites on IQ tests because they are treated as inferiors and are told they are stupid. As an answer to this position, he pointed out that women historically have received a similar message yet they score the same as men on standardized IQ tests. The implication, of course, was that a person's or group's position in society is not reflected by lower IQ scores. My professor either had no idea about how IQ tests are constructed, or he was consciously deceiving his students. Had I known anything about the subject in those years I would have answered that women score about the same as men because the tests are *designed* to produce this result, and that blacks score lower than whites for precisely the same reason.

Even if IQ tests were a valid measure of 'intelligence,' separate norms for groups experiencing vastly different social environments (such as blacks and whites in the Jim Crow South!) could have been created. This was acknowledged by Wechsler in 1944, although he claimed that it was not 'possible to do this at present.'[40] In fact, it was quite possible to create separate norms but it was undesirable from the standpoint of those who wanted to use the tests to provide scientific evidence for racial inequality. Had Wechsler decided to create separate norm groups he would have undermined one of the original purposes of American IQ tests, which was – literally – to demonstrate the genetic superiority of 'the Nordic race group.' That aspects of this 'original purpose' lives on is shown by the fact that contemporary IQ tests are seen as valid if they correlate well with the original Stanford-Binet. To his credit, Wechsler found it necessary to state that 'our norms cannot be used for the colored population of the United States.'[41] Naturally, this did not prevent subsequent commentators from citing racial differences in Wechsler IQ scores as evidence of black people's genetic inferiority.

From the standpoint of the economically powerful classes, IQ tests helped justify the inequalities of capitalism. They could argue, as Goddard argued, that genetic differences between classes precluded the egalitarian society advocated by socialists. This position is embraced by most contemporary IQ hereditarians. 'The tests,' wrote Mensh and Mensh in *The IQ Mythology*, 'do what their construction dictates; they correlate a group's mental worth with its place in the social hierarchy.'[42]

In response to Jensen's claim that the tests were not designed to discriminate between social groups, Mensh and Mensh responded, 'In reality — which is precisely the opposite of what Jensen claims it to be — test discrimination among individuals *within* any group is the incidental by-product of tests constructed to discriminate *between* groups.'[43] This position requires qualification. The eugenics movement, which played a major role in promoting IQ testing, was interested in individual as well as group differences. Even among the American white population, the 'chronic pauper stocks' (such as the 'Jukes' and the 'Kallikaks') were targeted for eugenic intervention. Generally speaking, and allowing for considerable overlap, eugenicists supported IQ tests as an instrument for identifying individuals who should be prevented from reproducing; the ruling elite wanted to scientifically legitimize the class-stratified society it had created; racists and segregationists saw IQ tests as evidence supporting their causes; and psychologists such as Terman, Brigham, Goddard, and others wanted all of these things. And all the while the interests of the poor, blacks, immigrants, native Americans, and the working class were ignored. After all, they were powerless.

This is not to deny that others (then and now) use IQ tests for more benign purposes. Yet it is important to understand why the tests were created and how present-day tests are based on many of the assumptions of the original Stanford-Binet, published in 1916. In this respect there are parallels between twin research and the American brand of IQ testing. Both were used (some might say invented) in order to identify people in need of eugenic intervention, which sometimes included compulsory sterilization or worse. Yet in spite of contemporary society's rejection of the eugenic program, IQ testing and twin research remain with us.

The fate of castelike minorities
An international perspective is often helpful in demystifying narrow national debates. Anthropologist John Ogbu has documented the fate of castelike minority groups such as the Maoris in New Zealand and the Burakumin of Japan.[44] He believed that the lower performance of minority children is the result of their status as members of a group relegated to the bottom of society, such as the case with blacks in the United States.

Expanding on this point, Fischer and colleagues documented the position of these and other groups. These include whites and aborigines in Australia, high caste and low caste in India, Jews and Arabs in Israel, English and Irish in Great Britain, and French and Flemish in Belgium. In each of these cases the latter group scores lower on standardized IQ tests.

Another example is South Africa between 1950 and 1990, where Afrikaaners (Dutch origin) scored from one-half to a full standard deviation lower on IQ tests than people of English descent. By the 1970s, however, the gap had disappeared: 'The convergence of Afrikaaner and English scores coincides with the rise of Afrikaaners to power in South Africa after generations of subordination to the English.'[45] Fischer and colleagues' thesis reads as follows: 'A racial or ethnic group's position in society determines its measured intelligence rather than vice versa.'[46]

Another example discussed by Fischer and associates are Koreans in Japan, who occupy a position in Japanese society analogous to blacks in the United States or West Indians in the UK. They also score lower than members of the dominant culture on IQ tests:

> Koreans, who are of the same 'racial' stock as Japanese and who in the United States do about as well academically as Americans of Japanese origin (that is, above average), are distinctly 'dumb' in Japan. The explanation cannot be racial, nor even cultural in any simple way. The explanation is that Koreans, whose nation was a colony of Japan for about a half-century, have formed a lower-caste group in Japan.[47]

Clearly, the United States is not the only country in which lower-caste groups score lower on standardized IQ tests.

Conclusion

The claims of behavior geneticists and others in support of important genetic influences on intelligence derive from their acceptance of many implausible assumptions. Several critics, such as Taylor and Richardson, have listed some of these assumptions.[48] If we include the false assumptions underlying behavior genetic methods such as twin and adoption studies, the case for important genetic influences on intelligence collapses completely.[49] As Richardson commented,

> I think it can be safely said that never before in any field of science have so many arbitrary assumptions been gathered together, *in full knowledge of their invalidity,* as the basis of substantive claims about the nature of people, with so many potentially dire consequences for them.[50]

Given the racial and class bias *built into* the most widely used IQ tests, claims by the authors of *The Bell Curve* and others about the innate cognitive inferiority of ethnic minorities and the working class are preposterous, to say the least. 'Nothing,' wrote Lewontin, Rose, and Kamin, 'demonstrates more clearly how scientific methodology and conclusions are shaped to fit ideological ends than the sorry story of the heritability of IQ.'[51]

Notes

[1] K. Richardson, 2000.

[2] Gould, 1981; Kamin, 1974; Chase, 1980; Tucker, 1994; Guthrie, 1976.

[3] Aly, 1994.

[4] Some of these objections are taken from Jensen's 1980 *Bias in Mental Testing*. Jensen, of course, argued against these objections.

[5] Neisser et al., 1996, p. 80.

[6] Fischer et al., 1996, p. 44.

[7] Ibid., p. 32.

[8] Simon, 1997, p. 203.

[9] Ibid., p. 204.

[10] Lewontin et al., 1984, p. 92.

[11] Ibid., p. 92.

[12] Cohen et al., 1992.

[13] Bouchard & McGue, 1981.

[14] For problems in meta-analytic research, see Begg, 1994; Joseph & Baldwin, 2000; R. Rosenthal, 1979; Simes, 1987.

[15] Kamin, in Eysenck vs. Kamin, 1981, p. 117.

[16] Ibid., p. 117. See Stoolmiller (1998) for further evidence of range restriction in IQ adoption studies.

[17] Stoolmiller, 1999. In a rejoinder to Stoolmiller, behavior genetic adoption researchers John Loehlin and Joseph Horn (2000, p. 245) acknowledged, 'Stoolmiller . . . argues that restriction of range is an important consideration in interpreting the results of adoption studies of IQ, such as the Texas Adoption Project (TAP). We agree, and have regularly pointed this out in our own publications . . .'

[18] Schiff & Lewontin, 1986, p. 179.

[19] Schiff et al., 1982.

[20] Plomin, Fulker, et al., 1997, p. 443. In a CAP study on children's television viewing habits, Plomin et al. concluded that there is a 'significant genetic influence on individual differences in children's television viewing' (Plomin, Corely, et al., 1990, p. 371).

[21] Stoolmiller, 1999, p. 395.

[22] In Jensen's 1998 book, *The g Factor*, (p. 476) he concluded, 'There is simply no good evidence that social environmental factors have a large effect on IQ, particularly in adolescence and beyond, except in cases of extreme environmental deprivation.' The evidence from adoption studies, however, flies in the face of this claim.

[23] Counterarguments to Jensen, and Herrnstein and Murray, can be found in many books, which include Block & Dworkin, 1976; Ceci, 1996; Chase, 1980; Devlin et al., 1997; Fischer et al., 1996; Fish, 2002; Fraser, 1995; Gould, 1981; Jacoby & Glauberman, 1995; Kamin, 1974; Lewontin et al., 1984.

[24] Quoted in Samelson, 1979, p. 105.

[25] Jensen, 1980, p. 71.

[26] Terman, 1916, p. 91.

[27] Ibid., p. 92.

[28] Brigham, 1923, p. xx.

[29] Ibid., p. xxi.

[30] Ibid., p. 207.

[31] Quoted in Kamin, 1974, p. 8.

[32] Terman & Merrill, 1937, p. 34.

[33] Wechsler, 1944, p. 107.

[34] Ibid., p. 144.

[35] Like Terman and Brigham, Wechsler (1958, p. 90) believed that racial differences in intelligence were real: '"Practical" handling of the problem does not, of course, imply an answer to the question of whether or not there are ethnic and cultural differences in intelligence. That such differences exist appear open to little doubt.'

[36] Jensen, 1980, p. 623.

[37] K. Richardson, 1998, p. 114.

[38] Mensh & Mensh, 1991.

[39] Ibid., p. 51.

[40] Wechsler, 1944, p. 107.

[41] Ibid., p. 107.

[42] Mensh & Mensh, 1991, p. 30.

[43] Ibid., p. 73.

[44] Ogbu, 1978.

[45] Fischer et al., 1996, p. 193.

[46] Ibid., p. 173.

[47] Ibid., p. 172.

[48] K. Richardson, 1998; Taylor, 1980.

[49] This observation applies to a recent brain-mapping twin study by Thompson et al. claiming important genetic influences on brain structure. The authors also claimed that frontal gray matter differences are correlated with Spearman's g (Thompson et al., 2001). According to Plomin and Craig, this study 'shows that g has a biological basis' (Plomin & Craig, 2002, p. 186).

[50] K. Richardson, 1998, p. 135.

[51] Lewontin et al., 1984, p. 100.

CHAPTER TEN

Molecular Genetic Research in Psychiatry and Psychology: An exercise in futility?

Psychiatry as a discipline has too often been characterized by many speculations based on few facts.

Psychiatric geneticists Edith Zerbin-Rüdin and Kenneth Kendler in 1996.[1]

Much of what passes for scientific advance in psychiatry is, in fact, rhetorical innovation.

Psychiatry critic Thomas Szasz in 1964.[2]

The search for the actual genes believed to contribute to psychiatric disorders and psychological trait differences has been underway for many years. In psychiatry, researchers attempt to identify genes for schizophrenia, attention-deficit hyperactivity disorder, and other diagnoses. Other investigators attempt to identify genes for intelligence, personality, and abnormal behavior. Molecular genetic research differs from the methods we looked at in the previous chapters in several respects. Most importantly, investigators must be able to identify actual genes. With twin and adoption studies, the existence of genes was inferred from observed or reported behavior, or test score correlations. To convince the public and professionals that these genes actually existed, proponents of the genetic position had merely to argue that their research was sound, and that their results pointed to genetic conclusions. In molecular genetic research, investigators attempt to identify specific genes contributing to a trait which, if found, must then be confirmed by replication studies.

There are two main reasons why we should not expect molecular genetic research in psychiatry and psychology to provide much useful information now or in the future. The first relates to the unwarranted belief that family, twin, and adoption studies have demonstrated the role of genetic factors. Most scientific articles reporting molecular genetic research begin by stating that these studies have demonstrated that genes are implicated. For example, Jönsson and colleagues began their 1997 schizophrenia molecular genetic study as follows: 'On the basis of family, twin and adoption studies, it has been proposed that genetic factors play a role in the aetiology of schizophrenia.'[3] And Tsuang and Faraone wrote in 2000, 'A century of genetic epidemiologic research shows that genes play a substantial role in the etiology of schizophrenia. This is the only reasonable conclusion we can draw from family, twin and adoption studies.'[4]

The search for genes is similar to the way that the existence of other phenomena are postulated and then looked for. In astronomy, the existence of Neptune was postulated on the basis of observing irregularities in Uranus's orbit. An astronomer was able to calculate the position of a heavy unknown object which could account for this irregularity, leading to the discovery of Neptune in 1846. Similar logic is observed in everyday life. When approaching a crossing road, a motorist will slow down when she sees the large vehicle in the next lane slowing down. Although the motorist cannot see the entire crossing road, the existence of a pedestrian is inferred from the behavior of the large vehicle's driver. For molecular geneticists, family, twin, and adoption studies are analogous to Uranus's orbit and the large vehicle slowing down at the crossing road. On the basis of their interpretation of the data produced by these studies, they reason that there must be a gene or genes for the condition or trait in question.

The second problem relates to the practical benefits of finding genes. Proponents of molecular genetic research argue that techniques such as gene therapy can be used to prevent or cure diseases influenced by identified genes. Although this is a laudable goal in the fight to eliminate physical diseases, does society wish to use this technology to alter 'intelligence genes,' or genes associated with unpleasant personalities? It might be argued that the discovery of relevant genes is hardly more enlightening than the 'discovery' of the original predisposition, since in both cases we might still choose to focus attention on environmental factors or triggers.

Segregation analysis

Psychiatric genetic researchers sometimes point to the results of *segregation analysis* in support of their positions. Segregation analysis bases itself on the assumption that the genetic basis of a condition has already been established. For example, Faraone and Tsuang have written that the mode of transmission is investigated 'after demonstrating that a disorder is influenced by genetic factors.'[5] Segregation analysis utilizes complex mathematical formulas and looks at patterns of familial transmission, for the purpose of determining which type of genetic transmission is operating. Segregation analyses often test a non-genetic or 'cultural transmission' model in addition to genetic models. However, researchers cannot know a priori what a purely environmental transmission might look like, or if it resembles a known genetic model. Thus segregation analysis might be useful in determining the type of genetic transmission of a *proven* genetic condition, but it is not a valid method for distinguishing genetic from environmental causation.

Like other methods in psychiatric genetics, segregation analysis is based on several questionable theoretical assumptions. For example, Faraone and Tsuang discussed the 'popular' segregation analysis model of Morton and MacLean.[6] In this model, according to Faraone and Tsuang, 'a genetic trait is assumed to be due to the influence of a major locus with mendelian transmission, a polygenic component, and random environmental effects.'[7] But a model *assuming* genetic causation cannot provide evidence in favor of genetic causation. Morton and MacLean based their model on several improbable assumptions: 'We have . . . assumed random mating, no gene-environment correlation, and no environment common to parents and children.'[8] In the real world, random mating for most traits is infrequent, there is plenty of gene-environment correlation, and parents and children usually share common environments.

As a demonstration of how a culturally transmitted trait can *simulate* a genetic model, in 1990 McGuffin and Huckle showed that the results of a complex segregation analysis of the families of British medical students, which tested for a gene *for attending medical school*, 'are compatible with a recessive-gene hypothesis.' The authors pointed out that a researcher could conclude 'that we have more consistent, and somewhat more persuasive, evidence of a major gene for attending medical school than for any of the neuropsychiatric disorders recently investigated in linkage studies.'[9]

Molecular genetic research in psychiatry

Schizophrenia

The search for 'schizophrenia genes' has been underway for many years. The most common methods in molecular genetic research are linkage and association studies. In a linkage study, investigators look for genetic markers linked with the putative disease gene among consanguineous family members. Findings are often represented as a logarithm of odds (LOD) score, which expresses the probability that the linkage occurred by chance. By convention, an LOD score higher than three (1000:1 odds in favor of linkage) is required in order to claim a significant linkage. Linkage studies are designed to identify areas of the chromosome where relevant genes might be located, but are unable to identify actual genes. Association studies compare the frequency of genetic markers among unrelated affected individuals and a control group.

Periodically, investigators claim to have found the location of genetic markers for schizophrenia. These studies are followed by the inevitable failures to replicate. The most well-known example is the study of Sherrington and colleagues in 1988, which was accompanied by an article in the same issue of *Nature* by Kennedy et al., who failed to replicate the findings.[10] Another highly publicized yet subsequently withdrawn claim (this time for bipolar disorder) was published by Egeland and associates in 1987.[11] Unfortunately, it is widely believed among the general public that genetic markers for these and other conditions have been located, when it is not the case. Furthermore, the finding of a marker is not the same as finding an actual gene; it merely points to an area where the gene might be located. Still, when the public hears of studies purporting to find markers they often assume that a gene has been found. In an analogy discussed by psychologist Ty Colbert, the finding of a marker is similar to miners coming across traces of gold in a river.[12] This 'marker' might indicate that a gold vein is nearby, or there could be other explanations. In any case, it would be wrong to conclude that a gold vein had been discovered in the absence of other evidence. Colbert noted in this example that we at least can be sure that gold actually exists, which certainly cannot be said about 'mental illness' genes.

In 1999 Williams and colleagues published a schizophrenia linkage study which looked at 196 sibling pairs diagnosed with DSM-IV schizophrenia. While the investigators found some evidence 'suggestive' of linkage, 'none approached the genome-wide significance of 0.05.'[13] Williams et al. concluded, 'Our results suggest that common genes of major effect . . . are unlikely to exist for schizophrenia.'[14] In the Spring of 2001 a new study was published by Meyer and colleagues which claimed to have found a significant association between a marker and catatonic schizophrenia.[15] However, there is little reason to believe that the results of this study will be replicated. More recently, a large 2002 internationally-based schizophrenia linkage study of 382 sibling pairs by DeLisi and associates failed to replicate reports of linkage in several previous studies. DeLisi et al. acknowledged that the 'most striking feature' of their results 'is the failure to confirm a number of earlier claims of positive findings,'[16] and wrote further that the 'present findings suggest that a critical reevaluation of the linkage approach is warranted.' Despite results of this type, the latter part of 2002 saw claims that susceptibility genes for schizophrenia had been discovered. For genetic researcher C. Robert Cloninger, the 2002 studies of Straub et al., Chumakov et al., and an additional unpublished study, led him to conclude that a 'watershed' event had occurred, and that for 'the first time, specific genes have been discovered that influence susceptibility to schizophrenia . . .' Cloninger's claim, however, is more the result of wishful thinking than objective scientific evaluation. In addition to the premature nature of his conclusions, how can 'schizophrenia susceptibility genes' (in reality, a mere association between 'schizophrenia' and genetic markers) be found when, as a leading American biological psychiatrist and schizophrenia researcher has admitted (see the beginning of Chapter 7), American psychiatry is unable to 'figure out who really has schizophrenia or what schizophrenia really is.'[17]

Claims to the contrary notwithstanding, molecular genetic studies have failed to find genes for schizophrenia.[18] Most leaders of the field now believe that many genes are involved (the polygenic theory), and have abandoned the single-gene approach. In 2000, psychiatric geneticists Tsuang and Faraone wrote, 'We can now conclusively reject the idea that there is one gene of major effect that causes schizophrenia.' They recommended that future researchers design studies 'to detect the many genes of small effect that each increase susceptibility to the disorder.'[19] The title of the paper in which these comments are found is, fittingly, 'The Frustrating Search for Schizophrenia Genes.'

Tsuang and Faraone noted that 'failures to replicate molecular genetic studies of schizophrenia might be interpreted to mean that schizophrenia genes do not exist,' and cautioned that such a conclusion would be 'premature' because the results from twin and adoption studies show that genes play an important role.[20] While it is true that the failure to find schizophrenia genes does not prove that such genes do not exist, the belief that twin and adoption studies show that genes are involved is erroneous. The irony here is that instead of confirming the results of schizophrenia twin and adoption studies, the failure to find schizophrenia genes may lead researchers to take a much needed second look at these greatly flawed and environmentally confounded twin and adoption studies. In 2000, schizophrenia genetic researcher Lynn DeLisi acknowledged that 'psychiatric genetics appears to be at a crossroads or crisis' as investigators continue to look for the 'elusive gene or genes' for schizophrenia.[21] DeLisi discussed 'those researchers who entered the new [*sic*] field of psychiatric genetics [who] feared most of all that they would screen the whole genome and find nothing.'[22] This is precisely the 'crisis' facing psychiatric genetics: its adherents are looking for genes that most likely do not exist.

Another reason that investigators believe they will find genes is the concurrent (though not necessarily related) view of schizophrenia as a brain disease. Here again, as Peter Breggin has convincingly argued, the evidence is weak.[23] On the other hand, real brain diseases such as Alzheimer's and Huntington's disease are put forward by biological psychiatrists as examples of what can be accomplished for schizophrenia. For example, Nancy Andreasen referred to Alzheimer's disease as 'one of the current "stars" in the molecular biology firmament.'[24] The problem with her reasoning is that Alzheimer's and Huntington's are proven diseases of the brain, which can be seen in post-mortem examinations. These conditions, therefore, are not 'behaviors' or 'mental illnesses.' Rather, they are bona-fide brain diseases. Schizophrenia and most other psychiatric disorders, on the other hand, are affective states or socially disapproved behaviors given the name 'disease' in the absence of convincing evidence in support of their biological bases. Why then do genetic researchers in psychiatry and psychology continue to point to Alzheimer's and Huntington's as examples of what they could find for conditions such as schizophrenia and attention-deficit hyperactivity disorder? The answer is that they have few other success stories to point to.

According to Thomas Szasz, when a psychiatric condition is shown to have a physical basis it leaves the domain of psychiatry and becomes a medical concern:

> As soon as a disease thought to be mental is proven to be physical, it is removed from the domain of psychiatry and placed in that of medicine, to be treated henceforth by internists, neurologists, or neurosurgeons. This is what happened with paresis, pellagra, epilepsy, and brain tumors. It is an ironic paradox, then, that while definitive proof that mental illnesses are brain diseases would destroy psychiatry's *raison d'être* as a medical specialty distinct and separate from neurology, the claim that mental illness is a brain disease has served, and continues to serve, as the psychiatrist's most effective justification for legitimacy as an independent medical discipline.[25]

Thus, the psychiatric domain consists of disorders that are claimed – but not proven – to have a biological basis. We might therefore say that psychiatry is where one will find *pseudo* brain diseases.

The current multinational effort to find genes for schizophrenia is based on a belief that it is a brain disease of genetic origin, yet there is little evidence in support of this belief. The best course of action for molecular genetic investigators would be to perform their own critical reanalysis of the original twin and adoption studies, upon which their current search is based. Only then will they discover the true reason that their search has turned up nothing. But suppose that they do eventually find something. One could still argue that, like phenylketonuria (PKU) and farsightedness, society's emphasis should be placed on environmental interventions. Thus, in the unlikely event that presumed genes for schizophrenia are discovered, the *decision* to emphasize genetic endowment or environmental triggers would still be related to one's beliefs. In any case, it is likely that psychiatric geneticists and molecular geneticists will one day give up on this thankless and futile endeavor, and researchers will focus on the environmental factors influencing the behaviors given the name 'schizophrenia.'

Attention-Deficit Hyperactivity Disorder

Another area of psychiatry where methodologically unsound research on twins and adoptees has led to the erroneous conclusion that genes must exist is attention-deficit hyperactivity disorder (ADHD). Even more than schizophrenia, ADHD exemplifies the way that socially disapproved behavior can be given a label and turned into a 'disease.' In Peter Breggin's words, the DSM-IV definition of ADHD is 'simply a list of behaviors that require extra attention from teachers.'[26] Like schizophrenia, the evidence supporting the brain disease theory of ADHD is weak.[27] The genetic evidence consists of family studies, twin studies, and adoption studies far more flawed than their schizophrenia counterparts. I have reviewed these studies in detail elsewhere.[28]

Predictably, textbooks usually report that ADHD has an important genetic component. Like schizophrenia, the validity of the equal environment assumption of the twin method is rarely challenged, and unsound ADHD adoption studies are often endorsed without analysis. Also like schizophrenia, the search for ADHD genes is underway. In 2000 Faraone and Biederman claimed that molecular genetic research, while still its 'infancy,' has 'already implicated several genes as mediating the susceptibility to ADHD.'[29] Faraone and Biederman cited several studies finding an association between ADHD and specific genes, as well as several studies failing to replicate these findings. This undoubtedly constitutes at best an inconclusive body of evidence, in spite of a meta-analysis cited by the authors which found a significant association between a specific gene and ADHD.[30] A disease-causing gene cannot be discovered by taking numbers from a body of inconclusive research and running them through a computer, as Faraone and Biederman seem to suggest.[31] Even in studies cited by Faraone and Biederman as finding a significant association, the results are treated with caution. For example, the authors of the 1999 Faraone et al. study linking ADHD to the 7-repeat allele of the DRD4 gene found a statistically significant association, but also noted that '58% of the subjects without the 7-repeat allele had ADHD. . . . suggest[ing] that the 7-repeat allele cannot be viewed as a necessary cause of ADHD.'[32]

Through the first part of 2002, and despite concerted efforts worldwide, no genes for ADHD have been discovered.[33] But the search goes on because psychiatric geneticists continue to place misguided faith in the equal environment assumption of the twin method. According to Faraone and Biederman, 'The twin literature about ADHD clearly indicates that ADHD is one of the most heritable of psychiatric disorders.'[34]

It is apparent from the evidence presented by Faraone and Biederman in 2000 that there is no proof that any specific genes cause ADHD.[35] As we have seen, the genetic literature is filled with claims of the discovery of specific genes for psychiatric conditions, which failed to be replicated and were soon forgotten. Given these and other highly publicized yet non-replicated claims, inconclusive evidence in favor of a specific association will not do; unequivocal proof of the operation of a specific gene or genes must be demonstrated. Like schizophrenia and other psychiatric conditions, it is extremely unlikely that genes for ADHD exist.

Why search for genes?

Tsuang and Faraone listed three potential benefits of identifying genes for schizophrenia: 'pharmacogenomics,' 'pharmacogenetics,' and the identification of high-risk children.[36] The first two words refer to producing psychotropic drugs with the aid of genetic information. The identification and 'treatment' of so-called high-risk children has already been proposed by some, in spite of the failure to find schizophrenia genes.

But suppose, for the sake of argument, that schizophrenia really does have a genetic basis (which would then be triggered by environmental factors), and investigators are able to identify the gene or genes responsible. The strategy of 'early intervention' would still be problematic.

The knowledge that a child is carrying schizophrenia genes could in itself be a life-altering event, affecting the way he or she is treated by parents and the social environment. Even Andreasen admitted that 'Treating young people who are not yet ill may adversely affect their self-esteem and self-image, perhaps creating a self-fulfilling prophecy that may lead them to eventually become ill.'[37] If Tienari and colleagues are correct that the predisposition manifests itself in people who grow up in disturbed homes, would this mean that predisposed children would be removed from these types of homes? And who would decide what constitutes 'disturbed'? The treatment proposed by psychiatric geneticists usually includes neuroleptic (also known as 'anti-psychotic') drugs. These drugs, however, merely modify behavior or affect. They do not treat physical illness or 'repair the brain.' Neuroleptics are known to cause troubling adverse effects and permanent neurological damage such as tardive dyskinesia. They can also be lethal (for example, they can cause neuroleptic malignant syndrome).[38] Furthermore, as seen in the experience of Mosher and colleagues' Soteria project, non-medicated people diagnosed with schizophrenia who enter a supportive home-like treatment program show as much improvement as others hospitalized and put on neuroleptics.[39] Thus, even in the unlikely event that schizophrenia genes are found, society might still decide that the best course of action would be to concentrate on non-medical interventions, and on eliminating the environmental triggers leading to the manifestation of the condition.

There is a potentially sinister aspect of identifying genes for psychiatric disorders, in that the information could be used for eugenic purposes. The identification of genes would have been warmly embraced by people like Rüdin, Luxenburger, Kallmann and other racial hygienists and eugenicists in their quest to rid the world of the 'schizophrenia genotype.' As a contemporary German psychiatric genetic researcher observed, 'doctors and scientists involved in the crimes of the Nazi period . . . would undoubtedly have welcomed the technical possibilities of present-day genetics.'[40] According to genetic researchers Moldin and Gottesman, however, the discovery of genes would have the effect of *reducing* stigma. They wrote that Sherrington and colleagues' (subsequently withdrawn) 1988 finding

> was welcomed with considerable optimism and insufficient scientific criticism not only because hopes were raised of identifying a factor that could finally provide clues to pathophysiology and ultimately new treatments, but also because localization of a locus was 'confirmation' that schizophrenia was in fact 'biological' and not a 'psychosocial' disorder. The long history of stigma associated with schizophrenia further reinforces the desire to see schizophrenia as a genetic or medical condition.[41]

Moldin and Gottesman have their history exactly backwards. The 'long history' of the stigmatization of people diagnosed with schizophrenia is directly related to their being viewed as the carriers of 'hereditary taint.' Even behavior geneticists Plomin and Rutter acknowledged in 1998 that the twentieth century provided 'chilling examples of compulsory sterilization in the name of eugenics and of genetic stigmatization.'[42] Although it is true that psychosocial theories can stigmatize families (or parents), there is a huge difference between the status of being a 'bad parent' and the stigma of being seen as the carrier of 'bad genes.' While psychoanalysts may have stigmatized 'schizophrenogenic' parents from time to time, the stigma of being a 'hereditary taint carrier' has led to the perpetration of crimes against people diagnosed with schizophrenia and other psychiatric or mental conditions. Thus, there is little to be gained by finding presumed schizophrenia genes, but the potential misuse of this information is enormous.

Looking for behavioral and IQ genes

In spite of the dreadful track record of genetic research in psychiatry, many people with psychiatric conditions suffer greatly and chronically. In the case of schizophrenia, young people are diagnosed in the beginning of their productive years and often behave in ways that bring great suffering to those who love and care about them. It is therefore understandable that researchers would be subjectively motivated to alleviate this suffering and to help restore people to productive lives.

The search for IQ genes and genes for normal variation in behavior is another matter. Even if we accept the behavior genetic argument that genes exert an important influence on behavioral and IQ differences among people, we could still question the necessity of identifying the genes producing these differences. Yet, leading behavior geneticists are currently searching for genes believed to influence psychological trait differences. According to Plomin and colleagues,

> Psychology is at the dawn of a new era in which molecular genetic techniques will revolutionize genetic research in psychology by identifying specific genes that contribute to genetic variance for complex dimensions and disorders. The quest is to find not *the* gene for a trait, but the multiple genes that affect the trait in a probabilistic rather than predetermined manner. The breathtaking pace of molecular genetics . . . leads us to predict that psychologists will routinely use DNA markers as a tool in their research to identify some of the relevant genetic differences among individuals.[43]

In a 1998 article in *Child Development*, Plomin and Rutter discussed and justified the search for 'behavior genes.'[44] They noted that while in the past researchers looked for single major genes influencing behavior, it is now recognized that many genes contribute to the appearance of a trait or to differences in continuously distributed traits (such as IQ). Markers for postulated genes contributing to these traits are called *quantitative trait loci* (or QTLs). Plomin and Rutter view the effects of genes on behavior as 'probabilistic,' as opposed to 'deterministic.' By this they mean that genes increase the likelihood of the appearance of a trait, but are not the sole cause. As examples of the discovery of behavioral genes, Plomin and Rutter cited studies claiming to have found genes associated with Alzheimer's disease and 'novelty seeking.' Why Alzheimer's disease is discussed in this context is unclear since, as we have seen, it is better understood as a brain disease and not as a behavior. Finding an association between the DRD4 gene and novelty seeking has seen many subsequent replication failures by other groups,[45] and in studies where replication was claimed, the findings are more likely the result of a 'statistical accident' than a real association.[46] Furthermore, it is difficult to imagine how a gene could influence the way people answer questions on personality tests. In any case, Plomin and Rutter admit that the DRD4 gene accounts for only 4% of the variance in novelty seeking, 'which may prove typical for complex quantitative traits.'[47]

Regarding the potential misuse of genetic information, Plomin and others frequently point out that policy decisions are based on values, which are allegedly separate from 'science.' Yet behavior genetics, like all social science, is unavoidably permeated with politics and values. For example, Plomin and associates have written that 'the basic message of behavioral genetics is that each of us is an individual. Recognition of, and respect for individual differences is essential to the ethic of individual worth.'[48] This is a social policy position, not a scientific finding. Plomin and colleagues' position on individuality would be rejected in Asian cultures where people are viewed more as members of their family than as individuals. There are many other ways that 'each of us' can view ourselves, such as members of a class, 'race,' political party, ethnic group, nationality, club, etc. The danger here, and this is a constant theme throughout the entire history of human genetic research, is the presentation of political and social programs disguised as objective, apolitical science. In any case, regardless of public statements by contemporary behavior geneticists, there is an enormous potential for the misuse of genetic information. Those who worry about this should also be aware that hard scientific evidence is not necessary — and never has been necessary — to bring about atrocities. Eugenic sterilization was flourishing in the United States before the twin method had even been invented, and the Nazis and their psychiatric genetic collaborators performed their deeds on the basis of what we would today consider very weak evidence.

Although behavior genetic researchers continue to look for 'IQ genes' (or QTLs correlated with *g*), the search has been unsuccessful. In 1995 Plomin and colleagues published the results of their study looking for QTLs for IQ, and were unable to find anything important: 'Although several significant associations were found in an original sample, only one association was replicated cleanly as an independent sample. This finding might well be a chance result, because 100 markers were investigated.'[49] Indeed, there is a particularly high risk for false positive results (Type I errors) in QTL studies, which can lead to spurious results.[50] One such fishing expedition was published by Plomin and Craig in 2001.[51] The investigators identified three markers (not actual genes) associated with general cognitive ability, whose association probability value, however, did not meet the conservative standards recommended by other genetic researchers.[52] Nonetheless, Plomin and Craig wrote later that 'the many hurdles that we set for acceptance of a quantitative trait locus association may have been too high . . .'[53] In addition, there are problems with the sampling procedures used by Plomin and Craig.[54] Although he has failed to find convincing evidence of QTLs for IQ, Plomin expects that when he does, it will show differences along class and racial lines:

> A more general concern involves group differences, such as average differences between classes and ethnic groups. As genes are found that are associated with differences among individuals within groups, the genes will inevitably be used to make comparisons between groups . . . The societal implications of such research need to be anticipated . . . Perhaps people will become less preoccupied with average differences between groups when DNA chips make it possible to focus on individuals.[55]

'Perhaps people will become less preoccupied' with alleged group differences . . . or perhaps they will become *more* preoccupied! In the United States, where race questions are a central obsession, these results would be explosive. The point is not whether group differences in 'intelligence genes' will be found — they won't — but rather the cavalier way in which investigators minimize the potential social impact of their research. Would Plomin have the same attitude if his research, for example, found that more Jews than non-Jews had QTLs associated with greed? Wouldn't there be a justifiable uproar that such research was even being done?[56]

In an article published in the same issue of *Science* containing the human genome sequencing paper (February 16, 2001), McGuffin, Riley, and Plomin could report only a few possible leads on genes for behavioral differences, acknowledging that 'progress so far has been slow.'[57] In fact, progress has been so slow that McGuffin et al. created an escape hatch for themselves in the likely event that they find no important results. They discussed the 'unsolved question of the distribution of effect sizes of QTL; some may involve effects so small or so complicated that they will never be detected.'[58] The search for QTLs, a prominent critic recently remarked, is a 'search for the proverbial needle in the haystack.'[59] Clearly, when Plomin and his associates give up looking for QTLs they will say that genes are important but are too 'complicated or small' to be found. Wouldn't it be far better to conclude that behavior genetics should take a hard second look at its core concepts, such as heritability and twin research?

Ken Richardson has provided an insightful analysis of the 'QTLs for *g*' fiasco:

> It doesn't surprise me in the slightest that, empirically, this enterprise is turning out to be rather hit-and-miss. What does surprise me is that such a search should be seriously conceived in the first place. One reason for my misgivings, of course, is the idea of trying to identify genes for a character that no one can define. Another is that, even if allelic association with IQ was found, this would not mean that it was causal. Such a correlation could arise from a number of possible connections other than genes for IQ. How these are to be eliminated are not mentioned.[60]

As Richardson points out, the search for IQ genes carries with it all of the problems and questionable assumptions of the previous IQ studies, such as the inability to agree on what intelligence is, whether IQ tests (which guide the search for QTLs) actually measure intelligence, whether general intelligence (*g*) has a physical reality, the fact that correlation does not mean cause, and a host of other problems. Still, like the failed attempts at finding genes for behavioral traits, the public has been led to believe that 'genes for IQ' have been found when, as Richardson reminds us, 'no such thing has happened.'[61]

Conclusions

Psychiatric genetic and molecular genetic researchers are, as DeLisi admitted, at a 'crossroads or crisis.' The failure to find genes for schizophrenia, ADHD, and other conditions, while not proving that these genes do not exist, is a testimony to the blind faith in the results of twin and adoption studies. The failure to find genes is glossed over by several researchers, such as Tsuang and colleagues' recent references to 'making great strides' in finding genes, and that recent technology has led to the 'implication of genes at several chromosomal loci.'[62] Gershon and associates claim that they are 'closing in on genes for manic-depressive illness and schizophrenia.'[63] More recently, Plomin wrote that according to geneticists, 'we will be awash with genes associated with complex traits including behaviour in the next few years.'[64] And in a 2001 testimony to the fact that no histopathology has ever been shown to cause schizophrenia, Andreasen spoke of the condition as being caused by an '"invisible lesion" that cannot be seen with the naked eye or under a microscope.'[65] (Perhaps the next step is to assert that schizophrenia is inherited through 'invisible genes'?) All of this is said to avoid having to admit to themselves the bitter truth that they have found nothing important, and will find nothing important. As Szasz once observed,

'Much of what passes for scientific advance in psychiatry is, in fact, rhetorical innovation.' Generally speaking, molecular genetic investigators use 'rhetorical innovation' in their *publications* in order to obscure the fact that they have found no genes in their *laboratories*. Ultimately, however, the use of language cannot eliminate the necessity of finding actual genes.

In the psychiatric genetic and molecular genetic literature, it is striking how little attention is paid to people's social environment as having anything to do with the causes of 'mental illness.' The causes are viewed as residing at the molecular level. Andreasen argues that it is important to understand how 'proteins are created through our DNA . . . since it will ultimately explain how mental illnesses are caused, treated, and prevented.'[66] This statement epitomizes the reductionistic view of human problems. In the past, such views were sustained because, in addition to their usefulness in absolving social conditions and political policies from causing human distress, the technology was not available to put reductionism to the test. Now that the technology *is* available, molecular geneticists in psychiatry will likely suffer through the failure to find genes for the non-existent physical basis of the non-existent 'diseases' they are investigating. The title of a 1992 article by psychiatric geneticist Michael Owen asked, 'Will schizophrenia become a graveyard for molecular geneticists?'[67] The answer, at least as it relates to psychiatric disorders, is yes.

Reductionistic views also can be found among those looking for genes for 'continuously distributed' traits such as personality and IQ. Although, as always, lip service is given to the environment, behavioral and performance differences are viewed as ultimately residing in the genes. Here again, nothing important will be found. Like the psychiatric geneticists, behavior geneticists have painted themselves into a corner, although as we have seen, the more thoughtful among them are already planning their escape. And escape they will, since the mistaken view that 'genes are destiny' is needed by powerful and well-connected interests promoting political, professional, and business agendas.

In short, behavior genetics and psychiatric genetics will have their Waterloo in molecular genetic research, even though this may not become apparent for quite some time. When it does, it will motivate investigators to re-examine the unsound research that inspired the search for genes in the first place.

Notes

[1] Zerbin-Rüdin & Kendler, 1996, p. 332.
[2] Szasz, 1964, p. 525.
[3] Jönsson et al., 1997, p. 414.
[4] Tsuang & Faraone, 2000, p. 1.
[5] Faraone & Tsuang, 1995, p. 93.

[6] Ibid., pp. 120–4.

[7] Ibid., p. 121.

[8] Morton & MacLean, 1974, p. 501.

[9] McGuffin & Huckle, 1990, p. 998.

[10] Sherrington et al., 1988; Kennedy et al., 1988.

[11] Egeland at al., 1987.

[12] Colbert, 2001.

[13] Williams et al., 1999, p. 1729.

[14] Ibid., p. 1729.

[15] Meyer et al., 2001.

[16] DeLisi et al., 2002, p. 808. In contrast to DeLisi and colleagues, Straub et al. (2002) found linkage between schizophrenia spectrum diagnoses and the 6p22 gene.

[17] DeLisi et al., 2002, p. 808. The Cloninger quote is from Cloninger, 2002, p. 13365. The three studies Cloninger referred to were, Straub et al., 2002; Chumakov et al., 2002; and an in-press publication by a Scandinavian group.

[18] See, for example, Bassett et al., 2001; DeLisi et al., 2002.

[19] Tsuang & Faraone, 2000, p. 1. Other examples of researchers' rejection of the single gene hypothesis include Portin & Alanen (1997), and Moldin & Gottesman (1997).

[20] Tsuang & Faraone, 2000, p. 1.

[21] DeLisi, 2000, p. 190.

[22] Ibid., p. 189.

[23] See Breggin, 1991.

[24] Andreasen, 2001, p. 121.

[25] Szasz, 1987, p. 70.

[26] Breggin, 2001, p. 203.

[27] See Breggin, 1998 for a critique of ADHD brain studies.

[28] See Joseph, 2000c, 2000e. Also see McMahon, 1980.

[29] Faraone & Biederman, 2000, p. 572.

[30] Faraone et al., 2001.

[31] For a critique of the Faraone et al. meta-analysis, see Pittelli, 2002b.

[32] Faraone, Biederman, et al., 1999, p. 770.

[33] Wilens et al., 2002.

[34] Faraone & Biederman, 2002, p. 497.

[35] See also Barr et al., 2001.

[36] Tsuang & Faraone, 2000.

[37] Andreasen, 2001, p. 330.

[38] See Breggin, 1991; Breggin & Cohen, 1999; Cohen, 1997; Cohen & McCubbin, 1990; Whitaker, 2002.

[39] Mosher & Menn, 1978; Mosher et al., 1995; Whitaker, 2002.

[40] Propping, 1992, p. 910.

[41] Moldin & Gottesman, 1997, p. 554.

[42] Plomin & Rutter, 1998, p. 1237.

[43] Plomin, DeFries, et al., 1997, p. 277.

[44] Plomin & Rutter, 1998.

[45] Barr, 2001; Bouchard & Loehlin, 2001; Prolo & Licinio, 2002.

[46] Beckwith & Alper, 2002, p. 321.

[47] Plomin & Rutter, 1998, p. 1226.

[48] Plomin, DeFries, et al., 1997, p. 279.

[49] Ibid., p. 153. For the original study, see Plomin et al., 1995.

[50] Wahlsten, 1999.

[51] Plomin & Craig, 2001.

[52] The need to establish more conservative p values for genome-wide scans was

discussed in Lander & Kruglyak, 1995, and Benjamini et al., 2001.

[53] Plomin & Craig, 2002, p. 186.

[54] Pittelli, 2002a, p. 186.

[55] Plomin, 1999, p. C29.

[56] This example is drawn from a similar one put forward by Breggin & Breggin in 1998, who argued against doing research on possible genetic influences on inner-city crime.

[47] McGuffin et al., 2001, p. 1232. In 2002, Plomin & Craig wrote that 'progress has been slow in identifying specific genes associated with most complex dimensions and disorders, probably because the effect sizes of individual genes is smaller than expected' (Plomin & Craig, 2002, p. 186).

[58] McGuffin et al., 2001, p. 1249.

[59] Wahlsten, 1999, p. 607.

[60] K. Richardson, 2000, p. 72.

[61] Ibid., p. 72.

[62] Tsuang et al., 2001, p. S18.

[63] Gershon et al., 1998, p. 233.

[64] Plomin, 2001, p. 138.

[65] Andreasen, 2001, p. 209.

[66] Ibid., p. 103.

[67] Owen, 1992.

CHAPTER ELEVEN

Where Do We Go from Here?

> Perhaps the most striking feature of the nature-nurture debate is the number of times it has ostensibly ended.
>
> *Historian of Science Diane B. Paul in 1998.*[1]

This book has challenged positions that are very well established and are defended by internationally respected researchers. Naturally, it would require thousands of pages to counter all of the arguments made in support of the genetic position. I therefore covered what I view as the most important topics. The arguments made in this book include the following points:

- The twin method is an invalid instrument for the detection of genetic influences on psychological trait differences and psychiatric conditions.
- So-called studies of twins reared apart are methodologically unsound, contain many pairs of twins who should not be viewed as having been reared apart, and failed to use the proper comparison group when evaluating the meaning of MZA correlations. Therefore, these studies provide no valid evidence in favor of genetic influences on psychological trait differences.
- The heritability statistic is wrongly used to estimate the magnitude of genetic influences on trait differences. Its use should be discontinued, except in cases where one is interested in the results of a selective breeding program.
- There is little evidence that schizophrenia (and most other psychiatric conditions) have a genetic basis.
- The schizophrenia adoption studies are (to varying degrees) methodologically unsound, and are unable to disentangle possible genetic and environmental influences on the condition.
- There is little evidence that socially undesirable behaviors such as criminality have a genetic basis.
- Possible genetic influences on intelligence cannot be determined through the use of standardized IQ tests, twin studies, or the heritability statistic.
- There is little evidence supporting the existence of specific genes for

psychiatric disorders, or differences in behavioral traits and intelligence.

- Concepts such as schizophrenia, intelligence, personality, and criminality are often difficult to define, and are controversial (even though human genetic researchers tend to ignore this). Attempts by human genetic researchers to categorize or quantify these concepts are more often means to the end of promoting personal, political, professional, or social agendas, than the result of scientific progress.

One of the running themes of this book has been that previous genetic theories were based on evidence considered overwhelming at the time, although *today* we can recognize that the evidence was actually quite weak. In Chapter 1 we saw how eugenicists Charles Davenport and Harry Laughlin could write, in the absence of twin studies, adoption studies, family studies, or even family pedigrees, that many physical, mental, and moral traits 'have been proved to have a hereditary basis.' They proposed that 'eugenical family studies' be performed as an additional method of studying the known hereditary factor. The conclusions of the early investigators should be seen more as statements of their beliefs than as objectively drawn from the evidence, and little, unfortunately, has changed since then.

The belief in important hereditary influences on psychiatric disorders has been with us for a long time. In his book on 'manic-depressive insanity,' originally published in English in 1921, the father of modern psychiatry, Emil Kraepelin, began his chapter on the causes of the condition as follows:

> *Hereditary taint.* I could demonstrate in about 80 per cent of the cases observed in Heidelberg. Walker found it in 73.4 per cent, Saiz in 84.7, Weygandt in 90, Albrecht in 80.6 per cent, and in the forms with numerous attacks still somewhat more frequently. Taint from the side of the parents he found in 36 per cent of the cases, in the last-named forms in 45 per cent.[2]

Although today the term 'hereditary taint' has been replaced by 'genetic predisposition,'[3] Kraepelin was as sure in his belief of the hereditary basis of manic-depression as contemporary psychiatric geneticists are for the various disorders they write about. This is also true as it relates to the 'inheritance of intelligence' question. The point is that today, as in the past, unsupported beliefs and prejudices are passed off as scientific facts.

Some proposals for the collection of data and the publication of research

A major aspect of this book has been the reanalysis of a large body of greatly flawed research. In the process, I do not want to leave readers with the impression that flawed research is confined to human genetic investigations. In fact, it is widespread. Criticism of the way research is carried out in all fields has been made for many years. In many areas of science there are documented cases of fraud and misconduct, and undoubtedly many more cases are never discovered.[4] Several years of reviewing bias in human genetic research has made it clear to me that many changes must be made in the way that scientific research is collected and reported. I would like to mention two of the many changes that need to be made.

The need for research registers

We saw in Chapter 3 that Farber called for a central register to house twin data. This idea should be expanded to cover all scientific research. Before initiating a study, researchers should be expected to submit a written description of how they will obtain participants, how they will define and measure the variables of interest, and how the comparison of groups will be performed. An internationally-based social science central register should be established to collect and store this information. The register would create a permanent public record of the intentions and methods of researchers before data collection and analysis. A research register has been defined as 'a database of research studies, either planned, active, or completed (or any combination of these), usually oriented around a common feature of the studies such as subject matter, funding source, or design.'[5]

Research registers are already used in some areas of medicine and should now be extended to the social sciences. Cooper and colleagues called for the creation of 'an international, standardized, computer accessible register of IRB [institutional review board] -approved research projects' for psychological research studies.[6] The most important purpose of an international social science register (ISSR) would be to record researchers' methods and definitions before the collection of data, a point made by Easterbrook:

> A further advantage [of a registry] is that through access to information on stated prior hypotheses, it would be possible to identify the use of multiple comparisons (that is whether positive results were based on prior hypotheses or posterior analyses). Such information could be helpful in the publication review process.[7]

A register of studies would help ensure that investigators analyze results according to their research design specifications. Precise design specifications would be completed and registered before data collection. Sometimes, as the research proceeds, changes are required in the design specifications. In these instances, a statement would be sent to ISSR explaining the protocol revisions. Ideally, information about research design sent to ISSR would be published in the form of a brief abstract before the publication of the final paper.

Registration of studies at ISSR would reduce the practice of 'data dredging,' that is, the purposeful posthoc analysis/manipulation of datasets in order to locate significant relationships that support the investigator's research hypothesis. To use one example, we saw in Chapter 8 that Bohman and colleagues were able to find some statistically significant correlations in their data, yet there was no indication that the comparisons of interest were determined in advance. It is well known that in a large set of data probability dictates that, by chance, some significant correlations will be found between variables having no causal or associative relationship. Information about methods submitted to ISSR would be kept confidential until publication. After publication, the information submitted to ISSR would become part of the public record.

Reduce the importance attached to researchers' conclusions
In reporting the results of genetic studies, considerable emphasis is placed on what researchers conclude about their data. Not surprisingly, most researchers conclude that their findings are consistent with their original hypothesis. A critic noted the 'pervasive . . . manner in which scientists can deliberately or, more often, unconsciously work in such a way that their conclusions are bound to support a particular position, policy, or action.'[8] Psychologist George Albee concluded that his early belief that social scientists discover facts in order to build theories was wrong, and that

> it is more accurate to say that people, and particularly social scientists, select theories that are consistent with their personal values, attitudes, and prejudices, and then go out into the world, or into the laboratory, to seek facts that validate their beliefs about the world and about human nature, neglecting or denying observations that contradict their personal prejudices.[9]

Undoubtedly, there is a high degree of correlation between what human genetic researchers believe in advance that their studies will show, and what they conclude after the data are collected and analyzed. However, the results of most studies are open to multiple interpretations. Thus, the conclusions reached by the original researchers about their data are not necessarily more accurate or valid than alternative explanations.

As I have argued throughout this book, the results of behavior genetic and psychiatric genetic research can be plausibly explained on the basis of methodological error, bias, and environmental factors. Still, the authors of secondary sources (such as textbooks) typically endorse the conclusions of the original researchers at the expense of critical analysis.[10] Not surprisingly, genetically oriented researchers tend to conclude that they found evidence supporting the genetic position. This leads to a reasonable proposal: society should be skeptical about the conclusions reached by human genetic researchers, since these conclusions are heavily influenced by their reductionistic views. Let them collect and publish data, but we should rely on others to help us determine what conclusions should be drawn from these data. According to Plomin and Rutter, 'Genetics is much too important a topic to be left to geneticists!'[11] This is true, although behavior geneticists and psychiatric geneticists should be included in this formulation.

In 1993 Bouchard complained that an 'entire industry has evolved up around the reanalysis of kinship data, particularly the large body of published data on identical twins reared apart.'[12] This remark captures the attitude of investigators who believe that they have a monopoly on drawing conclusions from the data they produce. The 'industry' Bouchard criticizes is nothing more than people pointing out methodological flaws and offering alternative interpretations of the data. If Bouchard's conclusions about the role of heredity are valid, then he and his colleagues should have little problem defending them. Similar displeasure with critics is seen in the works of the popularizers of genetic research, where 'armchair critics' are portrayed as seeking to tear down the work of 'scientists.' The implication is that those not directly carrying out research have little right to offer alternative explanations of results. However, it is not only a right but a duty for others to critique research, particularly in cases where it has important social consequences.

The danger of eugenics

In 1994 genetic researchers Garver and Garver wrote that 'a eugenic mentality has existed in the United States during the entire twentieth century,' although it was 'more apparent in the first 45 years.'[13] From this perspective the post-World-War-II era's rejection of the eugenic program was a mere aberration, or an interval. Regardless of the motivations of contemporary behavior geneticists and psychiatric geneticists, the acceptance of their results could pave the way for a rebirth of eugenics. One of the reasons that I have documented the historical relationship between genetic research and eugenics is to warn readers of the possibility that history can repeat itself. The United States was once the world leader in eugenics. Over 30 states had eugenic sterilization laws, and many also had eugenically inspired 'anti-miscegenation' laws.

Today, we are frequently reminded by behavior geneticists and psychiatric geneticists that nature and nurture work together and that the environment is important. While these statements are intended to allay public and professional fears about genetic determinism, there is nothing new about genetic researchers recognizing the importance of the environment.[14] Similar statements were made by prominent eugenicists in the first half of the twentieth century. For example, in 1935 Harry Laughlin wrote, 'In any good eugenics exhibit there must always be a section for "Heredity and Environment." Such an exhibit must bring out clearly how both heredity and environment are important—each a vital factor in the development of the individual.'[15] Frederick Osborn wrote in the second (1951) edition of his *Preface to Eugenics*, 'Which is more important, heredity or environment? Such a question lacks reality, because when two sets of factors are necessary to produce a given result — such as a living human being — there can be no question of the relative importance of either set of factors; both are essential to the final result.'[16] According to Popenoe and Johnson in *Applied Eugenics*, 'Nature and nurture are indissolubly associated in the achievements of every man and woman . . . in daily life they are interdependent.'[17] And even Rüdin, in an exchange with a critic in 1930, stated, 'Dr. Myerson has said that environment plays a great role in the production of a disease. Of course it does.'[18] Statements such as these did not prevent Laughlin, Popenoe, and Rüdin from being strident advocates of compulsory eugenic sterilization. Nor did it prevent Osborn from issuing the following chilling opinion in 1951:

> The aggregate of human suffering caused by defective genes will be enormously diminished if the destruction of the defective genes takes place *immediately after their appearance*. Human suffering will be further reduced if the destruction of the defective genes takes place by failure of reproduction rather than by the death of the individual carrying the defective genes [emphasis added].[19]

The warning I issue has nothing to do with the motivations of contemporary genetic researchers. However, people in the future who use their research to promote eugenic goals will not be concerned with the views and caveats of today's investigators. They will use the studies for their own purposes. As historian Robert Proctor observed in 1992,

> If there is a disconcerting continuity between genomics and eugenics, it is the fact that both have taken root in a climate where many people believe that the large part of human talents and disabilities are heritable through the genes.[20]

While the danger Proctor warned of is real, some current political and social policies are already influenced by eugenic ideas. I have in mind examples such as 'ending welfare as we know it' in the U.S., certain aspects of genetic counseling, and the continuing practice of surreptitiously sterilizing minorities and the poor. Eugenic sterilization was performed in Sweden until the mid-1970s, while Japan began a eugenic sterilization program in 1948 which resulted in the sterilization of over 800,000 people, and continued well into the 1990s.[21]

A new approach

Given the lack of evidence supporting behavior genetic and psychiatric genetic positions, I would like to propose alternative ways that the topics covered in this book should be approached.

Family studies

It may be helpful in epidemiological research to determine if a condition or trait is familial, provided that it is understood that 'familial' does not mean 'genetic.' Family studies may be useful for identifying physical conditions showing a Mendelian pattern of inheritance, although family pedigrees can often perform this function.

Twin research

The twin method should, at long last, be relegated to its proper place alongside the discarded pseudosciences of bygone eras, such as phrenology, alchemy, and craniometry. Given the appalling history of fraud, bias, and unwarranted claims, one is tempted to place reared-apart twin studies in this category. However, a valid study using reared-apart twins is theoretically possible, although it would be carried out far differently than the studies published to date (as outlined in Chapter 4). One area of twin research that could benefit humanity is the much neglected method of studying identical twins *discordant* for various conditions. This method is unpopular with most twin researchers because its purpose is to discover relevant *environmental* factors.

Adoption research

Like reared-apart twin studies, it is theoretically possible to perform a valid adoption study. The problem, as always, lies in the fact that studies of this type are fraught with problems related to the social and psychological aspects of the human condition. The investigators in most adoption studies, and the psychiatric studies in particular, tended to see their experiments as being only minimally affected by these problems. Future adoption researchers must (1) determine in advance, and publish or submit to a register, the specific hypotheses, methods, definitions, and comparison groups used in their studies; (2) make a serious attempt to come to grips with problems such as selective placement and range restriction; (3) publish or place with a registry raw case history information of those under study; (4) choose as subjects only those adoptees who were placed into their adoptive homes shortly after birth; and (5) ensure that all interviews, diagnoses, or ratings are performed blind.

Heritability

Simply put, the heritability statistic should be discarded in human genetics, psychology, and psychiatry.

Schizophrenia

It must be recognized that there is little evidence supporting a genetic basis for schizophrenia. Furthermore, the discovery of a genetic predisposition would have little meaning, because we could still choose to focus attention on discovering and eliminating the environmental triggers. Schizophrenia and most other psychiatric diagnoses describe behaviors and mental states, not diseases. In fact, some critics have called for discarding the schizophrenia concept entirely. There is no qualitative difference between 'schizophrenics,' 'manic-depressives,' and people suffering from other emotional problems. These conditions merely fall on different points of a continuum marking the way that people react to unhealthy familial, social, and political environments.

Crime

Criminal behavior is a social problem, and what constitutes 'crime' is fraught with political implications. While some aspects of criminal behavior may involve psychological factors, it is not a 'disorder' in need of 'treatment.' We could, however, say that the society producing the conditions leading to crime (such as poverty, racism, and income disparity) is in need of 'treatment.'

IQ Tests

In 1922 the American journalist Walter Lippmann wrote that if IQ tests become 'a sort of last judgment on the child's capacity, that they reveal "scientifically" his predestined ability, then it would be a thousand times better if all the intelligence testers and all their questionnaires were sunk without warning in the Sargasso Sea.'[22] This statement should be qualified in two ways. First, the testers should be spared. Second, warning should be provided. Otherwise, the total experience of nearly 100 years of IQ testing has been overwhelmingly negative. The few positive uses of the tests could be easily incorporated into a new type of test lacking any claim to measure 'intelligence.'

Molecular genetic research in psychology and psychiatry

If researchers in this area choose to continue the search for genes that probably do not exist, that is their right. Still, they would do much better to re-examine the body of literature upon which they have based their search. In the end, molecular genetic research in psychiatry will prove to be a gigantic waste of time, energy, and money. The same can be said for the search for behavioral and IQ genes.

Towards a positive look to the future

A major theme of several recent books popularizing behavior genetic and psychiatric genetic research is that genetic theories about the origins of psychological trait differences and psychiatric conditions are new and revolutionary. As we have seen, these theories are in fact quite old. We are being sold an old product in new packaging. Genetic researchers often call attention to human differences at the expense of emphasizing how much people have in common. They also tend to be interested in people's limits as opposed to their potential for achievement and growth. The behavior genetic culture, as Richardson observed, is truly 'pessimistic.'[23]

When, in the future, individuals and families come into psychotherapy seeking guidance and understanding, some have envisioned a 'quick swab of a client's cheek' as a part of the psychological diagnostic procedure.[24] To imagine that human psychological suffering can be boiled down to what one finds under a microscope is both reductionistic and untrue. A 'Gattaca' type world will never come about, because genes *are not* destiny. 'The genome is not "the very essence" of what it means to be human,' wrote Proctor, 'any more than sheet music is the essence of a concert performance.'[25] Whether or not people are born as 'blank slates,' the social, political, and psychological environment is the main factor shaping who we are. The evidence presented by behavior genetics and psychiatric genetics represents little more than a house of cards that falls down in the face of critical examination.

Why then have genetic and eugenic ideas flourished in the United States, the UK, and elsewhere for several generations? It would be a mistake to see the history of human genetic research merely as the actions of isolated individuals or groups. The 'scientific racism' chronicled in Alan Chase's *The Legacy of Malthus* was financed and promoted by the wealthy elite of Western societies. Families such as the Carnegies, Rockefellers, and Harrimans financed eugenic research in the United States, Germany, Denmark, and elsewhere.[26] And even today, academic researchers promoting biological explanations for racial and class differences are financed and cultivated by organizations such as the Pioneer Fund and other agencies of the wealthy elite.[27] Thus, in the absence of political change, scientific racism and biological explanations of human differences will continue to find fertile soil.

In contrast to the bleak view of humans and their future laid out by behavior genetics and psychiatric genetics, there exists a radically different perspective. Human psychological distress, to the extent that it goes beyond people's normal reactions to life events, is primarily the result of well-known and well-documented psychologically traumatic environments and events, and social conditions such as racism, sexism, homophobia, unemployment, economic inequality, war, and societal alienation. Future societies lacking these conditions will see a dramatic reduction of human suffering, as well as a flourishing of human ability and innovation, and the possible role of genetic influences in shaping human differences will be of interest only to historians.

Notes

[1] Paul, 1998, p. 82.
[2] Kraepelin, 1976, p. 165.
[3] I thank Alvin Pam for bringing this to my attention.
[4] For documentation of fraud in science see Altman & Hernon, 1997; Bell, 1992; Broad & Wade, 1982; Gould, 1981; LaFollette, 1992.
[5] Dickersin, 1994, p. 72.
[6] Cooper et al., 1997, p. 452.
[7] Easterbrook, 1992, p. 346.

8 Savan, 1988, p. 26.
9 Albee, 1982, p. 5.
10 Leo & Joseph, 2002.
11 Plomin & Rutter, 1998, p. 1238.
12 Bouchard, 1993a, p. 43.
13 Garver & Garver, 1994, p. 151.
14 See Paul, 1998.
15 Laughlin, 1935, p. 161.
16 Osborn, 1951, p. 81.
17 Popenoe & Johnson, 1933, p. 18.
18 Rüdin, 1932, p. 494.
19 Osborne, 1951, p. 26.
20 Proctor, 1992, p. 83.
21 Yamaguchi, 1997.
22 Quoted in Kamin, in Eysenck vs. Kamin, 1981, p. 90.
23 K. Richardson, 1998, p. 178.
24 Efran et al., 1998, p. 37.
25 Proctor, 1992, pp. 83–4.
26 See Paul, 1998; Kevles, 1985.
27 For a description of the Pioneer Fund's activities, see Sedgwick, 1995. According
   to Sedgwick (p. 147), Bouchard stated in regards to the MISTRA, 'We couldn't
   have done this project without the Pioneer Fund.' Bouchard was also in agreement
   with Herrnstein & Murray's conclusions in *The Bell Curve* about the genetic origins
   of racial differences in intelligence (Bouchard, 1995).

REFERENCES

Abrams, R., and Taylor, M. A. (1983). The genetics of schizophrenia: A reassessment using modern criteria. *American Journal of Psychiatry, 140,* 171–175.

Ainslie, R. C. (1985). *The psychology of twinship.* Lincoln: University of Nebraska Press.

Albee, G. W. (1982). The politics of nature and nurture. *American Journal of Community Psychology, 10,* 4–36.

Allen, G. E. (2001). The biological basis of crime: An historical and methodological study. *Historical Studies of the Physical and Biological Sciences, 31,* 183–222.

Allen, M. G., Cohen, S., and Pollin, W. (1972). Schizophrenia in veteran twins: A diagnostic review. *American Journal of Psychiatry, 128,* 939–45.

Allport, G. W. (1961). *Pattern and growth in personality.* New York: Holt, Rinehart, and Winston.

Altman, E., and Hernon, P. (Eds.). (1997). *Research misconduct: Issues, implications, and strategies.* Greenwich, CT: Ablex Publishing Corporation.

Aly, G. (1994). Medicine against the useless. In G. Aly, P. Chroust, and C. Pross (Eds.), *Cleansing the fatherland* (pp. 22–98). Baltimore: Johns Hopkins Press.

American Psychiatric Association. (1965). *Diagnostic and statistical manual of mental disorders.* Washington, DC: Author. (Originally published in 1952)

American Psychiatric Association. (1968). *Diagnostic and statistical manual of mental disorders* (2nd ed.). Washington, DC: Author.

American Psychiatric Association. (1980). *Diagnostic and statistical manual of mental disorders* (3rd ed.). Washington, DC: Author.

American Psychiatric Association. (1987). *Diagnostic and statistical manual of mental disorders* (3rd rev. ed.). Washington, DC: Author.

American Psychiatric Association. (1994). *Diagnostic and statistical manual of mental disorders* (4th ed.). Washington, DC: Author.

Andreasen, N. C. (1998). Understanding schizophrenia: A silent Spring? *American Journal of Psychiatry, 155,* 1657–9.

Andreasen, N. C. (2000). Schizophrenia: The fundamental questions. *Brain Research Reviews, 31,* 106–12.

Andreasen, N. C. (2001). *Brave new brain.* Oxford: Oxford University Press.

Anonymous. (1925). *Eugenical News, 10,* 69–71.

Arieti, S. (1974). *Interpretation of schizophrenia* (2nd ed.). New York: Basic Books.

Ash, M. G. (1998). From 'positive eugenics' to behavioral genetics: Psychological twin research under Nazism and since. *Paedagogica Historica (Supplementary Series, Vol. III),* 335–58.

Bucur, M. (2002). *Eugenics and modernization in interwar Romania.* Pittsburgh: University of Pittsburgh Press.

Bailey, J. M., and Pillard, R. C. (1993). Reply to Lidz's 'Reply to 'a genetic study of male sexual orientation' '[Letter to the editor]. *Archives of General Psychiatry,*

50, 240–1.

Barkley, R. A. (1998). *Attention-deficit hyperactivity disorder: A handbook for diagnosis and treatment* (2nd ed.). New York: The Guilford Press.

Barr, C. L. (2001). Genetics of childhood disorders: XXII. ADHD, Part 6: The dopamine D4 receptor gene. *Journal of the American Academy of Child and Adolescent Psychiatry, 40,* 118–21.

Barr, C., Swanson, J., and Kennedy, J. (2001). Molecular genetics of ADHD. In F. Levy and D. Hay (Eds.), *Attention, genes and ADHD* (pp. 173–95). Hove, UK: Brunner-Routledge.

Bassett, A. S., Chow, E.W.C., O'Neill, S., and Brzustowicz, L.M. (2001). Genetic insights into the neurodevelopmental hypothesis of schizophrenia. *Schizophrenia Bulletin, 27,* 417–30.

Bateson, G., Jackson, D. D., Haley, J., and Weakland, J. (1956). Toward a theory of schizophrenia. *Behavioral Science, 1,* 251–64.

Beckwith, J., and Alper, J. S. (2002). Genetics of human personality: Social and ethical implications. In J. Benjamin, R. Ebstein, & R. Belmaker, (Eds.). *Molecular genetics and the human personality* (pp. 315–31). Washington, DC: American Psychiatric Press.

Beckwith, J., Geller, L., and Sarkar, S. (1991). IQ and heredity [Letter to the editor]. *Science, 252,* 191.

Begg, C. B. (1994). Publication bias. In H. Cooper and L. Hedges (Eds.), *The handbook of research synthesis* (pp. 399–409). New York: Russell Sage Foundation.

Bell, A. E. (1977). Heritability in retrospect. *Journal of Heredity, 68,* 297–300.

Bell, R. (1992). *Impure science: Fraud, compromise, and political influence in scientific research.* New York: John Wiley and Sons.

Benjamin, L. S. (1976). A reconsideration of the Kety and associates study of genetic factors in the transmission of schizophrenia. *American Journal of Psychiatry, 133,* 1129–33.

Benjamini, Y., Drai, D., Elmer, G., Kafkafi, N., Goloni, I. (2001). Controlling the false discovery rate in behavior genetics research. *Behavioural Brain Research, 125,* 279–84.

Bernstein, V. H. (1945, August 21). 'Created Nazi Science of Murder. Meet 'gentle' Prof. Rudin, Theorist of 'Aryanism''*PM Daily,* 5.

Billings, P. R., Beckwith, J., and Alper, J. S. (1992). The genetic analysis of human behavior: A new era? *Social Science and Medicine, 35,* 227–38.

Bleuler, E. (1950). *Dementia praecox or the group of schizophrenias.* New York: International Universities Press. (Originally published in 1911)

Bleuler, M. (1955). Research and changes in concepts in the study of schizophrenia, 1941–1950. *Bulletin of the Isaac Ray Medical Library, 3,* 1–132.

Bleuler, M. (1978). *The schizophrenic disorders: Long-term patient and family disorders.* New Haven: Yale University Press.

Block, N. J., and Dworkin, G. (Eds.). (1976). *The IQ controversy.* New York: Pantheon.

Bohman, M. (1971). A comparative study of adopted children, foster children and children in their biological environment born after undesired pregnancies. *Acta Paediatrica Scandinavica* (Suppl. 221), 1–38.

Bohman, M. (1978). Some genetic aspects of alcoholism and criminality. *Archives of General Psychiatry, 35,* 269–76.

Bohman, M., Cloninger, C. R., Sigvardsson, S., and von Knorring, A. (1982). Predisposition to petty criminality in Swedish adoptees. *Archives of General Psychiatry, 39,* 1233–41.

Bohman, M., and Sigvardsson, S. (1980). A prospective, longitudinal study of children registered for adoption. *Acta Psychiatrica Scandinavica, 61,* 339–55.

Bouchard, T. J., Jr. (1982). Identical twins reared apart: Reanalysis or pseudo-analysis? *Contemporary Psychology, 27,* 190–1.

Bouchard, T. J., Jr. (1983). Do environmental similarities explain the similarity in intelligence of identical twins reared apart? *Intelligence, 7*, 175–84.

Bouchard, T. J., Jr. (1984). Twins reared together and apart: What they tell us about human diversity. In S. Fox (Ed.), *Individuality and determinism: Chemical and biological bases* (pp. 147–84). New York: Plenum Press.

Bouchard, T. J., Jr. (1993a). The genetic architecture of human intelligence. In P. Vernon (Ed.), *Biological approaches to the study of human intelligence* (pp. 33–93). Norwood, NJ: Ablex Publishing Corporation.

Bouchard, T. J., Jr. (1993b). Genetic and environmental influences on adult personality: Evaluating the evidence. In J. Hettema and I. Deary (Eds.), *Basic issues in personality* (pp. 15–44). Dordrecht, The Netherlands: Kluwer Academic Publishers.

Bouchard, T. J., Jr. (1995). Breaking the last taboo. Review of 'The Bell Curve.'*Contemporary Psychology, 40*, 415–21.

Bouchard, T. J., Jr. (1996). Behaviour genetic studies of intelligence, yesterday and today: The long journey from plausibility to proof. *Journal of Biosocial Science, 28*, 527–55.

Bouchard, T. J., Jr. (1997a). The genetics of personality. In K. Blum and E. Noble (Eds.), *Handbook of psychiatric genetics* (pp. 273–96). Boca Raton, FL: CRC Press.

Bouchard, T. J., Jr. (1997b). IQ similarity in twins reared apart: Findings and responses to critics. In R. Sternberg and E. Grigorenko (Eds.), *Intelligence, heredity, and environment* (pp. 126–60). New York: Cambridge University Press.

Bouchard, T. J., Jr. (1997c, September/October). Whenever twain shall meet. *The Sciences, 37*, 52–7.

Bouchard, T. J., Jr. (1999). Foreword. In N. Segal, *Entwined lives*. New York: Dutton.

Bouchard, T. J., Jr., and Loehlin, J. C. (2001). Genes, evolution, and personality. *Behavior Genetics, 31*, 243–73.

Bouchard, T. J., Lykken, D. T., McGue, M., Segal, N. L., and Tellegen, A. (1990). Sources of human psychological differences: The Minnesota Study of Twins Reared Apart. *Science, 250*, 223–8.

Bouchard, T. J., Jr., Lykken, D. T., McGue, M., Segal, N. L., and Tellegen, A. (1991). Response [Letter to the editor]. *Science, 252*, 191–2.

Bouchard, T. J., Jr., Lykken, D. T., Segal, N. L., and Wilcox, K. J. (1986). Development in twins reared apart: A test of the chronogenetic hypothesis. In A. Demirjian (Ed.), *Human growth: A multidisciplinary review* (pp. 299–310). London: Taylor and Francis.

Bouchard, T. J., Jr., and McGue, M. (1981). Familial studies of intelligence: A review. *Science, 212*, 1055–9.

Bouchard, T. J., and McGue, M. (1990). Genetic and rearing environmental influences on adult personality: An analysis of adopted twins reared apart. *Journal of Personality, 58*, 263–92.

Bouchard, T. J., Jr., McGue, M., Hur, Y., and Horn, J. M. (1998) A genetic and environmental analysis of the California Psychological Inventory using adult twins reared together and apart. *European Journal of Personality, 12*, 307–20.

Bouchard, T. J., Jr., and Pedersen, N. (1999). Twins reared apart: Nature's double experiment. In M. LaBuda and E. Grigorenko (Eds.), *On the way to individuality: Methodological issues in behavioral genetics* (pp. 71–93). Commack, NY: Nova Science.

Boyle, M. (1990). *Schizophrenia: A scientific delusion?* New York: Routledge.

Boyle, M. (1999) Diagnosis. In C. Newnes, G. Holmes, and C. Dunn (Eds.) *This is Madness: A critical look at psychiatry and the future of mental health services* (pp. 75–90). Ross-on-Wye: PCCS Books.

Bradburn, N. M., Rips, L. J., and Shevell, S. K. (1987). Answering autobiographical questions: The impact of memory and inference on surveys. *Science, 236*, 157–61.

Bradley, W., and Schaefer, K. (1998). *The uses and misuses of data and models.* Thousand Oaks, CA: Sage Publications.

Breggin, P. R. (1991). *Toxic psychiatry.* New York: St. Martin's Press.

Breggin, P. R. (1998). *Talking back to Ritalin.* Monroe, ME: Common Courage Press.

Breggin, P. R. (2001). Empowering social work in the era of biological psychiatry. *Ethical Human Sciences and Services, 3,* 197–206.

Breggin, P. R., and Breggin, G. R. (1998). *The war against children of color.* Monroe, ME: Common Courage Press.

Breggin, P. R., and Cohen, D. (1999). *Your drug may be your problem.* Reading, MA: Perseus.

Brennan, P., Mednick, S. A., and Gabrielli, W. F. (1991). Genetic influences and criminal behavior. In M. Tsuang, K. Kendler, and M. Lyons (Eds.), *Genetic issues in psychosocial epidemiology* (pp. 231–46). New Brunswick, NJ: Rutgers University Press.

Brigham, C. C. (1923). *A study of American intelligence.* Princeton, NJ: Princeton University Press.

Broad, W. J., and Wade, N. (1982). *Betrayers of the truth.* New York: Simon and Schuster.

Broberg, G., and Roll-Hansen, N. (Eds.). (1996). *Eugenics and the welfare state: Sterilization policy in Denmark, Sweden, Norway, and Finland.* East Lansing, MI: Michigan State University Press.

Broberg, G., and Tydén, M. (1996). Eugenics in Sweden: Efficient care. In G. Broberg and N. Roll-Hansen (Eds.), *Eugenics and the welfare state: Sterilization policy in Denmark, Sweden, Norway, and Finland* (pp. 77–149). East Lansing, MI: Michigan State University Press.

Brunner, H. G., Nelen, M., Breakefield, X. O., Ropers, H. H., and van Oost, B. A. (1993). Abnormal behavior associated with a point mutation in the structural gene for monoamine oxidase. *Science, 262,* 578–80.

Bulmer, M. (1999). The development of Francis Galton's ideas on the mechanism of heredity. *Journal of the History of Biology, 32,* 263–92.

Burbridge, D. (2001). Francis Galton on twins, heredity and social class. *British Journal for the History of Science, 34,* 323–40.

Burleigh, M. (1994). *Death and deliverance.* Cambridge, UK: Cambridge University Press.

Byerley, W., and Coon, H. (1995). Strategies to identify genes in schizophrenia. In J. Oldham and M. Riba (Eds.), *Review of psychiatry* (Vol. 14; pp. 361–81). Washington, DC: American Psychiatric Press.

Cadoret, R. J. (1978). Psychopathology in adopted-away offspring of biologic parents with antisocial behavior. *Archives of General Psychiatry, 35,* 176–84.

Cannon, T. D., Kaprio, J., Lönnqvist, J., Huttunen, M., and Koskenvuo, M. (1998). The genetic epidemiology of schizophrenia in a Finnish twin cohort. *Archives of General Psychiatry, 55,* 67–74.

Cardon, L. R., and Fulker, D. W. (1993). Genetics of specific cognitive abilities. In R. Plomin and G. McClearn (Eds.), *Nature, nurture, and psychology* (pp. 99–120). Washington, DC: American Psychological Association.

Carey, G. (1992). Twin imitation for antisocial behavior: Implications for genetic and family environmental research. *Journal of Abnormal Psychology, 101,* 18–23.

Carey, G., and DiLalla, D. L. (1994). Personality and psychopathology: Genetic perspectives. *Journal of Abnormal Psychology, 103,* 32–43.

Carter, H. D. (1940). Ten years of research on twins: Contributions to the nature-nurture problem. *National Society for the Study of Education Yearbook, 39,* 235–55.

Cassill, K. (1982) *Twins: Nature's amazing mystery.* New York: Atheneum.

Cassou, B., Schiff, M., and Stewart, J. (1980). Génétique et schizophrénie: Réévaluation d'un consensus [Genetics and schizophrenia: Re-evaluation of a consensus]. *Psychiatrie de l'Enfant, 23,* 87–201.

Ceci, S. J. (1996). *On intelligence* (expanded edition). Cambridge, MA: Harvard University Press.

Chase, A. (1975, February). The great pellagra cover-up. *Psychology Today, 8 (9),* pp. 82–6.

Chase, A. (1980). *The legacy of Malthus: The social costs of the new scientific racism.* Urbana, IL/Chicago: University of Illinois Press. (Originally published in 1977)

Cheverud, J. M. (1990). Inheritance and the additive genetic model. *Behavioral and Brain Sciences, 13,* 124.

Christiansen, K. O. (1977a). A preliminary study of criminality among twins. In S. Mednick and K. Christiansen (Eds.), *Biosocial bases of criminal behavior* (pp. 89–108). New York: Gardner Press.

Christiansen, K. O. (1977b). A review of studies of criminality among twins. In S. Mednick K. Christiansen (Eds.), *Biosocial bases of criminal behavior* (pp. 45–88). New York: Gardner Press.

Chumakov, I., Blumenfeld, M., Guerassimenko, O., Cavarec, L., Palicio, M., Abderrahim, H., Bougueleret, L., Barry, C., Tanaka, H., La Rosa, P., et al. (2002). Genetic and physiological data implicating the new human gene G72 and the gene for D-amino acid oxidase in schizophrenia. *Proceedings of the National Academy of Sciences, 99,* 13675–80.

Cloninger, C. R. (2002). The discovery of susceptibility genes for mental disorders. *Proceedings of the National Academy of Sciences, 99,* 13365-7.

Cloninger, C. R., and Gottesman, I. I. (1987). Genetic and environmental factors in antisocial behavior disorders. In S. Mednick, T. Moffitt, and S. Stack (Eds.), *The causes of crime: New biological approaches* (pp. 92–109). New York: Cambridge University Press.

Cohen, D. (1997). A critique of the use of neuroleptic drugs in psychiatry. In S. Fisher and R. Greenberg (Eds.), *From placebo to panacea: Putting psychotropic drugs to the test* (pp. 173–228). New York: John Wiley and Sons.

Cohen, D., and Cohen, H. (1986). Biological theories, drug treatments, and schizophrenia: A critical assessment. *Journal of Mind and Behavior, 7,* 11–36.

Cohen, D., and McCubbin, M. (1990). The political economy of tardive dyskinesia: Asymmetries in power and responsibility. *Journal of Mind and Behavior, 11,* 465–88.

Cohen, D. J., Dibble, E., Grawe, J. M., and Pollin, W. (1973). Separating identical from fraternal twins. *Archives of General Psychiatry, 29,* 465–9.

Cohen, R. J., Swerdlik, M. E., and Smith, D. K. (1992). *Psychological testing and assessment* (2nd ed.). Mountain View, CA: Mayfield.

Colbert, T. C. (2001). *Blaming our genes.* Tustin, CA: Kevco.

Cooper, H., DeNeve, K., and Charlton, K. (1997). Finding the missing science: The fate of studies submitted for review by a human subjects committee. *Psychological Methods, 2,* 447–52.

Coryell, W., and Zimmerman, M. (1988). The heritability of schizophrenia and schizoaffective disorders. *Archives of General Psychiatry, 45,* 323–7.

Craike, W. H., and Slater, E. (1945). Folie a deux in uniovular twins reared apart. *Brain, 68, Part III,* 213–21.

Crow, J. F. (1990). How important is detecting interaction? *Behavioral and Brain Sciences, 13,* 126–7.

Crowe, R. R. (1972). The adopted offspring of women criminal offenders. *Archives of General Psychiatry, 27,* 600–3.

Crowe, R. R. (1974). An adoption study of antisocial personality. *Archives of General Psychiatry, 31,* 785–91.

Crusio, W. E. (1990). Estimating heritabilities in quantitative behavior genetics: A station passed. *Behavioral and Brain Sciences, 13,* 127–8.

Dalgard, O. S., and Kringlen, E. (1976). A Norwegian twin study of criminality. *British Journal of Criminology, 16,* 213–32.

Davenport, C. B. (1911). *Heredity in relation to eugenics.* New York: Henry Holt and Company.

Davenport, C. B. (1916). The hereditary factor in pellagra. *Eugenics Record Office Bulletin No. 16.* Cold Spring Harbor, New York.

Davenport, C. B. (1920). Heredity of constitutional mental disorders. *Psychological Bulletin, 17,* 300–10.

Davenport, C. B., and Laughlin, H. H. (1915). How to make a eugenical family study. *Eugenics Record Office Bulletin No. 13.* Cold Spring Harbor, New York.

DeGrandpre, R. (1999). *Ritalin nation.* New York: W.W. Norton.

DeLisi, L. E. (2000). Critical overview of current approaches to genetic mechanisms in schizophrenia research. *Brain Research Reviews, 31,* 187–92.

DeLisi, L. E., Shaw, S. H., Crow, T. J., Shields, G., Smith, A. B., Larach, V. W., Wellman, N., Loftus, J., Nanthakumar, B., Razi, K., Stewart, J., Comazzi, M., Vita, A., Heffner, T., & Sherrington, R. (2002). A genome-wide scan for linkage to chromosomal regions in 382 sibling pairs with schizophrenia or schizoaffective disorder. *American Journal of Psychiatry, 159,* 803–12.

Detterman, D. K. (1990). Don't kill the ANOVA messenger for bearing bad interaction news. *Behavioral and Brain Sciences, 13,* 131–3.

Devlin, B., Fienberg, S. E., Resnick, D. P., & Roeder, K. (1997). *Intelligence, genes, and success: Scientists respond to The Bell Curve.* New York: Springer Verlag.

Dibble, E., Cohen, D. J., and Grawe, J. M. (1978). Methodological issues in twin research: The assumption of environmental equivalence. In *Twin research: Psychology and methodology* (pp. 245–51). New York: Alan R. Liss.

Dickersin, K. (1994). Research registers. In H. Cooper and L. Hedges (Eds.), *The handbook of research synthesis* (pp. 71–83). New York: Russell Sage Foundation.

Dugdale, R. L. (1910). *The Jukes: A study in crime, pauperism, disease, and heredity* (4th ed.). New York: G. P. Putnam's Sons. (Originally published in 1877)

Dusek, V. (1987). Bewitching science. *Science for the People, 19,* 19–22.

Duster, T. (1990). *Backdoor to eugenics.* New York: Routledge.

Easterbrook, P. J. (1992). Directory of registries of clinical trials. *Statistics in Medicine, 11,* 345–59.

Ebstein, R. P., Novick, O., Umansky, R., Priel, B., Osher, Y., Blaine, D., Bennett, E. R., Nemanov, L., Katz, M., and Belmaker, R. H. (1995). Dopamine D4 receptor (D4DR) exon III polymorphism associated with the human personality trait of novelty seeking. *Nature Genetics, 12,* 78–80.

Eckert, E. D., Bouchard, T. J., Jr., Bohlen, J., and Heston, L. L. (1986). Homosexuality in monozygotic twins reared apart. *British Journal of Psychiatry, 148,* 421–25.

Efran, J. S., Greene, M. A., and Gordon, D. E. (1998, March/April). Lessons of the new genetics. *Family Therapy Networker,* 27–32, 35–41.

Egeland, J. A., Gerhard, D. S., Pauls, D. L., Sussex, J. N., and Kidd, K. K. (1987). Bipolar affective disorders linked to DNA markers on chromosome 11. *Nature, 325,* 783–7.

Eley, T. C., Lichtenstein, P., and Stevenson, J. (1999). Sex differences in the etiology of aggressive and nonaggressive antisocial behavior: Results from two twin studies. *Child Development, 70,* 155–68.

Ellinger, T. U. H. (1942). On the breeding of Aryans. *Journal of Heredity, 33,* 141–3.

Essen-Möller, E. (1941). Psychiatrische Untersuchungen an einer Serie von Zwillingen [Psychiatric investigations on a series of twins]. *Acta Psychiatrica et Neurologica*

(Suppl. 23). Copenhagen: Munksgaard.

Essen-Möller, E. (1963). Twin research in psychiatry. *Acta Psychiatrica Scandinavica, 39*, 65–77.

Essen-Möller, E. (1970). Twenty-one psychiatric cases and their MZ cotwins. *Acta Geneticae Medicae et Gemellologiae, 19*, 315–17.

Eysenck, H. J., versus Kamin, L. J. (1981). *The intelligence controversy*. New York: John Wiley and Sons. (Published in the UK as *Intelligence: The battle for the mind*).

Falconer, D. S. (1965). The inheritance of liability to certain diseases, estimated from the incidence among relatives. *Annals of Human Genetics, 29*, 51–76.

Falconer, D. S., and Mackay, T. F. C. (1996). *Introduction to quantitative genetics* (4th ed.). Harlow, UK: Longman.

Faraone, S. V., and Biederman, J. (2000). Nature, nurture, and attention deficit hyperactivity disorder. *Developmental Review, 20*, 568–81.

Faraone, S. V., & Biederman, J. (2002). Reply to "Meta-analysis and Psychiatric genetics" [Letter to the editor]. *American Journal of Psychiatry, 159*, 496–7.

Faraone, S. V., Biederman, J., Weiffenbach, B., Keith, T., Chu, M. P., Weaver, A., Spencer, T. J., Wilens, T. E., Frazier, J., Cleves, M., and Sakai, J. (1999). Dopamine D4 gene 7-repeat allele and attention deficit hyperactivity disorder. *American Journal of Psychiatry, 156*, 768–70.

Faraone, S. V., Doyle, A. E., Mick, E., and Biederman, J. (2001). Meta-Analysis of the association between the 7-repeat allele of the dopamine D_4 receptor gene and attention deficit hyperactivity disorder. *American Journal of Psychiatry, 158*, 1052–7.

Faraone, S. V., and Tsuang, M. T. (1995). Methods in psychiatric genetics. In M. Tsuang, M. Tohen, and G. Zahner (Eds.), *Textbook in psychiatric epidemiology* (pp. 81–134). New York: Wiley-Liss.

Faraone, S. V., Tsuang, M. T., and Tsuang, D. W. (1999). *Genetics of mental disorders*. New York: The Guilford Press.

Farber, S. L. (1981). *Identical twins reared apart: A reanalysis*. New York: Basic Books.

Feldman, M. W., and Lewontin, R. C. (1975). The heritability hang-up. *Science, 190*, 1163–8.

Fink, A. E. (1938). *The causes of crime: Biological theories in the United States, 1800–1915*. New York: A. S. Barnes and Company.

Fischer, C. S., Hout, M., Sánchez Jankowski, M., Lucas, S. R., Swidler, A., and Voss, K. (1996). *Inequality by design*. Princeton, NJ: Princeton University Press.

Fischer, M. (1971). Psychoses in the offspring of schizophrenic monozygotic twins and their normal co-twins. *British Journal of Psychiatry, 118*, 43–51.

Fischer, M. (1973). *Genetic and environmental factors in schizophrenia*. Copenhagen: Munksgaard.

Fish, J. M. (Ed.). (2002). *Race and intelligence: Separating science from myth*. Mahwah, New Jersey: Lawrence Erlbaum.

Fisher, R. A. (1919). The genesis of twins. *Genetics, 4*, 489–99.

Fletcher, R. (1991). *Science, ideology, and the media: The Cyril Burt scandal*. New Brunswick, NJ: Transaction Publishers.

Frangos, E., Athanassenas, G., Tsitourides, S., Katsanou, N., and Alexandrakou, P. (1985). Prevalence of DSM-III schizophrenia among the first-degree relatives of schizophrenic probands. *Acta Psychiatrica Scandinavica, 72*, 382–6.

Franzek, E., and Beckmann, H. (1998). Different genetic background of schizophrenia spectrum diagnoses: A twin study. *American Journal of Psychiatry, 155*, 76–83.

Fraser, S. (Ed.). (1995). *The Bell Curve wars*. New York: Basic Books.

Fraser-Roberts, J. A. (1934). Twins. *Eugenics Review, 17*, 25–32.

Fuller, J. L., and Thompson, W. R. (1960). *Behavior genetics*. New York: John Wiley and Sons.

Fuller, J. L., and Thompson, W. R. (1978). *Foundations of behavior genetics.* St. Louis: C. V. Mosby.

Gabrielli, W. F., and Mednick, S. A. (1983). Genetic correlates of criminal behavior. *American Behavioral Scientist, 27,* 59–74.

Galton, F. (1875). The history of twins as a criterion of the relative powers of nature and nurture. *Journal of the Anthropological Institute of Great Britain and Ireland, 5,* 391–406.

Galton, F. (1881). *Hereditary genius.* New York: D. Appleton. (Originally published in 1869)

Galton, F. (1883). *Inquiries into human faculty and its development.* London: Macmillan and Company.

Galton, F. (1995). Hereditary talent and character. In R. Jacoby and N. Glauberman (Eds.), *The Bell Curve debate* (pp. 393–409). New York: Times Books. (Originally published in 1865)

Gardner, I. C., and Newman, H. H. (1940). Mental and physical traits of identical twins reared apart. *Journal of Heredity, 31,* 119–26.

Garver, K. L., and Garver, B. (1994). The Human Genome Project and eugenic concerns. *American Journal of Human Genetics, 54,* 148–58.

Gedda, L. (1961). *Twins in history and science.* Springfield, IL: Charles C. Thomas.

Gershon, E. S., Badner, J. A., Goldin, L. R., Sanders, A. R., Cravchik, A., and Detera-Wadleigh, S. D. (1998). Closing in on genes for manic-depressive illness and schizophrenia. *Neuropsychopharmacology, 18,* 233–42.

Gesell, A. L. (1942). The method of co-twin control. *Science, 95,* 446–8.

Gesell, A. L., and Thompson, H. (1929). Learning and growth in identical infant twins. *Genetic Psychology Monographs, 6,* 5–120.

Gillie, O. (1976). Crucial data was faked by eminent scientist. *The Sunday Times,* 24 October, London.

Goddard, H. H. (1927). *The Kallikak family: A study in the heredity of feeble-mindedness.* New York: The Macmillan Company. (Originally published in 1912)

Goldsmith, H. H., and Gottesman, I. I. (1996). Heritable variability and variable heritability in developmental psychopathology. In M. Lenzenweger and J. Haugaard (Eds.), *Frontiers of developmental psychopathology* (pp. 5–43). New York: Oxford University Press.

Goodman, R., and Stevenson, J. (1989). A twin study of hyperactivity – II. The etiological role of genes, family relationships and perinatal adversity. *Journal of Child Psychology and Psychiatry, 30,* 691–709.

Goodnight, C. J. (1990). On the relativity of quantitative genetic variance components. *Behavioral and Brain Sciences, 13,* 134–5.

Gosden, R. (2001). *Punishing the patient.* Melbourne: Scribe Publications.

Gottesman, I. I. (1978). Schizophrenia and genetics: Where are we? Are you sure? In L. Wynne, R. Cromwell, and S. Matthysse (Eds.), *The nature of schizophrenia: New approaches to research and treatment* (pp. 59–69). New York: John Wiley and Sons.

Gottesman, I. I. (1982). Identical twins reared apart: A reanalysis [Book review]. *American Journal of Psychology, 95,* 350–2.

Gottesman, I. I. (1991). *Schizophrenia genesis.* New York: W. H. Freeman and Company.

Gottesman, I. I. (1996). Blind men and elephants. In L. Hall (Ed.), *Genetics and mental illness: Evolving issues for research and society* (pp. 51–77). San Francisco: Freeman.

Gottesman, I. I., and Bertelsen, A. (1989). Confirming unexpressed genotypes for schizophrenia. *Archives of General Psychiatry, 46,* 867–72.

Gottesman, I. I., and Bertelsen, A. (1990). Reply to Torrey [Letter to the editor]. *Archives of General Psychiatry, 47,* 977–8.

Gottesman, I. I., and Bertelsen, A. (1996). Legacy of German psychiatric genetics: Hindsight is always 20/20. *American Journal of Medical Genetics (Neuropsychiatric Genetics), 67,* 317–22.

Gottesman, I. I., and Shields, J. (1966a). Contributions of twin studies to perspectives on schizophrenia. In B. Maher (Ed.), *Progress in experimental personality research* (Vol. 3; pp. 1–84). New York: Academic Press.

Gottesman, I. I., and Shields, J. (1966b). Schizophrenia in twins: 16 years' consecutive admissions to a psychiatric clinic. *British Journal of Psychiatry, 112,* 809–18.

Gottesman, I. I., and Shields, J. (1972). *Schizophrenia and genetics: A twin study vantage point.* New York: Academic Press.

Gottesman, I. I., and Shields, J. (1976a). A critical review of recent adoption, twin, and family studies of schizophrenia: Behavioral genetics perspectives. *Schizophrenia Bulletin, 2,* 360–401.

Gottesman, I. I., and Shields, J. (1976b). Rejoinder: Toward optimal arousal and away from original din. *Schizophrenia Bulletin, 2,* 447–53.

Gottesman, I. I., and Shields, J. (1982). *Schizophrenia: The epigenetic puzzle.* New York: Cambridge University Press.

Gottfredson, M. R., and Hirschi, T. (1990). *A general theory of crime.* Stanford, CA: Stanford University Press.

Gould, S. J. (1981). *The mismeasure of man.* New York: W. W. Norton and Company.

Grove, W. M., Eckert, E. D., Heston, L. L., Bouchard, T. J., Jr., Segal, N. L., and Lykken, D. T. (1990). Heritability of substance abuse and antisocial behavior: A study of monozygotic twins reared apart. *Biological Psychiatry, 27,* 1293–1304.

Gunderson, J. G., and Siever, L. J. (1985). Relatedness of schizotypal to schizophrenic disorders: Editor's introduction. *Schizophrenia Bulletin, 11,* 532–7.

Guthrie, R. V. (1976). *Even the rat was white.* New York: Harper and Row.

Gütt, A., Rüdin, E., and Ruttke, F. (1934). *Gesetz zur Verhütung erbkranken Nachwuchses* [Law for the prevention of genetically diseased offspring]. Munich: J. F. Lehmanns.

Häfner, H., and der Heiden, W. (1986). The contribution of European case registers to research on schizophrenia. *Schizophrenia Bulletin, 12,* 26–51.

Haier, R. J., Rosenthal, D., and Wender, P. H. (1978). MMPI assessment of psychopathology in the adopted-away offspring of schizophrenics. *Archives of General Psychiatry, 35,* 171–5.

Hall, W. S. (1914). The relation of crime to adolescence. *Bulletin of the American Academy of Medicine, 15,* 86–95.

Haller, M. H. (1963). *Eugenics: Hereditarian attitudes in American thought.* New Brunswick, NJ: Rutgers University Press.

Halverson, C. F. (1988). Remembering your parents: Reflections on the retrospective method. *Journal of Personality, 56,* 435–49.

Hamer, D., and Copeland, P. (1988). *Living with our genes.* New York: Anchor Books.

Hansen, B. (1996). Something rotten in the state of Denmark: Eugenics and the ascent of the welfare state. In G. Broberg and N. Roll-Hansen (Eds.), *Eugenics and the welfare state: Sterilization policy in Denmark, Sweden, Norway, and Finland* (pp. 9–76). East Lansing, MI: Michigan State University Press.

Hansen, S. (1925). Eugenics in Denmark. *Eugenical News, 10,* 81–2.

Harris, J. R. (1998). *The nurture assumption: Why children turn out the way they do.* New York: The Free Press.

Harvald, B., and Haugue, M. (1965). Hereditary factors elucidated by twin studies. In J. Neel, M. Shaw, and W. Schull (Eds.) *Genetics and the epidemiology of chronic diseases* (pp. 61–76). Washington, DC: Public Health Service Publication No. 1163.

Hearnshaw, L. (1979). *Cyril Burt: Psychologist.* Ithaca, NY: Cornell University Press.

Hemminki, E., Rasimus, A., and Forssas, E. (1997). Sterilization in Finland: From eugenics to contraception. *Social Science and Medicine, 45,* 1875–84.

Herrnstein, R. J., and Murray, C. (1994). *The bell curve.* New York: The Free Press.

Heston, L. L. (1966). Psychiatric disorders in foster home reared children of schizophrenic mothers. *British Journal of Psychiatry, 112,* 819–25.

Heston, L. L. (1970). The genetics of schizophrenic and schizoid disease. *Science, 167,* 249–56.

Heston, L. L. (1981/November). Family reunion. *The Sciences, 21,* 26–8.

Heston, L. L. (1988). What about environment? In D. Dunner, E. Gershon, and J. Barrett (Eds.), *Relatives at risk for mental disorder* (pp. 205–13). New York: Raven Press.

Heston, L. L., and Denny, D. D. (1968). Interactions between early life experience and biological factors in schizophrenia. In D. Rosenthal and S. Kety (Eds.), *The transmission of schizophrenia* (pp. 363–76). New York: Pergamon Press.

Heston, L. L., Denney, D. D., and Pauly, I. B. (1966). The adult adjustment of persons institutionalized as children. *British Journal of Psychiatry, 112,* 1103–10.

Hietala, M. (1996). From race hygiene to sterilization: The eugenics movement in Finland. In G. Broberg and N. Roll-Hansen (Eds.), *Eugenics and the welfare state: Sterilization policy in Denmark, Sweden, Norway, and Finland* (pp. 195–258). East Lansing, MI: Michigan State University Press.

Hill, D. (1983). *The politics of schizophrenia.* Lanham, MD: University Press of America.

Hirsch, J. (1970). Behavior-genetic analysis and its biosocial consequence. *Seminars in Psychiatry, 2,* 89–105.

Hirsch, J. (1990). A nemesis for heritability estimation. *Behavioral and Brain Sciences, 13,* 137–8.

Hirsch, J. (1997). Some history of heredity-vs-environment, genetic inferiority at Harvard (?), and The (incredible) Bell Curve. *Genetica, 99,* 207–24.

Hirsch, N. D. (1926). A study of natio-racial mental differences. *Genetic Psychology Monographs, 1,* 231–406.

Hirsch, N. D. (1930). *Twins: Heredity and environment.* Cambridge, MA: Harvard University Press.

Hoffer, A., and Pollin, W. (1970). Schizophrenia in the NAS-NRC panel of 15,909 veteran twin pairs. *Archives of General Psychiatry, 23,* 469–77.

Hoffman, L. W. (1991). The influence of the family environment on personality: Accounting for sibling differences. *Psychological Bulletin, 110,* 187–203.

Holmberg, D., and Holmes, J. G. (1994). Reconstruction of relationship memories: A mental models approach. In N. Schwarz and S. Sudman (Eds.) *Autobiographical memory and the validity of retrospective reports* (pp. 267–88). New York: Springer Verlag.

Holmes, S. J. (1930, September). Nature versus nurture in the development of the mind. *The Scientific Monthly, 31,* 245–52.

Holden, C. (1980). Identical twins reared apart. *Science, 207,* 1323–8.

Holzman, P. S. (2001). Seymour S. Kety and the genetics of schizophrenia. *Neuropsychopharmacology, 25,* 299–304.

Horgan, J. (1993). Eugenics revisited. *Scientific American, 268,* 122–31.

Hubbard, R., and Wald, E. (1993). *Exploding the gene myth.* Boston: Beacon Press.

Hurwitz, S., and Christiansen, K. O. (1983). *Criminology.* Boston: George Allen and Unwin.

Hutchings, B., and Mednick, S. A. (1975). Registered criminality in the adoptive and biological parents of registered male criminal adoptees. In R. Fieve, D. Rosenthal, and H. Brill (Eds.), *Genetic research in psychiatry* (pp. 105–16). Baltimore: The Johns Hopkins Press.

Ingraham, L. J., and Kety, S. S. (1988). Schizophrenia spectrum disorders. In M. Tsuang and J. Simpson (Eds.), *Handbook of schizophrenia, Vol. 3: Nosology, epidemiology and genetics* (pp. 117–37). New York: Elsevier Science Publishers.

Inouye, E. (1961). Similarity and dissimilarity of schizophrenia in twins. *Proceedings of The Third World Congress of Psychiatry* (Vol. 1; pp. 524–30). Montreal: University of Toronto Press.

Inouye, E. (1965). Similar and dissimilar manifestations of obsessive-compulsive neurosis in monozygotic twins. *American Journal of Psychiatry, 121,* 1171–5.

Irvine, R. T. (1903). The congenital criminal. *Medical News, 82,* 749–52.

Jackson, D. D. (1960). A critique of the literature on the genetics of schizophrenia. In D. Jackson (Ed.), *The etiology of schizophrenia* (pp. 37–87). New York: Basic Books.

Jackson, Donald Dale. (1980, October). Reunion of identical twins, raised apart, reveals some astonishing similarities. *Smithsonian,* pp. 48–56.

Jacobs, D. (1994). Environmental failure — oppression is the only cause of psychopathology. *Journal of Mind and Behavior, 15,* 1–18.

Jacoby, R., and Glauberman, N. (Eds.). (1995). *The Bell Curve debate.* New York: Times Books.

Jacquard, A. (1983). Heritability: One word, three concepts. *Biometrics, 39,* 465–77.

Jensen A. R. (1970). IQs of identical twins reared apart. *Behavior Genetics. 1,* 133–48.

Jensen, A. R. (1972). How much can we boost IQ and scholastic achievement? In *Genetics and Education.* London: Methuen and Co. (Originally published in 1969)

Jensen, A. R. (1974). Kinship correlations reported by Sir Cyril Burt. *Behavior Genetics, 4,* 1–28.

Jensen, A. R. (1980). *Bias in mental testing.* New York: Free Press.

Jensen, A. R. (1998). *The g factor.* Westport, CT: Praeger.

Jinks, J. L., and Fulker, D. W. (1970). Comparison of the biometrical genetic, MAVA, and classical approaches to the analysis of human behavior. *Psychological Bulletin, 73,* 311–49.

Jönsson, E., Brene, S., Zhang, X. R., Nimgaonkar, V. L., Tylec, A., Schalling, M., and Sedvall, G. (1997). Schizophrenia and neurotrophin-3 alleles. *Acta Psychiatrica Scandinavica, 95,* 414–19.

Joseph, J. (1998a). *A critical analysis of the genetic theory of schizophrenia.* Unpublished doctoral dissertation, California School of Professional Psychology, Alameda.

Joseph, J. (1998b). The equal environment assumption of the classical twin method: A critical analysis. *Journal of Mind and Behavior, 19,* 325–58.

Joseph, J. (1999a). A critique of the Finnish Adoptive Family Study of Schizophrenia. *Journal of Mind and Behavior, 20,* 133–54.

Joseph, J. (1999b). The genetic theory of schizophrenia: A critical overview. *Ethical Human Sciences and Services, 1,* 119–45.

Joseph, J. (2000a). A critique of the spectrum concept as used in the Danish-American schizophrenia adoption studies. *Ethical Human Sciences and Services, 2,* 135–60.

Joseph, J. (2000b). Inaccuracy and bias in textbooks reporting psychiatric research: The case of the schizophrenia adoption studies. *Politics and the Life Sciences, 19,* 89–99.

Joseph, J. (2000c). Not in their genes: A critical view of the genetics of attention-deficit hyperactivity disorder. *Developmental Review, 20,* 539–67.

Joseph, J. (2000d). Potential confounds in psychiatric genetic research: The case of pellagra. *New Ideas in Psychology, 18,* 83–91.

Joseph, J. (2000e). Problems in psychiatric genetic research: A reply to Faraone and Biederman. *Developmental Review, 20,* 582–93.

Joseph, J. (2001a). The Danish-American Adoptees' Family studies of Kety and associates: Do they provide evidence in support of the genetic basis of schizophrenia? *Genetic, Social, and General Psychology Monographs, 127,* 241–78.

Joseph, J. (2001b). Don Jackson's 'A critique of the literature on the genetics of schizophrenia' – A reappraisal after 40 years. *Genetic, Social, and General Psychology Monographs, 127,* 27–57.

Joseph, J. (2001c). Is crime in the genes? A critical review of twin and adoption studies of criminality and antisocial behavior. *Journal of Mind and Behavior, 22,* 179–218.

Joseph, J. (2001d). Separated twins and the genetics of personality differences: A critique. *American Journal of Psychology, 114,* 1–30.

Joseph, J. (2002). Twin studies in psychiatry and psychology: Science or pseudoscience? *Psychiatric Quarterly, 73,* 71–82.

Joseph, J., and Baldwin, S. (2000). Four editorial proposals to improve social sciences research and publication. *International Journal of Risk and Safety in Medicine, 13,* 117–27.

Joynson, R. B. (1989). *The Burt affair.* London: RKP.

Juel-Nielsen, N. (1980). *Individual and environment: Monozygotic twins reared apart* (rev. ed.). New York: International Publishers. (Originally published in 1965)

Kallmann, F. J. (1938a). *The genetics of schizophrenia: A study of heredity and reproduction in the families of 1,087 schizophrenics.* New York: J. J. Augustin.

Kallmann, F. J. (1938b). Heredity, reproduction, and eugenic procedure in the field of schizophrenia. *Eugenical News, 13,* 105–13.

Kallmann, F. J. (1938c). Wilhelm Weinberg. *Journal of Nervous and Mental Disease, 87,* 263–4.

Kallmann, F. J. (1946). The genetic theory of schizophrenia: An analysis of 691 schizophrenic twin index families. *American Journal of Psychiatry, 103,* 309–22.

Kallmann, F. J. (1953). *Heredity in health and mental disorder.* New York: W. W. Norton.

Kallmann, F. J. (1958). The uses of genetics in psychiatry. *Journal of Mental Science, 104,* 542–52.

Kamin, L. J. (1974). *The science and politics of I.Q.* Potomac, MD: Lawrence Erlbaum Associates.

Kamin, L. J. (1985). Criminality and adoption [Letter to the editor]. *Science, 227,* 983.

Kamin, L. J. (1986). Is crime in the genes? The answer may depend on who chooses the evidence. *Scientific American, 254,* 22–7.

Kamin, L. J., and Goldberger, A. S. (2002). Twin studies in behavioral research: A skeptical view. *Theoretical Population Biology, 61,* 83–95.

Kandel, E., Mednick, S. A., Kirkegaard-Sorensen, L., Hutchings, B., Knop, J., Rosenberg, R., and Schulsinger, F. (1988). IQ as a protective factor for subjects at high risk for antisocial behavior. *Journal of Consulting and Clinical Psychology, 56,* 224–6.

Karon, B. P. (1999). The tragedy of schizophrenia. *General Psychologist, 34,* 1–12.

Karon, B. P., and VandenBos, G. R. (1981). *Psychotherapy of schizophrenia: The treatment of choice.* New York: Jason Aronson.

Keefe, R. S. E., and Harvey, P. D. (1994). *Understanding schizophrenia.* New York: The Free Press.

Kempthorne, O. (1978). Logical, epistemological and statistical aspects of nature-nurture data interpretation. *Biometrics, 34,* 1–23.

Kempthorne, O. (1990). How does one apply statistical analysis to our understanding of the development of human relationships. *Behavioral and Brain Sciences, 13,* 138–9.

Kendler, K. S. (1983). Overview: A current perspective on twin studies of schizophrenia. *American Journal of Psychiatry, 140,* 1413–25.

Kendler, K. S. (1986). Genetics of schizophrenia. In A. Frances and R. Hales (Eds.), *American psychiatric association annual review* (Vol. 5; pp. 25–41). Washington DC: American Psychiatric Press.

Kendler, K. S. (1987). The genetics of schizophrenia: A current perspective. In H. Meltzer (Ed.), *Psychopharmacology: The third generation of progress* (pp. 705–13). New York: Raven Press.

Kendler, K. S. (1988). The genetics of schizophrenia: An overview. In M. Tsuang and J. Simpson (Eds.), *Handbook of schizophrenia, Vol. 3: Nosology, epidemiology and genetics* (pp. 437–62). New York: Elsevier Science Publishers.

Kendler, K. S. (1993). Twin studies of psychiatric illness: Current status and future directions. *Archives of General Psychiatry, 50,* 905–15.

Kendler, K. S. (1995). Genetic epidemiology in psychiatry: Taking both genes and environment seriously. *Archives of General Psychiatry, 52,* 895–99.

Kendler, K. S. (1997). Reply to Gejman, Gershon, and Lerer and Segman [Letter to the editor]. *American Journal of Medical Genetics (Neuropsychiatric Genetics), 74,* 461–3.

Kendler, K. S. (2000). Schizophrenia: Genetics. In B. Sadock and V. Sadock (Eds.) *Kaplan and Sadock's comprehensive textbook of psychiatry* (7th ed., Vol. 1; pp. 1147–58). Philadelphia: Lippincott, Williams, and Wilkins.

Kendler, K. S. (2001). Twin studies of psychiatric illness: An update. *Archives of General Psychiatry, 58,* 1005–14.

Kendler, K. S., and Diehl, S. R. (1993). The genetics of schizophrenia: A current, genetic-epidemiologic perspective. *Schizophrenia Bulletin, 19,* 261–85.

Kendler, K. S., and Diehl, S. R. (1995). Schizophrenia: Genetics. In H. Kaplan and B. Sadock (Eds.), *Comprehensive textbook of psychiatry* (6th ed., Vol. 1; pp. 942–57). Baltimore: Williams and Wilkins.

Kendler, K. S., and Gruenberg, A. M. (1984). An independent analysis of the Danish adoption study of schizophrenia. *Archives of General Psychiatry, 41,* 555–64.

Kendler, K. S., Gruenberg, A. M., and Kinney, D. K. (1994). Independent diagnoses of adoptees and relatives as defined by DSM-III in the provincial and national samples of the Danish adoption study of schizophrenia. *Archives of General Psychiatry, 51,* 456–68.

Kendler, K. S., Neale, M. C., Kessler, R. C., Heath, A. C., and Eaves, L. J. (1994). Parental treatment and the equal environment assumption in twin studies of psychiatric illness. *Psychological Medicine, 24,* 579–90.

Kendler, K. S., Neale, M. C., Kessler, R. C., Heath, A. C., and Eaves, L. J. (1993). A test of the equal-environment assumption in twin studies of psychiatric illness. *Behavior Genetics, 23,* 21–7.

Kendler, K. S., and Robinette, C. D. (1983). Schizophrenia in the national academy of sciences-national research council twin registry: A 16-year update. *American Journal of Psychiatry, 140,* 1551–63.

Kennedy, J. L., Giuffra, L. A., Moises, H. W., Cavalli-Sforza, L. L., Pakstis, A. J., Kidd, J. R., Castiglione, C. M., Sjogren, B., Wetterberg, L., and Kidd, K. K. (1988). Evidence against linkage of schizophrenia to markers on chromosome 5 in a northern Swedish pedigree. *Nature, 336,* 167–70.

Kety, S. S. (1959). Biochemical theories of schizophrenia, part II. *Science, 129,* 1590–6.

Kety, S. S. (1966). Research programs of the major mental illnesses: 1. *Hospital and Community Psychiatry, 17,* 226–32.

Kety, S. S. (1970). Genetic-environmental interactions in the schizophrenic syndrome. In R. Cancro (Ed.), *The schizophrenic reactions* (pp. 233–44). New York: Brunner/Mazel.

Kety, S. S. (1974). From rationalization to reason. *American Journal of Psychiatry, 131,* 957–63.

Kety, S. S. (1976). Studies designed to disentangle genetic and environmental variables in schizophrenia: Some epistemological questions and answers. *American Journal of Psychiatry, 133,* 1134–7.

Kety, S. S. (1978). Heredity and environment. In J. Shershow (Ed.), *Schizophrenia: Science and practice* (pp. 47–68). Cambridge, MA: Harvard University Press.

Kety, S. S. (1983a). Mental illness in the biological and adoptive relatives of schizophrenia adoptees: Findings relevant to genetic and environmental factors in etiology. *American Journal of Psychiatry, 140,* 720–7.

Kety, S. S. (1983b). Observations on genetic and environmental influences in the etiology of mental disorder from studies on adoptees and their relatives. In S. Kety, L. Rowland, R. Sidman, and S. Matthysse (Eds.), *Genetics of neurological and psychiatric disorders* (pp. 105–14). New York: Raven Press.

Kety, S. S. (1985a). The concept of schizophrenia. In M. Alpert (Ed.), *Controversies in schizophrenia: Changes and constancies – Proceedings of the 74th annual meeting of the psychopathological association, New York, March 1–3, 1984* (pp. 3–11). New York: Guilford Press.

Kety, S. S. (1985b). Schizotypal personality disorder: An operational definition of Bleuler's latent schizophrenia? *Schizophrenia Bulletin, 11,* 590–4.

Kety, S. S. (1987). The significance of genetic factors in the etiology of schizophrenia: Results from the national study of adoptees in Denmark. *Journal of Psychiatric Research, 21,* 423–9.

Kety, S. S., and Ingraham, L. J. (1992). Genetic transmission and improved diagnosis of schizophrenia from pedigrees of adoptees. *Journal of Psychiatric Research, 26,* 247–55.

Kety, S. S., Rosenthal, D., and Wender, P. H. (1978). Genetic relationships within the schizophrenia spectrum: Evidence from adoption studies. In R. Spitzer and D. Klein (Eds.), *Critical issues in psychiatric diagnosis* (pp. 213–23). New York: Raven Press.

Kety, S. S., Rosenthal, D., Wender, P. H., and Schulsinger, F. (1968). The types and prevalence of mental illness in the biological and adoptive families of adopted schizophrenics. In D. Rosenthal and S. Kety (Eds.), *The transmission of schizophrenia* (pp. 345–62). New York: Pergamon Press.

Kety, S. S., Rosenthal, D., Wender, P. H., and Schulsinger, F. (1976). Studies based on a total sample of adopted individuals and their relatives: Why they were necessary, what they demonstrated and failed to demonstrate. *Schizophrenia Bulletin, 2,* 413–27.

Kety, S. S., Rosenthal, D., Wender, P. H., Schulsinger, F., and Jacobsen, B. (1975). Mental illness in the biological and adoptive families of adopted individuals who have become schizophrenic: A preliminary report based on psychiatric interviews. In R. Fieve, D. Rosenthal, and H. Brill (Eds.), *Genetic research in psychiatry* (pp. 147–65). Baltimore: The Johns Hopkins Press.

Kety, S. S., Rosenthal, D., Wender, P. H., Schulsinger, F., and Jacobsen, B. (1978). The biologic and adoptive families of adopted individuals who became schizophrenic: Prevalence of mental illness and other characteristics. In L. Wynne, R. Cromwell, and S. Matthysse (Eds.), *The nature of schizophrenia* (pp. 25–37). New York: John Wiley and Sons.

Kety, S. S., Wender, P. H., Jacobsen, B., Ingraham, L. J., Jansson, L., Faber, B., and Kinney, D. K. (1994). Mental illness in the biological and adoptive relatives of schizophrenic adoptees: Replication of the Copenhagen study to the rest of Denmark. *Archives of General Psychiatry, 51,* 442–55.

Kevles, D. J. (1985). *In the name of eugenics.* Berkeley, CA: University of California Press.

Koch, H. L. (1966). *Twins and twin relations.* Chicago: The University of Chicago Press.

Kolb, L., and Brodie, H. K. (1982). *Modern clinical psychiatry* (10th ed.). Philadelphia: W. B. Saunders Company.

Koskenvuo, M., Langinvainio, H., Kaprio, J., Lönnqvist, J., and Tienari, P. (1984). Psychiatric hospitalization in twins. *Acta Geneticae Medicae et Gemellologiae,*

33, 321–32.

Kraepelin, E. (1976). *Manic-depressive insanity and paranoia*. New York: Arno Press. (Originally published in English in 1921)

Kranz, H. (1936). *Lebensschicksale krimineller zwillinge* [The life destiny of criminal twins]. Berlin: Julius Springer Verlag.

Kringlen, E. (1967a). *Heredity and environment in the functional psychoses: An epidemiological-clinical study*. Oslo: Universitetsforlaget.

Kringlen, E. (1967b). *Heredity and environment in the functional psychoses: Case histories*. Oslo: Universitetsforlaget.

Kringlen, E. (1976). Twins—still our best method. *Schizophrenia Bulletin, 2*, 429–33.

Kringlen, E. (1987). Contributions of genetic studies on schizophrenia. In H. Häfner and W. Gattaz (Eds.), *Search for the causes of schizophrenia* (pp. 123–42). New York: Springer Verlag.

Kringlen, E., and Cramer, G. (1989). Offspring of monozygotic twins discordant for schizophrenia. *Archives of General Psychiatry, 46*, 873–7.

Kuhn, T. S. (1996). *The structure of scientific revolutions* (3rd ed.). Chicago: The University of Chicago Press. (Originally published in 1962)

LaFollette, M. C. (1992). *Stealing into print: Fraud, plagiarism, and misconduct in scientific publishing*. Berkeley, CA: University of California Press.

Laing, R. D. (1967). *The politics of experience*. New York: Pantheon Books.

Laing, R. D. (1981). A critique of Kallmann's and Slater's genetic theory of schizophrenia. In R. Evans, *Dialogue with R. D. Laing* (pp. 97–156). New York: Praeger.

Lander, E., and Kruglyak, L. (1995). Genetic dissection of complex traits: Guidelines for interpreting and reporting linkage results. *Nature Genetics, 11*, 241–7.

Lang, J. S. (1987, April 13th). How genes shape personality. *U.S. News and World Report*.

Lange, J. (1930). *Crime and destiny*. New York: Charles Boni.

Langinvainio, H., Kaprio, J., Koskenvuo, M., and Lönnqvist, J. (1984). Finnish twins reared apart III: Personality factors. *Acta Geneticae Medicae et Gemellologiae, 33*, 259–64.

Langinvainio, H., Koskenvuo, M., Kaprio, J., and Sistonen, P. (1984). Finnish twins reared apart II: Validation of zygosity, environmental dissimilarity and weight and height. *Acta Geneticae Medicae et Gemellologiae, 33*, 251–8.

Largent, M. A. (in press). 'The greatest curse of the race': Eugenic sterilization in Oregon, 1909–83. *Oregon Historical Quarterly*.

Laughlin, H. H. (1935). The eugenics exhibit at Chicago. *Journal of Heredity, 26*, 155–162.

Lauterbach, C. E. (1925). Studies in twin resemblance. *Genetics, 10*, 525–79.

Lehtonen, J. (1994). From dualism to psychobiological interaction: A comment on the study by Tienari and his co-workers. *British Journal of Psychiatry, 164* (Suppl. 23), 27–8.

Leo, J., and Joseph, J. (2002). Schizophrenia: Medical students are taught it's all in the genes, but are they hearing the whole story? *Ethical Human Sciences and Services, 4*, 17–30.

Lerer, B., and Segman, R. H. (1997). Correspondence regarding German psychiatric genetics and Ernst Rüdin [Letter to the Editor]. *American Journal of Human Genetics (Neuropsychiatric Genetics), 74*, 459–60.

Lerner, I. M. (1958). *The genetic basis of selection*. New York: John Wiley and Sons.

Lewontin, R. C. (1974). The analysis of variance and the analysis of causes. *American Journal of Human Genetics, 26*, 400–11.

Lewontin, R. C. (1976). Race and intelligence. In N. Block and G. Dworkin (Eds.), *The IQ controversy* (pp. 78–92). New York: Pantheon Books. (Originally published in 1970)

Lewontin, R. C. (1987). The irrelevance of heritability. *Science for the People, 19*, 23, 32.

Lewontin, R. C. (1991). *Biology as ideology*. New York: Harper Perennial.

Lewontin, R. C., Rose, S., and Kamin, L. J. (1984). *Not in our genes*. New York: Pantheon.

Lidz, T. (1976). Commentary on a critical review of recent adoption, twin, and family studies of schizophrenia: Behavioral genetics perspectives. *Schizophrenia Bulletin, 2*, 402–12.

Lidz, T., and Blatt, S. (1983). Critique of the Danish-American studies of the biological and adoptive relatives of adoptees who became schizophrenic. *American Journal of Psychiatry, 140*, 426–35.

Lidz, T., Blatt, S., and Cook, B. (1981). Critique of the Danish-American studies of the adopted-away offspring of schizophrenic parents. *American Journal of Psychiatry, 138*, 1063–8.

Lifton, R. J. (1986). *The Nazi doctors*. New York: Basic Books.

Loehlin, J. C. (1981). Identical twins reared apart: A reanalysis [Book review]. *Acta Geneticae Medicae et Gemellologiae, 30*, 297–8.

Loehlin, J. C., and Horn, J. M. (2000). Stoolmiller on restriction of range in adoption studies: A comment. *Behavior Genetics, 30*, 245–7.

Loehlin, J. C., and Nichols, R. C. (1976). *Heredity, environment, and personality*. Austin: University of Texas Press.

Lombroso, C. (1968). *Crime: Its causes and remedies*. Montclair, NJ: Patterson Smith. (Originally published in 1911)

Lowing, P. A., Mirsky, A. F., and Pereira, R. (1983). The inheritance of schizophrenia spectrum disorders: A reanalysis of the Danish adoptee study data. *American Journal of Psychiatry, 140*, 1167–71.

Ludmerer, K. M. (1972). *Genetics and American society: A historical appraisal*. Baltimore: Johns Hopkins University Press.

Lush, J. L. (1945). *Animal breeding plans*. Ames, IA: Collegiate Press.

Lush, J. L. (1949). Heritability of quantitative characteristics in farm animals. *Hereditas* (Suppl.), G. Bonnier and R. Larsson, Eds., 356–75.

Luxenburger, H. (1928). Vorläufiger Bericht über psychiatrische Serienuntersuchungen an Zwillingen [Provisional report on a series of psychiatric investigations of twins]. *Zeitschrift fur die Gesamte Neurologie und Psychiatrie, 116*, 297–347.

Luxenburger, H. (1931a). Möglichkeiten und Notwendigkeiten für die psychiatrisch-eugenische Praxis [Possibilities and necessities for the psychiatric-eugenic practice]. *Münchener Medizinische Wochenschrift, 78*, 753–8.

Luxenburger, H. (1931b). Psychiatrische Erbprognose und Eugenik. [Psychiatric genetic prognosis and eugenics]. *Eugenik, 1*, 117–24.

Luxenburger, H. (1931c). Psychische Hygiene und psychiatrische eugenik [Psychic hygiene and psychiatric eugenics]. *Eugenik, 2*, 49–56.

Luxenburger, H. (1934a). Paranoia und Gesetz zur Verhütung erbkranken Nachwuchses [Paranoia and the law for the prevention of hereditarily ill offspring]. *Der Erbarzt, 1*, 33–6.

Luxenburger, H. (1934b). Rassenhygienisch wichtige Probleme und Ergebnisse der Zwillingspathologie [Important racial hygienic problems and results of twin pathology]. In E. Rüdin (Ed.). *Erblehre und rassenhygiene im völkischen staat* [Genetics and racial hygiene in the völkish state] (pp. 303–16). Munich: J. F. Lehmanns.

Luxenburger, H. (1936a). Die rassenhygienische Bedeutung der Lehre von den Manifestationsschwankungen schwanfungen erblicher Krankheiten [The racial hygienic meaning of the theory of variations in the manifestation of hereditary illness]. *Der Erbarzt, 3*, 33–6.

Luxenburger, H. (1936b). Die wichtigsten neueren Ergebnisse der Empirischen Erbprognose und der Zwillingsforschung in der Psychiatrie [The most important new results of the empirical genetic prognosis and twin research in psychiatry]. *Der Erbarzt, 3,* 129–33.

Luxenburger, H. (1939). Kritische Besprechungen und Referate [Review of G. Banu L'Hygiène de la race (Racial Hygiene)]. *Archiv für Rassen- und Gesellschaftsbiologie, 33,* 433–8.

Luxenburger, H. (1940a). Erbpathologie der Schizophrenie [Genetic pathology of schizophrenia] In A. Gütt (Ed.), *Handbuch der erbkrankheiten* (Vol. 2; pp. 191–294). Leipzig: Georg Thieme.

Luxenburger, H. (1940b). Schizophrenie und manisch-depressives Irresein [Schizophrenia and manic-depressive psychosis]. *Fortschritte der Erbpathologie und Rassenhygiene, 4,* 239–59. Abstract obtained from *Psychological Abstracts (1942), 16, (4),* 168.

Lykken, D. T. (1978). The diagnosis of zygosity in twins. *Behavior Genetics, 8,* 437–73.

Lykken, D. T. (1995). *The antisocial personalities.* Hillsdale, NJ: Lawrence Erlbaum Associates.

Lykken, D. T., McGue, M., Tellegen, A., and Bouchard, T. J., Jr. (1992). Emergenesis: Genetic traits that may not run in families. *American Psychologist, 47,* 1565–77.

Lyons, M. J. (1996). A twin study of self-reported criminal behavior. In G. Bock and J. Goode (Eds.), *Genetics of criminal and antisocial behavior* (pp. 61–75). New York: John Wiley and Sons.

Lyons, M. J., and Bar, J. L. (2001). Is there a role for twin studies in the molecular genetics era? *Harvard Review of Psychiatry, 9,* 318–23.

Lyons, M. J., Kendler, K. S., Provet, A., and Tsuang, M. T. (1991). The genetics of schizophrenia. In M. Tsuang, K. Kendler, and M. Lyons (Eds.), *Genetic issues in psychosocial epidemiology* (pp. 119–52). New Brunswick, NJ: Rutgers University Press.

Lyons, M. J., True, W. R., Eisen, S. A., Goldberg, J., Meyer, J. M., Faraone, S. V., Eaves, L. J., and Tsuang, M. T. (1995). Differential heritability of adult and juvenile antisocial traits. *Archives of General Psychiatry, 52,* 906–15.

Marchese, F. J. (1995). The place of eugenics in Arnold Gesell's maturation theory of child development. *Canadian Psychology, 36,* 89–114.

Marshall, R. (1990). The genetics of schizophrenia: Axiom or hypothesis? In R. Bentall (Ed.), *Restructuring schizophrenia* (pp. 89–117). London: Routledge.

Martin, J. D., Blair, G. E., Dannenmaier, W. D., Jones, P. C., and Asako, M. (1981). Relationship of scores on the California Psychological Inventory to age. *Psychological Reports, 49,* 151–4.

McClearn, G. E. (1964). The inheritance of behavior. In L. Postman (Ed.), *Psychology in the making* (pp. 144–252). New York: Alfred A. Knopf.

McCourt, K., Bouchard, T. J., Jr., Lykken, D. T., Tellegen, A., and Keyes, M. (1999). Authoritarianism revisited: Genetic and environmental influences in twins reared apart and together. *Personality and Individual Differences, 27,* 985–1014.

McGue, M., and Bouchard, T. J., Jr. (1984). Adjustment of twin data for the effects of age and sex. *Behavior Genetics, 14,* 325–43.

McGue, M., and Bouchard, T. J., Jr. (1989). Genetic and environmental determinants of information processing and special mental abilities: A twin analysis. In R. Sternberg (Ed.), *Advances in the psychology of human intelligence* (Vol. 5; pp. 7–45). Hillsdale, NJ: Lawrence Erlbaum.

McGuffin, P., and Katz, R. (1990). Who believes in estimating heritability as an end in itself? *Behavioral and Brain Sciences, 13,* 141–2.

McGuffin, P., and Huckle, P. (1990). Simulation of Mendelism revisited: The recessive gene for attending medical school. *American Journal of Human Genetics, 46,* 994–99.

McGuffin, P., Riley, B., and Plomin, R. (2001). Towards behavioral genomics. *Science*, *291*, 1232–49.

McGuire, T. R., and Hirsch, J. (1977). General intelligence (g) and heritability (H2, h2). In I. Uzgiris and F. Weitzmann (Eds.), *The structuring of experience* (pp. 25–72). New York: Plenum Press.

McMahon, R. C. (1980). Genetic etiology in the hyperactive child syndrome: A critical review. *American Journal of Orthopsychiatry*, *50*, 145–50.

Mednick, S. A. (1996). General discussion III. In G. Bock and J. Goode (Eds.), *Genetics of criminal and antisocial behavior* (pp. 129–37). New York: John Wiley and Sons.

Mednick, S. A., Brennan, P., and Kandel, E. (1988). Predisposition to violence. *Aggressive Behavior*, *14*, 25–33.

Mednick, S. A., Gabrielli, W. F., and Hutchings, B. (1984). Genetic influences in criminal convictions: Evidence from an adoption cohort. *Science*, *224*, 891–4.

Mednick, S. A., Gabrielli, W. F., and Hutchings, B. (1987). Genetic factors in the etiology of criminal behavior. In S. Mednick, T. Moffitt, and S. Stack (Eds.). *The causes of crime: New biological perspectives* (pp. 75–91). New York: Cambridge University Press.

Mednick, S. A., and Hutchings, B. (1977). Some considerations in the interpretation of the Danish adoption studies in relation to asocial behavior. In S. Mednick and K. Christiansen (Eds.), *Biosocial bases of criminal behavior* (pp. 159–64). New York: Gardner Press.

Mednick, S. A., and Kandel, E. S. (1988). Congenital determinants of violence. *Bulletin of the American Academy of Psychiatry and the Law*, *16*, 101–9.

Mednick, S. A., and Volavka, J. (1980). Biology and crime. In N. Morris and M. Tonry (Eds.), *Crime and justice: An annual review of research* (Vol. 2; pp. 85–158). Chicago: The University of Chicago Press.

Mensh, E., and Mensh, H. (1991). *The IQ mythology*. Carbondale, IL: Southern Illinois Press.

Merriman, C. (1924). The intellectual resemblance of twins. *Psychological Monographs*, *33*, (5), 1–58.

Meyer, J., Huberth, A., Ortega, G., Syagailo, Y. V., Jatzke, S., Mössner, R., Strom, T. M., Ulzheimer-Teuber, I., Stöber, G., Schmitt, A., and Lesch, K. P. (2001). A missense mutation in a novel gene encoding a putative cation channel is associated with catatonic schizophrenia in a large pedigree. *Molecular Psychiatry*, *6*, 302–6.

Mischel, W. (1968). *Personality and assessment*. New York: Wiley.

Mittler, P. (1971). *The study of twins*. Middlesex, UK.: Penguin Books.

Moldin, S. O., and Gottesman, I. I. (1997). At issue: Genes, experience, and chance in schizophrenia — positioning for the 21st century. *Schizophrenia Bulletin*, *23*, 547–61.

Moore, D. S. (2001). *The dependent gene: The fallacy of 'nature vs. nurture'*. New York: Times Books.

Moos, R. H., and Moos, B. S. (1986). *Family Environment Scale manual* (2nd ed.). Palo Alto, CA: Consulting Psychologists Press.

Morris-Yates, A., Andrews, G., Howie, P., and Henderson, S. (1990). Twins: A test of the equal environments assumption. *Acta Psychiatrica Scandinavica*, *81*, 322–6.

Morton, N. E., and MacLean, C. J. (1974). Analysis of family resemblance. III. Complex segregation of quantitative traits. *American Journal of Human Genetics*, *26*, 489–503.

Mosher, L. R. (1975). Unpublished letter to Gottesman, dated 10/14/75.

Mosher, L. R., and Menn, A. Z. (1978). Community residential treatment for schizophrenia: Two-year follow up. *Hospital and Community Psychiatry*, *29*, 715–23.

Mosher, L. R., Vallone, R., and Menn, A. (1995). The treatment of acute psychosis without neuroleptics: Six-week psychopathology outcome data from the Soteria project. *International Journal of Social Psychiatry, 41,* 157–73.

Muller, H. J. (1925). Mental traits and heredity. *Journal of Heredity, 16,* 433–48.

Müller-Hill, B. (1998). *Murderous science.* Plainview, NY: Cold Spring Harbor Laboratory Press. (Original English version published in 1988)

Munk-Jørgensen, P. (1985). The schizophrenia diagnosis in Denmark. *Acta Psychiatrica Scandinavica, 72,* 266–73.

Murray, R. M., and Reveley, A. M. (1986). Genetic aspects of schizophrenia: Overview. In A. Kerr and P. Snaith (Eds.), *Contemporary issues in schizophrenia* (pp. 261–7). London: Gaskell.

Myerson, A. (1925). *The inheritance of mental diseases.* Baltimore: Williams and Wilkins.

Myerson, A. (1932). Discussion in E. Rüdin, 'The significance of eugenics and genetics for mental hygiene.'In F. Williams (Ed.), *Proceedings of the First International Congress on Mental Hygiene* (pp. 471–95). New York: International Committee for Mental Hygiene.

Neale, J. M., and Oltmanns, T. F. (1980). *Schizophrenia.* New York: John Wiley and Sons.

Neel, J. V., and Schull, W. J. (1954). *Human heredity.* Chicago: University of Chicago Press.

Neisser, U., Boodoo, G., Bouchard, T. J., Jr., Boykin, A. W., Brody, N., Ceci, S. J., Halpern, D. F., Loehlin, J. C., Perloff, R., Sternberg, R. J., Urbina, S. (1996). Intelligence: Knowns and unknowns. *American Psychologist, 51,* 77–101.

Newman, D. L., Tellegen, A., and Bouchard, T. J., Jr. (1998). Individual differences in adult ego development: Sources of influences in twins reared apart. *Journal of Personality and Social Psychology, 74,* 985–95.

Newman, H. H. (1917). *The biology of twins (mammals).* Chicago: University of Chicago Press.

Newman, H. H. (1932). *Evolution, genetics, and eugenics* (3rd ed.). Chicago: University of Chicago Press.

Newman, H. H., Freeman, F. N., and Holzinger, K. J. (1937). *Twins: A study of heredity and environment.* Chicago: The University of Chicago Press.

Ochsner, A. J. (1899). Surgical treatment of habitual criminals. *Journal of the American Medical Association, 32,* 867–8.

Ogbu, J. U. (1978). *Minority education and caste.* New York: Academic Press.

Olson, C. P. (Ed.). (1920). *Oregon laws: Showing all the laws of a general nature in force in the state of Oregon* (Vol. 2). San Francisco: The Bancroft-Whitney Company.

Onstad, S., Skre, I., Torgersen, S., and Kringlen, E. (1991). Twin concordance for DSM-III-R schizophrenia. *Acta Psychiatrica Scandinavica, 83,* 395–401.

Osborn, F. (1951). *Preface to eugenics* (rev. ed.). New York: Harper and Brothers.

Owen, M. J. (1992). Will schizophrenia become a graveyard for molecular geneticists? *Psychological Medicine, 22,* 289–93.

Paikin, H., Jacobsen, B., Schulsinger, F., Godtfredsen, K., Rosenthal, D., Wender, P. H., and Kety, S. S. (1974). Characteristics of people who refused to participate in a social and psychopathological study. In S. Mednick, F. Schulsinger, J. Higgins, and B. Bell (Eds.), *Genetics, environment and psychopathology* (pp. 293–322). New York: American Elsevier Publishing Company.

Pam, A. (1995). Biological psychiatry: Science or pseudoscience? In C. Ross and A. Pam (Eds.), *Pseudoscience in biological psychiatry: Blaming the body* (pp. 7–84). New York: John Wiley and Sons.

Pam, A., Kemker, S. S., Ross, C. A., and Golden, R. (1996). The 'equal environment assumption'in MZ-DZ comparisons: An untenable premise of psychiatric genetics? *Acta Geneticae Medicae et Gemellologiae, 45,* 349–60.

Paul, D. B. (1985). Textbook treatments of the genetics of intelligence. *Quarterly Review of Biology, 60,* 317–26.

Paul, D. B. (1998). *The politics of heredity: Essays on eugenics, biomedicine, and the nature-nurture debate.* Albany, NY: State University of New York Press.

Pedersen, N. L., McClearn, G. E., Plomin, R., and Nesselroade, J. R. (1992). Effects of early rearing environment on twin similarity in the last half of the life span. *British Journal of Developmental Psychology, 10,* 255–67.

Pedersen, N. L., Plomin, R., Nesselroade, J. R., and McClearn, G. E. (1992). A quantitative genetic analysis of cognitive abilities during the second half of the life span. *Psychological Science, 3,* 346–53.

Pedersen, N. L., Plomin, R., McClearn, G. E., and Friberg, L. (1988). Neuroticism, extroversion, and related traits in adult twins reared apart and reared together. *Journal of Personality and Social Psychology, 55,* 950–7.

Penrose, L. S. (1973). *Outline of human genetics* (3rd ed.). New York: Crane, Russak and Company.

Peters, U. H. (2001). On Nazi psychiatry. *Psychoanalytic Review, 88,* 295–309.

Pittelli, S. J. (2002a). Genetic research on cognitive ability [Letter to the editor]. *British Journal of Psychiatry, 180,* 186.

Pittelli, S. J. (2002b). Meta-analysis and Psychiatric genetics [Letter to the editor]. *American Journal of Psychiatry, 159,* 496.

Plomin, R. (1990). Trying to shoot the messenger for his message. *Behavioral and Brain Sciences, 13,* 144.

Plomin, R. (1994). The role of inheritance in behavior. *Science, 248,* 183–8.

Plomin, R. (1996). Beyond nature versus nurture. In L. Hall (Ed.), *Genetics and mental illness* (pp. 29–49). New York: Plenum Press.

Plomin, R. (1999). Genetics and cognitive ability. *Nature, 402,* (Suppl.), C25–C29.

Plomin, R. (2001). Genetics and behavior. *The Psychologist, 14,* 134–9.

Plomin, R., Corley, R., Caspi, A., Fulker, D. W., and DeFries, J. C. (1998). Adoption results for self-reported personality: Evidence for nonadditive genetic effects? *Journal of Personality and Social Psychology, 75,* 211–18.

Plomin, R., Corely, R., DeFries, J. C., and Fulker, D. W. (1990). Individual differences in television viewing in early childhood: Nature as well as nurture. *Psychological Science, 1,* 371–7.

Plomin, R., and Craig, I. (2001). Genetics, environment and cognitive abilities: Review and work in progress towards a genome scan for qualitative trait locus associations using DNA pooling. *British Journal of Psychiatry, 178,* (Suppl. 40), s41–s48.

Plomin, R., and Craig, I. (2002). Authors' reply [Letter to the editor]. *British Journal of Psychiatry, 180,* 186.

Plomin, R., and Daniels, D. (1987). Why are children in the same family so different from one another? *Behavioral and Brain Science, 10,* 1–16.

Plomin, R., and DeFries, J. C. (1985). *Origins of individual differences in infancy: The Colorado Adoption Project.* Orlando, FL: Academic Press.

Plomin, R., and DeFries, J. C. (1998, May). The genetics of cognitive abilities and disabilities. *Scientific American,* 62–9.

Plomin, R., DeFries, J. C., and Loehlin, J. C. (1977). Genotype-environment interaction and correlation in the analysis of human behavior. *Psychological Bulletin, 84,* 309–22.

Plomin, R., DeFries, J. C., and McClearn, G. E. (1990). *Behavioral genetics: A primer* (2nd ed.). New York: W. H. Freeman and Company.

Plomin, R., DeFries, J. C., McClearn, G. E., and Rutter, M. (1997). *Behavioral genetics* (3rd ed.). New York: W. H. Freeman and Company.

Plomin, R., Fulker, D. W., Corley, R., and DeFries, J. C. (1997). Nature, Nurture, and cognitive development from 1 to 16 years: A parent-offspring adoption study?

Psychological Science, 8, 442–7.

Plomin, R., McClearn, G. E., Smith, D. L., Vignetti, S., Chorney, M. J., Chorney, K., Venditti, C. P., Kasarda, S., Thompson, L. A., Detterman, D. K., Daniels, J., Owen, M., and McGuffin, P. (1995). DNA markers associated with high versus low IQ: The IQ quantitative trait loci (QTL) project. *Behavior Genetics, 24,* 107–18.

Plomin, R., and Rutter, M. (1998). Child development, molecular genetics, and what to do with genes once they are found. *Child Development, 69,* 1223–42.

Pope, H. G., Jonas, J. M., Cohen, B. M., and Lipinski, J. F. (1982). Failure to find evidence of schizophrenia in first-degree relatives of schizophrenic probands. *American Journal of Psychiatry, 139,* 826–8.

Popenoe, P. (1936). Twins and criminals. *Journal of Heredity, 27,* 388–90.

Popenoe, P., and Johnson, R. (1933). *Applied eugenics* (rev. ed.). New York: Macmillan.

Portin, P., and Alanen, Y. O. (1997). A critical review of genetic studies of schizophrenia. II. Molecular genetic studies. *Acta Psychiatrica Scandinavica, 95,* 73–80.

Price, B. (1950). Primary biases in twin studies. *American Journal of Human Genetics, 2,* 293–352.

Proctor, R. N. (1988). *Racial hygiene: Medicine under the Nazis.* Cambridge, MA: Harvard University Press.

Proctor, R. N. (1992). Genomics and eugenics: How fair is the comparison? In G. Annas and S. Elias (Eds.), *Gene mapping* (pp. 57–93). New York: Oxford University Press.

Prolo, P., and Licinio, J. (2002). DRD4 and novelty seeking. In J. Benjamin, R. Ebstein, & R. Belmaker, (Eds.). *Molecular genetics and the human personality* (pp. 91–107). Washington, DC: American Psychiatric Press.

Propping, P. (1992). Abuse of genetics in Nazi Germany [Letter to the editor]. *American Journal of Human Genetics, 51,* 909–10.

Rafter, N. H. (Ed.). (1988). *White trash: The eugenic family studies 1877–1919.* Boston: Northeastern University Press.

Raine, A. (1993). *The psychopathology of crime.* San Diego: Academic Press.

Raine, A., Brennan, P., and Mednick, S. A. (1997). Interaction between birth complications and early maternal rejection in predisposing individuals to adult violence: Specificity to serious, early-onset violence. *American Journal of Psychiatry, 154,* 1265–71.

Rainer, J. (1966). The contributions of Franz Josef Kallmann to the genetics of schizophrenia. *Behavioral Science, 11,* 413–37.

Ratner, C. (1982). Do the studies on identical twins prove that schizophrenia is genetically inherited? *International Journal of Social Psychiatry, 28,* 175–8.

Reilly, P. R. (1991). *The surgical solution: A history of involuntary sterilization in the United States.* Baltimore: The Johns Hopkins University Press.

Rende, R. D., Plomin, R., and Vandenberg, S. G. (1990). Who discovered the twin method? *Behavior Genetics, 20,* 277–85.

Reuband, K. (1994). Reconstructing social change through retrospective questions: Methodological problems and prospects. In N. Schwarz and S. Sudman (Eds.), *Autobiographical memory and the validity of retrospective reports* (pp. 305–11). New York: Springer Verlag

Rhee, S. H., and Waldman, I. D. (2002). Genetic and environmental influences on antisocial behavior: A meta-analysis of twin and adoption studies. *Psychological Bulletin, 128,* 490–529.

Richardson, K. (1998). *The origins of human potential.* London: Routledge.

Richardson, K. (2000). *The making of intelligence.* New York: Columbia University Press.

Richardson, L. F. (1913). The measurement of mental 'nature' and the study of adopted children. *Eugenics Review, 4,* 391–4.

Robbins, L. C. (1963). The accuracy of parental recall of aspects of child development and of child rearing practices. *Journal of Abnormal and Social Psychology, 66,* 261–70.

Rosanoff, A. J. (1938). *Manual of psychiatry and mental hygiene* (7th ed., rewritten and enlarged). New York: John Wiley and Sons.

Rosanoff, A. J., Handy, L. M., Plesset, I. R., and Brush, S. (1934). The etiology of so-called schizophrenic psychoses. *American Journal of Psychiatry, 91,* 247–86.

Rosanoff, A. J., Handy, L. M., and Rosanoff, I. A. (1934). Criminality and delinquency in twins. *Journal of the American Institute of Criminal Law and Criminology, 24,* 923–34.

Rosanoff, A. J., and Orr, F. I. (1911). A study of heredity in insanity in the light of the mendelian theory. *Eugenics Record Office, Bulletin No. 5.* Cold Spring Harbor, New York.

Rose, R. J. (1982). Separated twins: Data and their limits. *Science, 215,* 959–60.

Rose, R. J. (1991).Twin studies and psychosocial epidemiology. In M. Tsuang, K. Kendler, and M. Lyons (Eds.), *Genetic issues in psychosocial epidemiology* (pp. 12–32). New Brunswick, NJ: Rutgers University Press.

Rosenthal, D. (1959). Some factors associated with concordance and discordance with respect to schizophrenia in monozygotic twins. *Journal of Nervous and Mental Disease, 129,* 1–10.

Rosenthal, D. (1960). Confusion of identity and the frequency of schizophrenia in twins. *Archives of General Psychiatry, 3,* 297–304.

Rosenthal, D. (1961). Sex distribution and the severity of illness among samples of schizophrenic twins. *Journal of Psychiatric Research, 1,* 26–36.

Rosenthal, D. (1962a). Familial concordance by sex with respect to schizophrenia. *Psychological Bulletin, 59,* 401–21.

Rosenthal, D. (1962b). Problems of sampling and diagnosis in the major twin studies of schizophrenia. *Journal of Psychiatric Research, 1,* 116–34.

Rosenthal, D. (1963). *The Genain quadruplets.* New York: Basic Books.

Rosenthal, D. (1967). An historical and methodological review of genetic studies of schizophrenia. In J. Romano (Ed.), *The origins of schizophrenia: Proceedings of the first Rochester international conference on schizophrenia, March 29–31, 1967* (pp. 15–26). New York: Excerpta Medica Foundation.

Rosenthal, D. (1968). The heredity-environment issue in schizophrenia: Summary of the conference and present status of our knowledge. In D. Rosenthal and S. Kety (Eds.), *The transmission of schizophrenia* (pp. 413–27). New York: Pergamon Press.

Rosenthal, D. (1970a). Genetic research in the schizophrenia syndrome. In R. Cancro (Ed.), *The schizophrenic reactions* (pp. 245–58). New York: Brunner/Mazel.

Rosenthal, D. (1970b). *Genetic theory and abnormal behavior.* New York: McGraw-Hill.

Rosenthal, D. (1971a). *Genetics of psychopathology.* New York: McGraw-Hill.

Rosenthal, D. (1971b). A program of research on heredity in schizophrenia. *Behavioral Science, 16,* 191–201.

Rosenthal, D. (1972). Three adoption studies of heredity in the schizophrenic disorders. *International Journal of Mental Health, 1,* 63–75.

Rosenthal, D. (1974a). The concept of subschizophrenic disorders. In S. Mednick, et al. (Eds.), *Genetics, environment, and psychopathology* (pp. 167–76). New York: American Elsevier.

Rosenthal, D. (1974b). The genetics of schizophrenia. In S. Arieti and E. Brody (Eds.), *American handbook of psychiatry* (2nd ed.; pp. 588–600). New York: Basic Books.

Rosenthal, D. (1975a). Discussion: The concept of schizophrenic disorders. In R. Fieve, D. Rosenthal, and H. Brill (Eds.), *Genetic research in psychiatry* (pp. 199–208).

Baltimore: The Johns Hopkins Press.

Rosenthal, D. (1975b). The spectrum concept in schizophrenic and manic-depressive disorders. In D. Freedman (Ed.), *Biology of the major psychoses* (pp. 19–25). New York: Raven Press.

Rosenthal, D. (1978). Eugen Bleuler's thoughts and views about heredity in schizophrenia. *Schizophrenia Bulletin, 4,* 476–7.

Rosenthal, D. (1979). Genetic factors in behavioural disorders. In M. Roth and V. Cowie (Eds.), *Psychiatry, genetics and pathography: A tribute to Eliot Slater* (pp. 22–33). London: Oxford University Press.

Rosenthal, D., and Kety, S. S. (Eds.). (1968). *The transmission of schizophrenia: Proceedings of the second research conference of the foundations' fund for research in psychiatry, Dorado, Puerto Rico, 26 June to 1 July 1967.* New York: Pergamon Press.

Rosenthal, D., Wender, P. H., Kety, S. S., Schulsinger, F., Welner, J., and Østergaard, L. (1968). Schizophrenics' offspring reared in adoptive homes. In D. Rosenthal and S. Kety (Eds.), *The transmission of schizophrenia* (pp. 377–91). New York: Pergamon Press.

Rosenthal, D., Wender, P. H., Kety, S. S., Welner, J., and Schulsinger, F. (1971). The adopted-away offspring of schizophrenics. *American Journal of Psychiatry, 128,* 307–11.

Rosenthal, R. (1979). The 'file drawer problem'and tolerance for null results. *Psychological Bulletin, 86,* 638–41.

Rosenthal, R., and Rosnow, R. L. (1975). *The volunteer subject.* New York: John Wiley and Sons.

Rowe, D. C. (1983). Biometrical genetic models of self-reported delinquent behavior: A twin study. *Behavior Genetics, 13,* 473–89.

Rowe, D. C. (1986). Genetic and environmental components of antisocial behavior: A study of 265 twin pairs. *Criminology, 24,* 513–532.

Rowe, D. C. (1994). *The limits of family influence: Genes, experience, and behavior.* New York: The Guilford Press.

Rowe, D. C. (2002). *Biology and crime.* Los Angeles: Roxbury.

Rowe, D. C., and Jacobson, K. C. (1999). In the mainstream. In R. Carson and M. Rothstein (Eds.), *Behavioral genetics: The clash of culture and biology* (pp. 12–34). Baltimore: The Johns Hopkins University Press.

Rüdin, E. (1916). *Zur Vererbung und Neuentstehung der Dementia praecox* [On the heredity and new development of Dementia praecox].Berlin: Springer Verlag OHG.

Rüdin, E. (1932). The significance of eugenics and genetics for mental hygiene. In F. Williams (Ed.), *Proceedings of the First International Congress on Mental Hygiene* (pp. 471–95). New York: International Committee for Mental Hygiene.

Rüdin, E. (1939). Bedeutung der Forschung und Mitarbeit von Neurologen und Psychiatrie im nationalsozialistischen Staat [The meaning of research and cooperation of neurologists and psychiatry in the National Socialist state]. *Zeitschrift für die Gesamte Neurologie und Psychiatrie, 165,* 7–17.

Rüdin, E. (1942). Zehn Jahre nationalsozialistischer Staat [Ten years of the National Socialist state]. *Archiv für Rassen- und Gesellschaftsbiologie, 36,* 321–2.

Rutter, M. (1996). Introduction: Concepts of antisocial behaviour, of cause, and of genetic influences. In G. Bock and J. Goode (Eds.), *Genetics of criminal and antisocial behavior* (pp. 1–20). New York: John Wiley and Sons.

Rutter, M., Silberg, J., and Simonoff, E. (1993). Wither behavioral genetics? – A developmental psychopathological perspective. In R. Plomin and G. McClearn (Eds.), *Nature, nurture, and psychology* (pp. 433–56). Washington, DC: American Psychological Association.

328 *The Gene Illusion*

Samelson, F. (1979). Putting psychology on the map. In A. Buss (Ed.), *Psychology in social context* (pp. 103–67). New York: Irvington Publishers.

Sanger, M. (1921, October). The eugenic value of birth control propaganda. *Birth Control Review, 5, (10),* 5.

Sarbin, T. R. (1990). Toward the obsolescence of the schizophrenia hypothesis. *Journal of Mind and Behavior, 11,* 259–83.

Sarbin, T. R., and Mancuso, J. C. (1980). *Schizophrenia: Medical diagnosis or moral verdict?* New York: Pergamon Press.

Savan, B. (1988). *Science under siege: The myth of objectivity in scientific research.* Montreal: CBC Enterprises.

Scarr, S. (1968). Environmental bias in twin studies. *Eugenics Quarterly, 15,* 34–40.

Scarr, S., and Carter-Saltzman, L. (1979). Twin method: Defense of a critical assumption. *Behavior Genetics, 9,* 527–42.

Schiff, M., Duyme, M., Dumaret, A., and Tomkiewicz, S. (1982). How much *could* we boost scholastic achievement and IQ scores? A direct answer from a French adoption study. *Cognition, 12,* 165–96.

Schiff, M., and Lewontin, R. C. (1986). *Education and class: The irrelevance of IQ genetic studies.* Oxford: Clarendon Press.

Schulsinger, F. (1977). Psychopathy: Heredity and environment. In S. Mednick and K. Christiansen (Eds.), *Biosocial bases of criminal behavior* (pp. 109–25). New York: Gardner Press. (Originally published in 1972)

Schulz, B. (1934). Rassenhygienische Eheberatung [Racial hygienic marriage counseling]. *Volk und Rasse, 9,* 138–43.

Schulz, B. (1939). Über die Beteutung der empirischen Erbvorhersageforschung [On the meaning of genetic empirical prognostic research]. *Der Erbarzt, 6, (4),* 43–4.

Schwesinger, G. C. (1952). The effect of differential parent-child relation on identical twin resemblance in personality. *Acta Geneticae Medicae et Gemellologiae, 1,* 40–7.

Sedgwick, J. (1995). Inside the Pioneer Fund. In R. Jacoby and N. Glauberman (Eds.), *The Bell Curve debate* (pp. 144–61). New York: Times Books.

Segal, N. L. (1999). *Entwined lives: Twins and what they tell us about human behavior.* New York: Dutton.

Sharp, H. C. (1909). Vasectomy as a means of preventing procreation in defectives. *Journal of the American Medical Association, 53,* 1897–902.

Sherrington, R., Brynjolfsson, J., Petursson, H., Potter, M., Duddleston, K., Barraclough, B., Wasmuth, J., Dobbs, M., and Gurling, H. (1988). Localization of a susceptibility locus for schizophrenia on chromosome 5. *Nature, 336,* 164–7.

Shields, J. (1954). Personality differences and neurotic traits in normal twin schoolchildren. *Eugenics Review, 45,* 213–46.

Shields, J. (1962). *Monozygotic twins brought up apart and brought up together.* London: Oxford University Press.

Shields, J. (1968). Summary of the genetic evidence. In D. Rosenthal and S. Kety (Eds.), *The transmission of schizophrenia* (pp. 95-126). New York: Pergamon Press.

Shields, J. (1978). MZA Twins: Their use and abuse. In *Twin research: Psychology and methodology* (pp. 79–93). New York: Alan R. Liss.

Shorter, E. (1997). *A history of psychiatry.* New York: John Wiley and Sons.

Siemens, H. W. (1924a). *Race hygiene and heredity.* New York: D. Appleton.

Siemens, H. W. (1924b). *Die zwillingspathologie* [Twin pathology]. Berlin: Springer Verlag.

Siemens, H. W. (1927). The diagnosis of identity in twins. *Journal of Heredity, 18,* 201–9.

Siemens, H. W. (1937). *Grundzüge der vererbungslehre, rassenhygiene und bevölkerungspolitik* [Foundations of genetics, racial hygiene, and population policy] (8th ed.). Munich and Berlin: J. F. Lehmanns Verlag.

Simes, R. J. (1987). Confronting publication bias: A cohort design for meta-analysis. *Statistics in Medicine, 6,* 11–29.

Simon, J. L. (1997). Four comments on The Bell Curve. *Genetica, 99,* 199–205.

Sims, V. M. (1931). The influence of blood relationship and common environment on measured intelligence. *Journal of Educational Psychology, 22,* 56–65.

Slater, E. (1953). Psychotic and neurotic illnesses in twins. *Medical Research Council Special Report Series No. 278.* London: Her Majesty's Stationary Office.

Slater, E. (1968). A review of earlier evidence on genetic factors in schizophrenia. In D. Rosenthal and S. Kety (Eds.), *The transmission of schizophrenia* (pp. 15–26). New York: Pergamon Press.

Slater, E. (1971). Autobiographical sketch. In J. Shields and I. Gottesman (Eds.), *Man, mind, and heredity: Selected papers of Eliot Slater on psychiatry and genetics* (pp. 1–23). Baltimore: Johns Hopkins Press.

Slater, E., and Cowie, V. (1971). *The genetics of mental disorders.* London: Oxford University Press.

Slater, E., Hare, E., and Price, J. (1971). Marriage and fertility of psychiatric patients compared with national data. *Social Biology, 18* (Suppl.), S60–S73.

Smith, C. (1974). Concordance in twins: Methods and interpretation. *American Journal of Human Genetics, 26,* 454–66.

Smith, R. T. (1965). A comparison of socioenvironmental factors in monozygotic and dizygotic twins, testing an assumption. In S. Vandenberg (Ed.), *Methods and goals in human behavior genetics* (pp. 45–61). New York: Academic Press.

Sofair, A. N., and Kaldjian, L. C. (2000). Eugenic sterilization and a qualified Nazi analogy: The United States and Germany, 1930–45. *Annals of Internal Medicine, 132,* 312–19.

Spitzer, R. L., and Endicott, J. (1979). Justification for separating schizotypal and borderline personality disorders. *Schizophrenia Bulletin, 6,* 95–104.

Stern, C. (1962). Wilhelm Weinberg. *Genetica, 47,* 1–5.

Stocks, P. (1930). A biometric investigation of twins and their brothers and sisters. *Annals of Eugenics, 4,* 49–108.

Stoltenberg, S. F. (1997). Coming to terms with heritability. *Genetica, 99,* 89–96.

Stoolmiller, M. (1998). Correcting estimates of shared environmental variance for range restriction in adoption studies using a truncated multivariate normal model. *Behavior Genetics, 28,* 429–41.

Stoolmiller, M. (1999). Implications of the restricted range of family environments for estimates of heritability and nonshared environment in behavior-genetic adoption studies. *Psychological Bulletin, 125,* 392–409.

Straub, R. E., Jiang, Y., MacLean, C. J., Ma, Y., Webb, B. T., Myakishev, M. V., Harris-Kerr, C., Wormley, B., Sadek, H., Kadambi, B., Cesare, A. J., Gibberman, A., Wang, X., O'Neill, F. A., Walsh, D., and Kendler, K. S. (2002). Genetic variation in the 6p22.3 gene DTNBP1, the human ortholog of the mouse dysbindin gene, is associated with schizophrenia. *American Journal of Human Genetics, 72* (electronically published July 3, 2002).

Strömgren, E. (1993). Fini Schulsinger's contribution to psychiatric research in genetic epidemiology. *Acta Psychiatrica Scandinavica* (Suppl. 370), 11–13.

Strömgren, E. (1994). Recent history of European psychiatry—Ideas, development, and personalities. *American Journal of Medical Genetics (Neuropsychiatric Genetics), 54,* 405–10.

Stumpfl, F. (1936). *Die Ursprunge des Verbrechen. Dargesttellt am Lebenslauf van Zwillingen* [The origins of crime as represented in the resume of twins.] Leipzig: Georg Thieme.

Szasz, T. S. (1961). *The myth of mental illness: Foundations of a theory of personal conduct.* New York: Hoeber-Harper.

Szasz, T. S. (1964). The moral dilemma of psychiatry: Autonomy or heteronomy? *American Journal of Psychiatry, 121,* 521–8.

Szasz, T. S. (1968). Hysteria. In D. Sills (Ed.), *International encyclopedia of the social sciences* (vol. 7; pp. 47–52). New York: Macmillan.

Szasz, T. S. (1970). *The manufacture of madness.* New York: Delta.

Szasz, T. S. (1976). *Schizophrenia: The sacred symbol of psychiatry.* New York: Basic Books.

Szasz, T. S. (1987). *Insanity: The idea and its consequences.* Syracuse, NY: Syracuse University Press.

Taylor, H. F. (1980). *The IQ game: A methodological inquiry into the heredity-environment controversy.* New Brunswick, NJ: Rutgers University Press.

Taylor, P. (1996). General discussion III. In G. Bock and J. Goode (Eds.), *Genetics of criminal and antisocial behavior* (pp. 129–37). New York: John Wiley and Sons.

Teasdale, T. W. (1979). Social class correlations among adoptees and their biological adoptive parents. *Behavior Genetics, 9,* 103–14.

Tellegen, A., Lykken, D. T., Bouchard, T. J., Jr., Wilcox, K. J., Segal, N. L., and Rich, S. (1988). Personality similarity in twins reared apart and together. *Journal of Personality and Social Psychology, 54,* 1031–9.

Terman, L. M. (1916) *The measurement of intelligence.* Boston: Houghton Mifflin.

Terman, L. M., and Merrill, M. A. (1937). *Measuring intelligence.* Boston: Houghton Mifflin.

Thompson, P. M., Cannon, T. D., Narr, K., van Erp, T., Poutanen, V. P., Huttunen, M., Lönnqvist, J., Standertksjöld-Nordenstam, C. G., Kaprio, J., Khaledy, M., Dail, R., Zoumalan, C., I., and Toga, A. W. (2001). Genetic influences on brain structure. *Nature Neuroscience, 12,* 1253–8.

Thorndike, E. L. (1905). Measurements of twins. *Archives of Philosophy, Psychology, and Scientific Methods, 1,* v–64.

Tienari, P. (1963). *Psychiatric illnesses in identical twins.* Copenhagen: Munksgaard.

Tienari, P. (1968). Schizophrenia in monozygotic male twins. In D. Rosenthal and S. Kety (Eds.), *The transmission of schizophrenia* (pp. 27–36). New York: Pergamon Press.

Tienari, P. (1971). Schizophrenia and monozygotic twins. *Psychiatria Fennica, 1971,* 97–104.

Tienari, P. (1975). Schizophrenia in Finnish male twins. *British Journal of Psychiatry Special Publication, No. 10.* M. Lader (Ed.) pp. 29–35.

Tienari, P. (1991). Interaction between genetic vulnerability and family environment: The Finnish adoptive family study of schizophrenia. *Acta Psychiatrica Scandinavica, 84,* 460–5.

Tienari, P. (1992). Implications of adoption studies on schizophrenia. *British Journal of Psychiatry, 161* (Suppl. 18), 52–8.

Tienari, P., Kaleva, M., Lahti, I., Läksy, K., Moring, J., Naarala, M., Sorri, A., Wahlberg, K. E., and Wynne, L. C. (1991). Adoption studies on schizophrenia. In C. Eggers (Ed.), *Schizophrenia and youth* (pp. 42–51). New York: Springer Verlag.

Tienari, P., Lahti, I., Sorri, A., Naarala, M., Wahlberg, K., Rönkkö, T., Moring, J., and Wynne, L. C. (1987). The Finnish adoptive family study of schizophrenia: Possible joint effects of genetic vulnerability and family interaction. In K. Halweg and M. Goldstein (Eds.), *Understanding major mental disorder: The contribution of family interaction research* (pp. 33–54). New York: Family Process Press.

Tienari, P., Sorri, A., Lahti, I., Naarala, M., Wahlberg, K., Moring, J., and Pohjola, J. (1985). Interaction of genetic and psychosocial factors in schizophrenia. *Acta Psychiatrica Scandinavica* (Suppl. No. 319, 71), 19–30.

Tienari, P., Sorri, A., Lahti, I., Naarala, M., Wahlberg, K., Moring, J., Pohjola, J., and Wynne, L. C. (1987). Genetic and psychosocial factors in schizophrenia: The Finnish adoptive family study. *Schizophrenia Bulletin, 13,* 477–84.

Tienari, P., Wynne, L. C., Moring, J., Lahti, I., Naarala, M., Sorri, A., Wahlberg, K., Saarento, O., Seitamaa, M., Kaleva, M., and Läksy, K. (1994). The Finnish adoptive family study of schizophrenia. *British Journal of Psychiatry, 164* (Suppl. 23), 20–6.

Tienari, P., Wynne, L. C., Moring, J., Läksy, K., Nieminen, P., Sorri, A., Lahti, I., Wahlberg, K. E., Naarala, M., Kurki-Suonio, K., Saarento, O., Koistinen, P., Tarvainen, T., Hakko, H., and Miettunen, J. (2000). Finish Adoptive Family Study: Sample selection and adoptee DSM-III-R diagnoses. *Acta Psychiatrica Scandinavica, 101,* 433–43.

Torrey, E. F. (1980). *Schizophrenia and civilization.* New York: Jason Aronson.

Torrey, E. F. (1990). Offspring of twins with schizophrenia [Letter to the editor]. *Archives of General Psychiatry, 47,* 976–7.

Torrey, E. F. (1995). *Surviving schizophrenia* (3rd ed.). New York: Harper Perennial.

Torrey, E. F., Bowler, A. E., Taylor, E. H., and Gottesman, I. I. (1994). *Schizophrenia and manic-depressive disorder: The biological roots of mental illness as revealed by the landmark study of identical twins.* New York: Basic Books.

Torrey, E. F., and Yolken, R. H. (2000). Familial and genetic mechanisms in schizophrenia. *Brain Research Reviews, 31,* 113–17.

Treasure, J., and Holland, A. (1995). Genetic factors in eating disorders. In G. Szmukler, C. Dare, and J. Treasure (Eds.), *Handbook of eating disorders* (pp. 65–81). New York: John Wiley and Sons.

Tsuang, M. T., and Faraone, S. V. (1990). *The genetics of mood disorders.* Baltimore: Johns Hopkins University Press.

Tsuang, M. T., and Faraone, S. V. (2000). The frustrating search for schizophrenia genes. *American Journal of Medical Genetics (Semin. Med Genet.), 97,* 1–3.

Tsuang, M. T., Stone, W. S., and Faraone, S. V. (2001). Genes, environment, and schizophrenia. *British Journal of Psychiatry, 178* (Suppl. 40), S18–S24.

Tucker, W. H. (1994). *The science and politics of racial research.* Urbana, IL: University of Illinois Press.

Turkheimer, E. (2000). Three laws of behavior genetics and what they mean. *Psychological Science, 9,* 160–4.

Van Dyke, J. L., Rosenthal, D., and Rasmussen, P. V. (1975). Schizophrenia: Effects of inheritance and rearing on reaction time. *Canadian Journal of Behavioral Science, 7,* 223–36.

von Bracken, H. (1934). Mutual intimacy in twins. *Character and Personality, 2,* 293–309.

von Verschuer, O. (1939). Twin research from the time of Francis Galton to the present-day. *Proceedings of the Royal Society, 128 (Series B),* 62–81.

Wahlberg, K. E., Wynne, L. C., Keskitalo, P., Nieminen, P., Moring, J., Läksy, K., Sorri, A., Koistinen, P., Tarvainen, T., Miettunen, J., and Tienari, P. (2001). Long-term stability of communication deviance. *Journal of Abnormal Psychology, 110,* 443–8.

Wahlberg, K. E., Wynne, L. C., Oja, H., Keskitalo, P., Anais-Tanner, H., Koistinen, P., Tarvainen, T., Hakko, H., Lahti, I., Moring, J., Naarala, M., Sorri, A., and Tienari, P. (2000). Thought Disorder Index of Finnish adoptees and communication deviance of their adoptive parents. *Psychological Medicine, 30,* 127–36.

Wahlberg, K. E., Wynne, L. C., Oja, H., Keskitalo, P., Pykäläinen, M., Lahti, I., Moring, J., Naarala, M., Sorri, A., Seitamaa, M., Läksy, K, Kolassa, J., and Tienari, P. (1997). Gene-environment interaction in vulnerability to schizophrenia: Findings from the Finnish Adoptive Family Study of Schizophrenia. *American Journal of Psychiatry, 154,* 355–62.

Wahlsten, D. (1990). Insensitivity of the analysis of variance to heredity-environment interaction. *Behavioral and Brain Sciences, 13,* 109–20.

Wahlsten, D. (1994). The intelligence of heritability. *Canadian Psychology, 35,* 244–59.

Wahlsten, D. (1999). Single-gene influences on brain and behavior. *American Review of Psychology, 50,* 599–624.

Walters, G. D., and White, T. W. (1989). Heredity and crime: Bad genes or bad research? *Criminology, 27,* 455–85.

Wasserman, D., and Wachbroit, R. (2001). Introduction: Methods, meanings, and morals. In D. Wasserman and R. Wachbroit (Eds.), *Genetics and criminal behavior* (pp. 1–21). Cambridge, UK: Cambridge University Press.

Watson, P. (1981). *Twins: An uncanny relationship?* New York: The Viking Press.

Weatherly, J., and Deabler, H. L. (1954). Schizophrenia in identical twins one of whom was lobotomized. *Journal of Nervous and Mental Disease, 120,* 262–7.

Weber, M. M. (1996). Ernst Rüdin, 1874–1952. *American Journal of Medical Genetics (Neuropsychiatric Genetics), 67,* 323-31.

Wechsler, D. (1944). *The measurement of adult intelligence* (3rd ed.). Baltimore: Williams and Wilkins.

Wechsler, D. (1958). *The measurement and appraisal of adult intelligence* (4th ed.). Baltimore: Williams and Wilkins.

Weindling, P. (1989). *Health, race, and German politics between national unification and Nazism, 1870–1945.* Cambridge, UK: Cambridge University Press.

Weindling, P. (1993). The survival of eugenics in 20th-Century Germany. *American Journal of Human Genetics, 52,* 643–9.

Weindling, P. (1999). International eugenics: Swedish sterilization in context. *Scandinavian Journal of History, 24,* 179–97.

Weingart, P. (1989). German eugenics between science and politics. *Osiris, 5 (second series),* 260–82.

Weingart, P. (1999). Science and political culture: Eugenics in comparative perspective. *Scandinavian Journal of History, 24,* 163–77.

Weinreich, M. (1946). *Hitler's professors.* New York: Yiddish Scientific Institute — Yivo.

Wender, P. H., and Klein, D. (1981). *Mind, mood, and medicine.* New York: Farrar, Straus, and Giroux.

Wender, P. H., Rosenthal, D., Kety, S. S., Schulsinger, F., and Welner, J. (1974). Crossfostering: A research strategy for clarifying the role of genetic and experiential factors in the etiology of schizophrenia. *Archives of General Psychiatry, 30,* 121–8.

Wetzell, R. F. (2000). *Inventing the criminal: A history of German criminology, 1880–1945.* Chapel Hill, NC: University of North Carolina Press.

Wheelan, L. (1951). Aggressive psychopathy in one of a pair of uniovular twins: A clinical and experimental study. *British Journal of Delinquency, 2,* 130–43.

Whitaker, R. (2002). *Mad in America: Bad science, bad medicine, and the enduring mistreatment of the mentally ill.* Cambridge, MA: Perseus.

Wiener, K. (1958). Preface. In H. W. Siemens, *General diagnosis and therapy of skin diseases.* Chicago: University of Chicago Press.

Wilens, T. E., Biederman, J., and Spencer, T. J. (2002). Attention-deficit/ hyperactivity disorder across the lifespan. *Annual Review of Medicine, 53,* 113–31.

Williams, N. M., Rees, M. I., Holmans, P., Norton, N., Cardno, A. G., Jones, L. A., Murphy, K. C., Sanders, R. D., McCarthy, G., Gray, M. Y., Fenton, I., McGuffin, P., and Owen, M. J. (1999). A two-stage genome scan for schizophrenia susceptibility genes in 196 affected sibling pairs. *Human Molecular Genetics, 8,* 1729–39.

Wilson, P. T. (1934). A study of twins with special reference to heredity as a factor determining differences in environment. *Human Biology, 6,* 324–54.

Wilson, J. Q., and Herrnstein, R. J. (1985). *Crime and human nature.* New York: Simon and Schuster.

Wingfield, A. H. (1928). *Twins and orphans: The inheritance of intelligence.* London: J. M. Dent and Sons.

Wingfield, A. H. (1930). The intelligence of twins. *Eugenics Review, 22,* 183–6.

Woodworth, R. S. (1941). *Heredity and environment: A critical survey of recently published material on twins and foster children.* New York: Social Science Research Center.

World Health Organization. (1966). The use of twins in epidemiological studies. *Acta Geneticae Medicae et Gemellologiae, 15,* 109–28.

Wright, L. (1997). *Twins: And what they tell us about who we are.* New York: John Wiley and Sons.

Wright, W. (1998). *Born that way.* New York: Alfred A. Knopf.

Wyatt, W. J. (1993). Identical twins, emergenesis and environments. (Letter to the editor.) *American Psychologist, 48,* 1294–5.

Wyatt, W. J., Posey, A., Welker, W., and Seamonds, C. (1984). Natural levels of similarities between identical twins and between unrelated people. *The Skeptical Inquirer, 9,* 62–6.

Wynne, L., Singer, M., and Toohey, M. (1976). Communication of the adoptive parents of schizophrenics. In J. Jorstad and E. Ugelstad (Eds.), *Schizophrenia 75 – psychotherapy family studies, research* (pp. 413–51). Oslo: University of Oslo Press.

Yamaguchi, M. (1997, December 27). Japan's victims of sterilization stop hiding. *The San Francisco Chronicle,* p. A12.

Yarrow, M. R., Campbell, J. D., and Burton, R. V. (1970). Recollections of childhood: A study of the retrospective method. *Monographs of the Society for Research in Child Development, 35* (5, Serial No. 138).

Zerbin-Rüdin, E. (1972). Genetic research and the theory of schizophrenia. *International Journal of Mental Health, 1,* 42–62.

Zerbin-Rüdin, E., and Kendler, K. S. (1996). Ernst Rüdin (1874–1952) and his Genealolgic-Demographic Department in Munich (1917–1986): An introduction to their family studies of schizophrenia. *American Journal of Medical Genetics (Neuropsychiatric Genetics), 67,* 332–7.

NAME INDEX

SUBJECT INDEX

THE JOURNAL OF CRITICAL PSYCHOLOGY, COUNSELLING AND PSYCHOTHERAPY

THIS NEW JOURNAL brings a strong focus on a compassionate critique of the counselling, psychology and psychotherapy worlds. Interviews, special issues, leading critics and new writing combine to ask some fundamental questions of psychology, psychotherapy and counselling. In it's previous incarnation as the critical Journal *Changes,* it broke new ground for almost twenty years, asking along the way, if all psychotherapy was mere wishful thinking, how to nurture the political while respecting the personal and, above all, if therapy, psychology and psychiatry could remain human in the endeavour to help people.

THE JOURNAL OF CRITICAL PSYCHOLOGY, COUNSELLING AND PSYCHOTHERAPY will continue to question and examine the assumptions inherent in therapeutic practices, to explore the implications of making the person central to psychological enquiry and to invite discussion and comment between people who believe that there is more to counselling than meets the scientific eye. It is the journal of the Psychology and Psychotherapy Association and will do its utmost to remind us that people are more important than theories and what we do is more important than what we say we do. It will speak out when the need arises and give a platform to new writers wanting to question the increasingly conservative world of psychology, psychotherapy and counselling. It might even make you laugh.

One Year [Four Issues]:
Individual [UK and Europe] £30.00. Individual [outside UK and Europe] £40.00. Institutional £125.00.

Sample issue FREE on application to PCCS Books.
Call +44 (0)1989 770 707